# LIFE

## TRUTH IN ITS VARIOUS PERSPECTIVES

# ANALECTA HUSSERLIANA

## THE YEARBOOK OF PHENOMENOLOGICAL RESEARCH

## VOLUME LXXVI



For sequel volumes see the end of this volume.

# LIFE

## TRUTH IN ITS VARIOUS PERSPECTIVES

## *Cognition, Self-Knowledge, Creativity, Scientific Research, Sharing-in-Life, Economics...*

*Edited by*

ANNA-TERESA TYMIENIECKA

*The World Phenomenology Institute*

Published under the auspices of
*The World Institute for Advanced Phenomenological Research and Learning*
A-T. Tymieniecka, President

**KLUWER ACADEMIC PUBLISHERS**
DORDRECHT / BOSTON / LONDON

Library of Congress Cataloging-in-Publication Data is available.

ISBN 978-90-481-5847-8

---

Published by Kluwer Academic Publishers,
P.O. Box 17, 3300 AA Dordrecht, The Netherlands.

Sold and distributed in North, Central and South America
by Kluwer Academic Publishers,
101 Philip Drive, Norwell, MA 02061, U.S.A.

In all other countries, sold and distributed
by Kluwer Academic Publishers,
P.O. Box 322, 3300 AH Dordrecht, The Netherlands.

*Printed on acid-free paper*

# TABLE OF CONTENTS

ACKNOWLEDGEMENTS vii
TOPICAL STUDY / Truth – The Ontopoietic Vortex of Life ix

SECTION I

ELDON C. WAIT / Do We See the Things Themselves? 3
SITANSU RAY / The Culmination of Reality: Man in the Universe 19
BEATA SZYMAŃSKA / An Experience of Pure Consciousness in Zen Buddhism 47
MARTA KUDELSKA / An Attempt at a Phenomenological Description of the Self-Knowledge Process in the Chandogya Upanishad 57
JORGE GARCÍA-GÓMEZ / Reflections on José Ortega y Gasset's Notion of *Àlétheia* 67
STEPHANIE GRACE SCHULL / Knowing Thyself: Paradox, Self-Deception, and Intersubjectivity 99
PEDRO LUIS BLASCO AZNAR / The Truth of the I and Its Intuitive Knowledge 117

SECTION II

DIANE G. SCILLIA / Blurring the Boundaries between Art and Life: Jan van Eyck's *Ghent Altarpiece* (1425–32); Allan Kaprow's *Apple Shrine* (1960) and *Eat* (1964) 135
CARMEN COZMA / Musical Art as Enlightenment and Understanding through Ethos: The Experience of the "Human" 151
ROBERT D. SWEENEY / Trace, Testimony, Portrait 159
RENATO PRADA OROPEZA / Phenomenology and Literary Aesthetics 171
GERALD NYENHUIS / Los Cuasi-Juicios 183
MIHAI PĂSTRĂGUŞ / Illusion and Truth in the Work of Art 195

## SECTION III

W. KIM ROGERS / Truthfulness in Science and Art 215
JULIO E. RUBIO / Phenomenology and Levels of Organization in Science 223
MILAN JAROS / Machinic Inscriptions of Fragment Objectness 233
JOSÉ LUIS BARRIOS LARA / Los Bordes Imaginarios del Asco y el Morbo: Una Fenomenología del Tiempo en las Fronteras de la Animalidad en el Cine de Pier Paolo Pasolini y David Cronenberg 247

## SECTION IV

LEONARDO SCARFÓ / On the Necessary Form of Philosophy in the Vital Determination of Every Beginning Thereof 265
TADEUSZ CZARNIK / Is Freedom a Condition of Responsibility? An Analysis based on Roman Ingarden's Notion of Freedom 281
J. J. VENTER / Economism: The Debate About the Universality Claims of Orthodox Economics 289

## SECTION V

PIERO TRUPIA / Peoplegram vs Organization Chart: The New Management of Human Resources 323
BRONISŁAW BOMBAŁA / The Autocreation of a Manager in the Process of Transformational Leadership 335
MAREK PYKA / Business and Ethics, a General Approach 347
JIM I. UNAH / Intellectuals and the Legitimation Crisis: A Phenomenological Ontology of Human Relations 355

INDEX OF NAMES 369

## ACKNOWLEDGEMENTS

From two different parts of the world, Krakow, Poland and Puebla, Mexico come to us the studies brought together in this volume. All have the same focus: truth. Some of them were presented at our Third International Congress on Life, which had the theme "Phenomenology/Philosophy and the Sciences of Life." This conference took place at the Jagiellonian University, Krakow, September 14–16, 1999 under the auspices of Professor Dr. Alexander Koj and Dean Rydzewski and the Institute of Philosophy at the University, to all of whom we owe our thanks. In two previous volumes of papers from this conference we expressed our thanks to the local organizing committee: Drs. T. Czarnik, I. Fiut, P. Mroz, Z. Zalewski, L. Pyra, and J. Handerek. The authors who contributed to that conference will find themselves most appropriately placed in the present volume dedicated to the theme of truth.

It was at the Universidad Benemérita of Puebla, Mexico that our conference on the theme of "Truth" took place, on May 20–21, 2000. We express our thanks to the Autonomous Benemérita University for its hospitality. We enormously enjoyed the vivacity of the Mexican spirit.

As usual, thanks go to Jeff Hurlburt and Louis Houthakker for helping with the preparation of the program as well as for the registration on site.

A-T. T.

The site of the congress in Puebla, the Benemérita Universidad Autónoma.

TOPICAL STUDY

# TRUTH – THE ONTOPOIETIC VORTEX OF LIFE

Although the definition of truth first proposed by Aristotle and maintained as the reference point for all succeeding views was cast in terms reflecting the intellective sphere of rationality/logos in the human unfolding, its validity, that is, the validity of the proposition framing it, and its verification reaches far below the logical sphere of statements. Truth's validity reverberates down from the intellective sphere of the mind's rationality into the spheres of sense that sustain it, within the multiple spheres of the network of the sense in which the logos of life projects its manifestation through living beings and whole world of life. To grasp the full significance of notions of truth we cannot stop at any one perspective or sphere, whether it be cognitive, intertextual, or pragmatic. To understand what "truth" means we should elucidate it in its origin and nature, that is, in its generative significance for the entire expanse of life and in its role within the logoic schema of its dynamic manifestation. I propose to outline this in a succinct way in what follows. Truth will emerge as a crucial logoic device, as the regulative vortex for the ontopoietic balancing out of life's forces in their constructive course.

## I. THE ROOTS OF TRUTH: THE NATURAL 'REALITY' OF LIFE

At the roots of the transcendental constitution of the lifeworld, Husserl saw the basic belief of the human being – belief in the natural world of life – as being prior to all the intentional differentiations of this world itself.

I corroborate this notion of a basic *existential trust* in the constancy of the world of life by shifting the focus from the world to *life*. I see it as our basic trust in the constancy of life and of ourselves as we incorporate it. With this trust our entire life progresses from day to day, from hour to hour, from instant to instant. It consists of our mute natural conviction of an indubitable *constant* background of our reality insofar as our life-individualizing process is simultaneously crystalizing the "outward" framework of our existence within the world and manifesting "inwardly" the entire spread of our vital existential and creative virtualities as they may unfold.

This conviction or belief differs essentially from any other type of what we call "belief." Each belief is suspended on a specific context, somewhat evident or presumed, from which it draws its significance and power of conviction. The existential trust that is here in question in contrast consists in the existential quintessence of our very *ontopoietic ingrownness* into our own sphere of life's subsistence within which our individualizing process enmeshes us in a mutual interaction with and adaptation to circumambient forces, on the one hand, and the universal system of life, on the other, inasmuch as we crystalize its constitutive rules through our self-individualizing becoming. This amounts to a mute but most powerful self-awareness in life that is rooted in our ontopoietic ingrownness within the life context. This is the way in which I will understand the terms "existential trust" in what follows. This basic trust – or self-awareness in life – incorporates our specific centered and outward expanding vital/existential system of propensities as well as our being activated and potentially (virtually) partaking with all our individual powers in the entire context of the life stream. Thus this basic trust in life is simultaneously a trust in ourselves as well as in the life system crystallized in our living world. Since we are subject to misjudging situations, illusions about matters of fact, errors in observation, and drawing false conclusions, etc., it is upon the ground of that world and our life scheme that we constantly check on the factual "real" status of all our concerns and it is over against this groundwork and its naturally presumed forthcoming expansion that this checking process proceeds. We constantly surmise certain states of fact to be owing to such and such, according to a "natural scheme of things." Upon these tacit assumptions the course of human existence proceeds smoothly, but since circumstances – organic as well as vital, psychic, or societal – may change, and since the perception of things, of the affairs, processes, feelings, attitudes, commitments, etc. involved in everyday life are also subject to natural changes of all sorts, we are constantly checking whether things be "so," not always attentively but with a doubting/assuming mechanism intrinsic to our trust in the world's constancy.

This "so" means that it falls into the "natural scheme of things," with what we would expect to hold "true." The "truth" of things is, then, first, the moment intrinsic in our basic trust that things are as they give themselves to be in our instantaneous experience of them, and simultaneously, as they belong, and as they – concurrently – "should be" or are to be expected to be in accord with the entire schema in which our life is involved.

## II. IDENTIFICATION, COMPARISON, DOUBT, ADEQUATION

Within basic trust in the unshakable presence of reality there lies the tacit re-cognition "at a glance" of the already experienced status of things, beings, facts, situations, etc. This "re-cognition" *identifies, discriminates, establishes adequation* between the originary sphere of experience and its "repetitive" occurrence. Here come two important points. First, this trust extends throughout the entire network of life with its innumerable entanglements of existential significance for living beings. Second, and what lies at the heart of our slowly unfolding argument, trust in the constancy of life, individual beings, and their world is not a prerogative of the specifically human being only. It extends down the evolutionary ladder to the entire animal kingdom relative to the different experiences or the "living" reactivity/receptivity of the different species of living beings. In different modalities there is a comparable "belief" in the constancy of life, the constancy of the world, the constancy of each living being itself as it runs through the animal spheres of the manifestation of the logos. Even the simplest living creature does not start its life over again each day, but proceeds upon the re-cognition of the data of the previous day. For the less complex creatures there is a sense of the "fitness" between their organic/vital system of individual life, the system of life, and the given reality in front of them. (See my monograph "The Moral Sense and the Human Person within the Fabric of Communal Life," in Anna-Teresa Tymieniecka (ed.), *The Moral Sense in the Communal Significance of Life*, Analecta Husserliana, Vol. XX [Dordrecht: Kluwer Academic Publishers, 1986], pp. 3–100). More complex animals, which master a sensing apparatus, re-cognize, identify, and discriminate among present data in repetitions of an originary re-cognition. They check on the external world to identify whether something is hay or fresh meat or a plastic bone. But it does not seem that they raise further questions, e.g., in just what way does their experience of today correspond to that of yesterday, or in what does this correspondence consist, etc. They do not search out the specifically human significance of life or the intellective sphere of the logos of life's deployment.

Advancing spheres of sense accompany the complexity of structure that allows the animal to select its food and seek it, to re-cognize its enemies and attempt to avoid danger. These advances are introduced in a move of "hesitation," on which follows a more sophisticated mode of re-cognition and discrimination that allows for sensed and identifiable "doubt." In verifying the presence of the constant core of "reality," the vitally significant fear that already senses fitness or danger, what is to be expected instinctively, acquires a modality of psychological-intellective re-cognition and identification, even an intrinsic psychological prototype of doubt. Doubt calling for verification of the state of affairs is present at every step of a pursuit. Doubt and verification is the intrinsic mechanism at hand for the seeking and selection of new ways to satisfy existential needs. With the highest animals, however, such as domestic animals or apes, there is doubt in the given data of life and the world, i.e., uncertainty as to whether the incoming data is what is to be expected or not, whether all is as it seems or appears deceptive. Then with us human beings there is full-blown doubt as to whether others' declarations of feeling are authentic and whether promises, agreements will be kept or not. This is a constant existential concern of individuals, societies, communities, nations, etc.

There is, indeed, along the entire evolutionary ladder extended by the logos of life an incessant play of "recognition-identification," "discrimination," "doubt," "verification," "assessment of adequation," etc.

## III. CONCERN WITH "TRUTH" – THE BALANCING DEVICE IN THE LOGOIC DYNAMIC SYSTEM OF THE MANIFESTATION OF LIFE AND OF THE WORLD

The ontopoietic unfurling of life's subjacent workings in the innumerable projects of the logos' constructive impetus is by no means a smooth, unquestioned flow of generative, developmental, growth and decay sequences. On the contrary, each of the steps of all individual progress in the self-individualizing process is wrung out of challenging obstacles to be overcome by adapting to present conditions, preparing to meet new obstacles, and advancing according to outlined intentions.... In this turmoil is located an active basic foothold of trust in the constancy of life and all that it entails. In fact, at each and every step there is an ongoing, ceaseless sequence of estimation, selection, adjustment. At the higher spheres, namely that of the *intellective logos of the specifically human mind*, there is the absolutely unavoidable use of judgment and decisions made upon it.

In sum there is a crucial, constructive logoic device built within the development of each system in the advancing and interlocking, fusing, intertwining spheres of sense. At their frontiers along the individual routes of self-unfolding, within the course of tacit sensing an either instinctive or intellectually formulated questioning of and checking on the status of everything, a quest after truth is carried on. This questioning lies at the heart of our expectations, needs, wishes, tendencies, life situation, valuations, decision making, etc. The adequation of our expectations and the actual state of affairs we call, in general, "truth."

This reference to truth is tacit too. It is a built-in crucial device of the constructive logos for bringing together the matching moments of its advancing course, for bridging disruptive moments, for easing unbearable tensions between opposed tendencies, for adjusting the seemingly unadjustable, in short, to serve as the constant point of recourse for salvaging actions within the merciless turmoil of the stream in which the living being strives to fashion a consistent course. This is concern with "truth," that of the stream of life, which is ever ready to throw up a submerging rift.

In short, *the search for truth is the constructive device intrinsic to the logos' ontopoietic manifestation in life.*

## IV. THE SEARCH FOR "TRUTH" IN OUR SELF-REALIZATION

It belongs to our ontopoietic, specifically human self-individualization that through the entire conundrum of our existential pursuit we direct our innermost – and not always clearly conscious – attention to "being ourselves." Whatever act, thought, emotion, judgment we perform, it is "our" act; it is through our acts that we "enact" our life and unfold our innermost self. Indeed, we identify ourselves with our acts by assuming their existential validity and also by feeling ourselves affirmed in them. Otherwise we deny to them this innermost personal adherence as self-expressions by a judgmental assertion that we did them "only for convenience's sake," declaring that we really did not believe in such actions but did them for some other reason. Briefly, all our acts express a reference to truth, to ourselves, to our identity. Not all of them – as a matter of fact, extremely few – allow us to take a clear stand on this referential significance. We enact our existence with such a velocity, and amid such a conundrum of indispensable momentary decision making, that we are at a loss to answer "Where are we going?" and to know "What are we really achieving?" But we are always poised in a critical situation in our deliberations: "What do we really want?"; "What do we really feel?"; "What is our 'true opinion' about such and such a matter?"

This questioning, if pursued, extends over the entire realm of our psychic, intellective, volitional, imaginative experience, reaching the unfathomable depth of our yearnings and dreams.

It is obvious that there can be no question of pursuing the truth of *direct* reference to the relevant data of our multisphered unfurling; that eludes us, withdrawing even further away when we seek causes, reasons, motivations, influences, etc. Yet in order to go on with our life enactment we have to make – and we always are making – provisory estimates of all these on the assumed grounds of a given pursuit of ours; we also project provisional conclusions that we have to believe conform to our state of fundamental ontopoietic ingrownness in life and the world around and within us. Indeed, the conundrum of intertextual relations between and among the spheres of sense does not allow for any clear-cut evidential adequation leading to a basic existential experience of our ontopoietic status itself. And yet our lives are led over against that status, as is corroborated through the innumerable lines and segments of sense in our life enactment, in which we find a tacit confirmation of reality, since from the incipient moment of our becoming this basic trust becomes progressively incarnated in our growing innermost rationale.

In the swing of the human spirit we launch ourselves beyond the world's frontiers in attempting to transcend it. The question of truth which was always running *sotto voce* through our life enactment as the "truth of ourselves," here, within the transcending elan leaves direct or extended reference to the world of life and assumes a specific life-transcending turn about which we will speak elsewhere.

## V. THE ALL-PERVADING QUEST AFTER BALANCE/CONSTANCY AND ITS VORTEX

All of the spheres of rationality – vital, Dionysian, and Apollonian – partake in and are sustained by their reference to ontopoietic "reality," whatever this expression may stand for and however "far" away the originary experiential evidence of the reality of life, the world, our root existence may be. As many as are the significant moments of life accumulated, as manifest themselves in innumerable modalities of the logos of life that carries them, so are the referential points of reality sharing these modalities differentiated and so are the essential forms of this relevance.

In the dynamisms of the logoic constructivism of sense in the ever more complex schemata that carry the progress of life in its vital, societal-sharing, and institutional systems and through intellective, judgmental, and creative aesthetic elevations, significant moments emerge across the spheres themselves that acquire specific sense in fusing, molding, interacting, criss-crossing the spheres and even singular senses already established in significant schemata.

Without reference, however weak, connecting with evidence of "reality" emphasized by the life system of the experiencing, acting, dreaming, creating person, the entire logoic system of human rational existence/life would float in the thin air and be a phantasmagoric play. The consistency of the ontopoietic individualizing course calls for the constancy of life's circuits. The world of existence and the living individual have to remain "the same." Indeed the display of logoic rationalities of the Dionysian and the Apollonian turns – that is, of the entire human dimensions of life – relies on the incessant conscious or just mutely experienced identification, verification, and confirmation of references to the "real" as being basically crystallized and evidenced in the core of our existential self-experience of our human ontopoietic course. It is in this evidential core of our existential self-experience that lies the "truth value" of our constancy in the world and of the world itself. There it is that our ontopoietic relevancies to the system of life – and beyond that to the laws of the earth, our life-maintaining planet, and of all the cosmos – are maintained. This relevancy for all the logoic spheres – or rational orders – of turbulent life is comparable to an umbilical cord.

## VI. THE ONTOPOIETIC VORTEX OF TRUTH AS THE GUARANTEE FOR THE EXISTENTIAL CONSTANCY OF THE DYNAMIC PLAY OF ALL THE SPHERES OF THE LOGOS OF LIFE, BRINGING THEM TOGETHER

Beginning with Parmenides and Plato, concern with truth has meant doubt and query into the truthfulness of reality as such. Plato's division of reality into two registers, that of the "true" and that of the merely "appearing," gave us the epistemologico-metaphysical perspective on truth. With Aristotle's concept of *adaequatio*, i.e., the conformity of a true proposition with its object, the concept of the truth was brought to its highest intellective level, that of logic and its rules. Bacon, in contrast, conceives of truth in a pragmatic fashion, seeing its validity as being proved in the success of an operation/action. Tarski extended the truthfulness of a single proposition to its place within discourse, in which context it receives its confirmation.

All of these conceptions of the truth and all of the numerous others deriving from them held by contemporary thinkers (e.g., the conception of Quine, which holds that concepts in general emerge and develop following the practical interests of human life) express the different and yet intimately conjoined perspectives of the representational, intellective, interactive, contextually interworldly, and utilitarian accomplishments and interests so crucial for the enactment of life. That is to say, the ontopoietic self-individualizing dynamic stream of life maintains its balancing powers within the turmoil of soliciting and rejecting forces through the logoic device of truth seeking, which runs through the entire spread of the interactive mesh of the advancing living being with its circumambient conditions, thus crystallizing the life schemata.

In short, although the constant search for "truth" or adequation reaches its highest intellective modality in the specifically human sphere of the cognitive logos, without which no course of individual – and a fortiori societal – enactment could be carried out, specifically human, cognitive, intellection being the clearest and strongest instrument of individual life enactment – this constant search for truth sustains the entire dynamic/constructive spread of the logos of life in its various spheres, using all the varied modalities of each.

Essentially, in pursuing the origins of the notion of truth within the ontopoietic deployment of life, we see it as the intrinsic device of the logos of life for its constructive enterprise of life's unfurling. It is of universal constructive significance. In its constant search for adequation, it presides over the singular self-individualizing process of living beings. There it plays its essential universal role by working out – through the attunements of individual existential quests for the adequation of present, at hand conditions with the past as well as with the universal schema of life – an interactive, shared platform of constancy in which interactions are balanced with other living beings amid the disruptive pulls of life forces. This balancing effort projecting a relative constancy in life's dynamic progress has to be worked out continuously. It has to be pursued in all the spheres of the logos of life as well as in all their interrelations and in all their perspectives; it has to be ceaselessly on the move within its dynamic transformations, which involve, in principle, all these perspectives on life's enactment and their significance (intellective, pragmatic, aesthetic, creative, etc.). All of them are intergenerative in some or other way, to a greater or lesser degree.

As we may see, taking into due consideration this fundamental generative notion of truth as being immersed in all spheres of sense and being appropriately qualified by them, none of the partial perspectives may claim a preponderant validity or claim precedence over the others. Each of the above-mentioned conceptions of truth – and others – may hold a claim to only partial validity. And only together can they, since they express the three main concerns of human life involvement – the intellective/cognitive conception, the contextual/interrelational conception, and the pragmatic/directional conception – adequately respond to the essential life situations of the human being from whom the question and quest for "truth" proceeds. Each of them plays its specific role in life situations within the sphere of sense that is in question.

It is clear from our analyses that the elucidation of the question of "truth" within the entire field of the phenomenology of the ontopoiesis of life undercuts any hasty, tunnel-visioned temptation to relativize this notion by reducing it to one perspective on life enactment with disregard for the others.

Only the consideration of all conceptions of truth may do justice to the full significance of "truth" in its innumerable modalities of manifestation as they come together in the operation of the crucial logoic device balancing life.

*Anna-Teresa Tymieniecka*

# SECTION I

Eldon Wait at the buffet.

ELDON C. WAIT

# DO WE SEE THE THINGS THEMSELVES?

Although Merleau-Ponty did not leave us a discursive argument proving that in perception we "reach" the things themselves, he did leave many indications as to how such an argument could be made. Our objective is to take up these indications, in order to develop such an argument, and finally to corroborate the conclusions of the argument with evidence from contemporary research on the psychology of perception.

For example, on various occasions Merleau-Ponty indicated that Husserlian transcendental idealism, although not the ultimate vantage point, is a "route that has to be followed".[1] We will follow that route by applying the phenomenological reduction, in the form of a retort argument. Applying the reduction will enable us to show how the traditional sceptical arguments of the rationalists and empiricists appear to presuppose, without ever recognising it, that the subject who takes up the arguments, is not an isolated consciousness who only "represents" to himself an external world, but rather one which is in "direct" contact with the world and with its transcendence, such that this world and this transcendence are in themselves what they are for him. In other words, by applying the reduction we can show how the reflecting philosopher can come to recognise himself as the constituting consciousness of transcendental idealism. The flaw of this transcendental idealism proves to be its inability to recognise other subjects with whom I share the same world. The recognition of this flaw enables us to argue that the ultimate subjectivity is not the constituting consciousness of transcendental phenomenology, but rather intersubjectivity, or more precisely intercorporeality.[2]

The reduction is generally presented as a suspension of an uncritical belief shared by both rationalists and empiricists that the world and the perceptual processes of the perceiver are "outside of", or numerically distinct from, consciousness, such that the subject has no direct access to the real world, or in fact to his own perceptual processes, such that he needs to represent these to himself, in the form of ideas or images. From the scientific perspective, for example, given the structure of the eye, the retina, and the effects of light on the retina, it is inconceivable that we could perceive the transcendent thing. Light passing through the lens produces only an image on the retina, and any experience of being directly present to the object, can only be an illusion. The perception of the object *must* involve at some level an interpretation or processing of retinal information.



*A.-T. Tymieniecka (ed.), Analecta Husserliana LXXVI*, 3–18.

The reduction is a suspension of the belief in this numerical, absolute distinction between consciousness and the world and would embrace therefore a suspension of the traditional uncritical attitude toward the scientific perspective, and in particular toward to the way in which the scientific perspective is assumed to prevail over all other perspectives. What the reduction teaches us is that the question of whether in perception we reach the world itself cannot be answered through a simple deduction from metaphysical assumptions about the nature of the subject and the world, or from what is deemed possible or impossible, that it can only be answered through an appeal to experience and evidence corroborating that experience.

## THE REDUCTION AS RETORT ARGUMENT

Husserl argued that this traditional way of conceiving reality was absurd, and that by suspending our belief in such a reality we loose nothing.[3] It is absurd because if it were metaphysically impossible for me to have a direct encounter with the real or with my own perceptual acts in the world, how could I ever know what it is for something to be real. How could I ever have interpreted correctly or incorrectly the alleged signs of "transcendence". And without this knowledge the problem of solipsism would not even be intelligible to me.[4]

If it were true, as Descartes argues, that I am unable to distinguish the dreamt world from the real world, from where could I derive my idea of the real? Perhaps I've only had dreams of dreaming and dreams of being awake and perhaps my understanding of the distinction was acquired in a dream, in which case I would only know the difference between dreamt reality and dreamt dreams. If I am to understand Descartes' claim, I need to understand the distinction between real dreams and real reality. The dream argument places me in a vicious circle because by accepting its conclusion I am accepting that I could never know what it was that I had accepted.

If ever I can claim to know the difference between what it is for something to be part of the dreamt world and for something to be part of the real world, it can only be because I have made contact with the real world, and that this contact did not itself presuppose "representations" and their "interpretation", a contact in which I come face to face with the transcendent world itself, such that at that moment it does not transcend me, but is in itself what it is for me. At that moment, I can no longer envisage a metaphysical distinction between what is "for me" and what is "in itself", and for me, the real and this contract would have to be whatever they are for me.

Similarly, in empiricism, if it were metaphysically impossible for me to have a direct encounter with anything other than subjective entities like representations, how could I ever know what it is for something to be "outside" of my mind, as Berkeley puts it? How could I ever be able to interpret the alleged signs of an object's being outside?[5] Could I picture to myself an object's lying outside my mind as somebody else could witness it? If I were always confined to representations of reality, all I could represent to myself would be representations. Before this "picture" could teach me what it is for something to lie outside of my mind, I would have to think of it as representing a possible situation in the world, a situation which lies outside of my mind. In other words, I need to represent to myself an outside world, in order to exploit my powers of representing an "outside" to myself.

## THE ARGUMENT FROM ILLUSIONS

In general the empiricist arguments from illusions presuppose that we can infer from a sequence of experiences that we do not perceive the world itself, that in some sense the real world must transcend the perceiver. But for a being which could only represent to itself a transcendent reality, even the past would have to be represented and at any particular moment I could have no direct contact with the entire sequence itself, but only with representations of the sequence. I will be unable to deduce anything from a present representation of a sequence of experiences if I could not think of them as being in a temporal sequence, and hence if I did not know what it was for something to be *past*. But if I am forever confined to present representations, what could "being in the past" be for me? It would be as difficult to know what it was for some event to be "past" as it would be to know what it was for something to be real. How could I know, if I did not have a contact with the past itself, if I were incapable of transcending my realm of present representations such that I could have a direct, non-inferential contact with the past event itself, where it is, in the past. The argument from illusions, rather than proving that I am confined to representations about a real world, presupposes that I am open to the very pastness of the past. The past therefore could not transcend me, it would have to be whatever it is for me.

Similarly, how can I accept Descartes' dream argument if the argument undermines any certainty that I have that I am awake? If I am not sure that I am awake, how can I be sure that the argument is valid? Perhaps the fact that last night, while I was lying in bed dreaming, I was convinced that I was awake and standing in front of the fireplace, and the fact that right now I am equally convinced that I am standing in front of the fireplace, might not at all imply that I could be dreaming now. Perhaps the conclusion of the argument (I could be dreaming now), only appears to follow the premises because I am asleep. Perhaps the relationship between premises and conclusion has been dreamt, produced by the dream. The dream argument has the strange ability to undermine itself, since I need to know that I am awake before I can trust my evaluation of its validity. Descartes' dream argument is either absurd or it surreptitiously exploits a certainty of being awake and in the world *which does not rely on judgements*, or *inferences from signs*, since the certainty that I am awake is presupposed in every act of judgement or inference. It must therefore be conceivable that I could be more sure of being awake, as a state in the world, than I am of the validity of any act of judgement or of the reliability of any "representation". It is conceivable therefore that my actual being awake and in the real world could be numerically distinguishable from what my being awake is for me. Any numerical distinction would oblige me to infer from the fact that as far as I am concerned I am awake, that I am in actual fact awake, or awake as far as others are concerned. I avoid the vicious circle only by recognising that I am a subject which has a direct contact with his own being awake, a contact unimaginable within the confines of the natural attitude. I avoid the vicious circle by accepting that this contact is whatever it is for me.

I can reveal what this contact must be in itself only in reflecting on what it is for me. I can develop a philosophy of perception by bracketing the assumptions of the natural attitude, by bracketing everything that science teaches me about perception, by bracketing any preconceived ideas about what is possible and what is not, and by returning to that contact with the world through which the "real" of the scientific world presents itself as the "real", and the past as the past. Whether or not a specific act of perception reaches the world itself cannot be deduced from metaphysical assumptions, it can only be decided on the basis of evidence.

## ARGUMENT AGAINST THE RETORT ARGUMENT

However, even though *I* cannot assume that the real world lies beyond me or is more than what it is for me without undermining my own understanding of what it is for something to lie beyond me, isn't it possible, in principle, that I could *in fact* be confined to my own internal states. Isn't it possible, for example, for an external observer to assume that all my perceptions are confined to private data and that even though *I* may not be able to make this assumption without calling into doubt my understanding of what is being assumed, the external observer is free to do so? Even though *I* cannot accept an argument that I am asleep at this moment without undermining the confidence I have in my ability to distinguish valid from invalid arguments, she could recognise its validity and conclude that, in actual fact, I have no genuine certainty that I am awake. Even though I cannot assume that I am confined to representations without having to concede that I might not know the difference between immanence and transcendence and hence just what it is that I have assumed, an external observer can make these assumptions about me without placing herself in a vicious circle. Isn't it conceivable that everything the phenomenologist has revealed about perception "from the inside" is merely a description of what his perception has to be for him, that it tells us nothing about what perception could be in itself, or what it could be for an external observer?[6] The argument against the retort argument claims therefore that, purely on the basis of internal consistency, I am not able to deduce anything about my relation with the world, and that the fact that the traditional concept of reality is absurd for me does not imply that it is absurd in itself.

## THE PRESUPPOSITIONS OF THE ARGUMENT AGAINST THE RETORT ARGUMENT

It is clear that in her description of what is possible, the external observer would not merely be describing what is possible *as far as she was concerned.* Essential to her claim is that this situation is a possible situation even for me. If this were not implied in her argument, she would have to concede that there could be a truth about perception from within, and a truth from without, and that what we are able to assume about perception from within has nothing to do with perception as this is revealed from without. She would have to concede that philosophical research could go no further, being unable to choose between two incompatible and *incommensurable* accounts of perception. Clearly this is not something she would accept, since she assumes that the two points of view are actually concerned with the same act of perception, for she argues that the same act which *I* cannot assume to be reducible to private images can be assumed to be such from the outside. She does not accept that there is a truth about my perception from within and another from without, such that I am free to develop a philosophy of perception from within unhindered by what she has revealed is possible from without. She introduces her external perspective as an argument against my internal perspective. In some way therefore, the external observer needs to know that her point of view prevails over mine, not only for her . . . but even for me. She needs to know that *I* can be brought to recognise that what she has revealed pertains to everything about my perception that I have been led to through the retort argument.

How can she know that this truth is relevant to my own view of myself? Since from her external point of view she can find in me nothing but nervous tissue, muscles and bones, how can she make any claims about what is "for me", about what I can recognise?

Can she represent to herself what her revelations, or what her world, must be for me? Could she by exploiting an argument by analogy with herself, prove to herself that the possibility she has revealed could be recognised by me, or could be "for me", such that I would abandon developing a philosophy of perception purely from the inside, unhindered by what she has revealed? Could she imagine *herself* in my position being made aware of this possibility? If her only contact with consciousness is with her own, and from the inside, she will find herself in the same position in which I found myself through the retort argument. She would be unable to accept that the real world lay beyond her and that she could only represent it to herself, because she would undermine her own confidence that she understood the difference between the subjective and the objective, the relative and the universal, and consequently that her "argument" by analogy could establish anything about what can be for me. Consequently she could not grasp in thought my recognition that the world lay beyond me, or is more than what it is for me.

If she is not able to "represent" this to herself, to constitute it in thought, we are left with two alternatives. Either she is simply unable to know that I must recognise the possibility that she has revealed, in which case she would have to concede that what she has revealed is only true for her and is not relevant to what I have revealed about myself, or else she must be able to derive her assurance that her world is for me from a "direct contact" with her world's "being for me". If we are to assume that such a contact is in principle inconceivable, she could never claim that what she has revealed is in any way relevant to what I have revealed about my perception from the inside, and the philosophy of perception will be forever stymied, being unable to choose between a philosophy of perception from the inside and a philosophy of perception from the outside.

But such a direct "contact", which she needs to have with her world's being for me, is indeed paradoxical. What is clear is that she cannot infer from anything in my behaviour that her world is for me, simply because she is unable to constitute or represent to herself the idea that her world is for me. But neither can the being of her world for me be for her what it is for her, because without any contact with my consciousness, from the realm of what is "for her", she cannot infer anything about what is "for me". In this contact she would have to overcome the distinction between what is for her and what is for me, and we will have to give up the ontology of consciousnesses being parallel to each other, but distinct from each other, each confined to what is for it. From this contact it will have to be possible for her to glean an intelligibility which she could not have grasped in thought or represented to herself, but an intelligibility nevertheless which would assure her that her world is for me, and that even for myself I am not free to develop a philosophy of perception purely from the inside, unhindered by her world.

Is there anything in our experience which corresponds to such a contact? Let us suppose that I have taken my friend Paula to a spot from which I know one can get a spectacular view of the Pilatus and its cable car. Unfortunately, the mountain is still hidden under a thick mist, and Paula buries herself in her guidebook reading out loud information about the dimensions and colours of the mountain. As she reads, I try to conjure up in my mind an image of what the mountain would look like from here, and I have some impression that she is doing the same. Then suddenly the mist lifts, and I cry out, "Look at the cable car . . . over there!", while pointing in its direction. She looks at the mountain, and then suddenly fixing her gaze on the cable car, she says, "Oh! *There* it is", at the same time pointing with her hand and nodding her head. If I believed with the empiricists and realists, that the real world lay beyond all experience, and that the perceptual event was a private event, immanent in consciousness, I would have to describe the scene more or less as follows: Light waves from the cable car strike Paula's retina, producing impulses which are then transported to her brain, where the information is processed. Her consciousness would take place in the hidden recesses of her brain or her mind, and would be inaccessible to me. Paula could not actually see *the same* cable car I see, for "her" cable car would be reconstructed from information gathered from her retina. Clearly this is not the way in which I experience looking at the mountain with her. For me, she sees *numerically the same* mountain that I do. Neither of us is confined to private images. We share the same world, and this not merely in a manner of speaking, which means that the same mountain emerges as being "for us".[7]

To experience her as being open to the same mountain which I see, to experience this mountain being for her, is not to experience her as having a reliable image of it, or as having good reasons for *believing* it is there and that it is such and such. It is to experience her as *seeing* the same mountain that I see, *there where it is* in front of us, such that she will be *more* sure of the presence and nature of the mountain than she is about any images she may have or about any judgements she may have made. My experience of sharing the mountain involves therefore in some way an experience of her perceptual consciousness "opening up" or "reaching out" beyond her private realm of representations, beyond the realm of what is for her, in order to embrace that which is "for us".

If the argument against the retort argument is going to be possible, if philosophy is to avoid being confronted by incommensurable accounts of perception, one from within and one from without, we cannot assume that my awareness of Paula's perceptual grasping of the mountain, of the mountain being "for us", must be illusory. Whether in a specific instance my experience of Paula's consciousness of the mountain is illusory or veridical can only be decided on the basis of evidence.

Firstly, what empirical evidence is there that this experience of others seeing the world itself is veridical, that others are not necessarily confined to private representations of the world and their reconstructions of an external, but for them unreachable, world? In spite of the widespread acceptance of empiricism, there is no unambiguous evidence from research on perception and recognition to suggest that the intuition and interpretation of "representations" plays any role in normal perception.

To experience Paula as reaching in perception the things themselves is to experience her as perceiving the actual size and actual shape of the object irrespective of her perspective. Perceiving from a distance diminishes the size of the retinal image, not the size of the object, perceiving from an angle distorts the retinal image and not the object. It is now generally accepted in psychological research on perception and recognition that it is easier for subjects to identify shapes of objects in the world because of their actual resemblance to a known shape, than to identify objects because of the resemblance of the shapes of their retinal images. Research has shown that very young infants hardly ever confuse objects which project similar retinal images but which are in themselves very different. Infants never confuse for example, a large ball far away with a small ball close up, even when their retinal images are identical. They only confuse objects which are *in actual fact* similar. Infants therefore appear to see the things themselves and there is nothing to suggest they are obliged to infer the nature, the size and the shape of objects from retinal images.

Irwin Rock for example, has carried out a number of fascinating experiments measuring our ability to recognise patterns from different perspectives. If subjects are exposed to a series of patterns projected onto a screen, and then later exposed to a new series of patterns in which some of the original patterns are included, Rock has shown that subjects recognise patterns they have seen before more quickly if these are presented upright, i.e., in the same orientation in which they were originally presented.

In the second part of his experiment subjects were asked to look at the screen from an inverted position. They were required to stand with their backs to the screen and bend over so as to look at the screen through their legs. Looking from an inverted position would result in an inversion of the retinal image. From an inverted position the retinal image of an inverted pattern would be identical to the retinal image of an upright pattern seen from an upright position.[8] If we invert an inverted image, we obtain an image which is upright. If we believed that all perception involves the intuition of images, we would expect subjects looking from an inverted position to recognise the inverted presentations of the patterns more quickly than the upright presentations. This however did not happen. Subjects continued to find it easier to recognise a pattern presented in an upright orientation even when they themselves looked from an inverted position. Rock concludes that what plays a role in the recognition of patterns is the orientation of the pattern *in the world*, not the orientation of the image on the retina.[9]

It is of course always possible to imagine hidden mechanisms which would account for Rock's results, without undermining empiricism. For example we could argue that whenever we look from an inverted position the brain automatically inverts all images, perhaps in its attempt to compensate for the fact that we are looking from an inverted position. But what other evidence is there for such hidden automatic processes? Is there anything in my experience which resembles such an automatic turning of the visual field? When I look at a picture from an inverted position my experience is not in any way like the experience of seeing the picture from an upright position plus some kinaesthetic impressions of being inverted. If I tilt my head to the left, such that my forehead is now closer to the left wall of my study, I certainly do not have an impression of the world tilting over to the right, even though the image of the ceiling now falls on that part of the retina on which previously the image of the right wall, had fallen. I continue to experience an upright world, even from an inclined position, and from this position it is easier for me to describe its appearance as an upright world, than to predict what a photograph taken by a camera tilted to the left, tilted in the same way that my head is tilted, would be like. But neither is there any impression of an image of the world turning with me so as to take into account the tilting of my head, as I could turn a book to make it easier to read from an inclined position. There is only one irreducible impression of my head tilting to the left *within a static world*, an irreducible impression of looking at an upright world from an inclined position. Could we argue that this irreducible impression has been produced by a mind which takes into account the changing orientation of the retinal image, and the kinaesthetic sensations of inclining one's head? Could we argue that there is a hidden mechanism through which data is processed and a representation of my situation in the world offered to consciousness? This would be to assume that perceptual experience was no more than a representation of a situation in the world. As we have seen, what the reduction teaches is that we are not free to assume that there is a numerical distinction between what my perception is for me, and what it is as an unconscious process through which "representations" are constructed and offered to consciousness, unless there is evidence for such a distinction. Once we have *assumed* there to be such a distinction, we would have to concede that we could never be in contact with the real world, and could consequently never know what it was for something to be real. I would never have contact with the "outside" of my perceptual acts. How would I ever know what it was for something to be decor, to lie outside? The theory of a hidden mechanism inverting the image of the world is introduced simply to save the hypothesis that all perception is the intuition of images, that the perceptual process must in some way still begin with the retinal image. If we are free to speculate about the existence of hidden mechanisms for which there is no other evidence than the fact that it is the orientation of the object in the world and not the orientation of the retinal image which plays a role in the recognition of the object, we place ourselves in an unassailable position, and the only reason for choosing the empiricist account could be a commitment to a metaphysical assumption, that we have argued cannot be made. The argument against the retort argument presupposes that it is possible to encounter the world's being for someone else. It must therefore be conceivable that others are open to the world itself and not merely to their own representations, and it must be possible to identify evidence which would enable us to argue one way or the other.

Secondly, what evidence is there that we are in contact with the other's direct perception of the world? Butterworth and Cochran (1980) have shown that infants are acutely aware of what others are looking at, long before they are able to examine in a mirror the movements of their own eyes when looking in a certain direction. It is thus impossible for them to have learnt how to infer what others are looking at from an observation of their own eye movements. Once again, we are not free to speculate about hidden mechanisms developed through evolution which would enable the child to interpret "signs" of what we are looking at, without any prior learning, unless there is other evidence for the existence of such mechanisms. We could introduce such occult mechanisms only if it could be assumed that there can be no direct contact with the other's perception and that the experience of such a contact is necessarily illusory. We have shown above that these assumptions cannot be made.

How are we to make sense of this direct contact that I appear to have with the conscious lives of others? Merleau-Ponty has argued in favour of an embodied or "incarnate" consciousness, which means that for him there is no metaphysical demarcation between the point of view from the inside and the point of view from the outside. Every point of view from the inside is inseparable from a point of view from the outside and vice versa.[10]

As we have seen, if, while looking at a pattern with my head tilted to one side, I attempt to confine myself to *what* I experience, ignoring what I may know about the angle of my head, confining myself purely to my experience, to that which is for me, I find it impossible to isolate *what* I see from the angle at which I see. I find it impossible to isolate a description from the inside, from a description of myself perceiving from the outside. I find that there is simply one irreducible phenomenon of "seeing an upright pattern from an inclined position", such that I will be more sure of looking at an upright pattern from an inclined position than I am sure about what a photograph of the pattern taken from an inclined position would be like.

In answer to Descartes' dream argument, Merleau-Ponty would say that if I tried to describe my experience of waking up, bracketing the question of whether I am actually waking up or not, I would find it impossible to distinguish between the consciousness of waking up and the waking up of consciousness. For me to wake up, is to have an experience which takes me beyond experience, such that I will be more sure of waking up *as an event in the world*, than I am of having the experience of waking up. I will be more sure that I am actually awake, than I am of what being awake means to me. Consequently, although I have a direct contact with my own being awake as an event in the world, it is not whatever it is for me.

Similarly, empiricists have claimed that when I see a man at a distance, he appears smaller than when seen close by. But there is no experiential evidence corroborating this claim. I can make the man at a distance appear smaller only by closing one eye and holding up an object like my thumb, and then by attempting to see the man *as if* he were at the same distance from me as my thumb. In doing so, however, I have flattened out my visual field and created an artificial perspective. As soon as I open both eyes the man "moves back" from me, a distance appears between us, which is not something for me, since it is invisible, but which is not an idea of depth inferred from signs. It is rather something *between* us, something through which I perceive. The man experienced in natural perception is neither smaller nor larger than the man seen close by, he is the same man *seen from further away*. "The same man seen from further away" is an irreducible phenomenon, which means that my perception of the man cannot be distinguished from a complementary perception of myself, as seen from the outside, situated in the world perceiving the man from a certain distance.[11] The more carefully I devote myself exclusively to what there is for me, the more I discover myself in the world, as seen from the outside. Ultimately, I am more sure that I *see the same man at a distance*, than I am sure about what size he appears to have for me, and I am generally surprised to see how small he appears in photographs taken from this distance. Similarly, as I look at a circular plate from an angle, I am more sure that I "perceive a circular plate from an angle" than I am able to predict what shape the image of the plate would have if photographed from this angle.

According to Merleau-Ponty the genius of Cezanne lies in his ability, through the use of colour and brush stroke, to make visible this invisible with which we are in direct contact, namely the "seeing from a distance" or the "seeing from an angle". In his still lifes he does not paint an ellipse, but "a plate seen at an angle", and in his many renditions of Mont Saint Victoire he has made visible "seeing a large mountain from a great distance".

What Merleau-Ponty argues is that this irreducible phenomenon of "seeing from a point of view", or of perceiving as an incarnate consciousness, cannot be explained psychologically. It is not the effect of an unconscious reconstruction in my mind of my situation in the external world. We have shown elsewhere that infants are aware of their spatial relations with objects around them long before they have any conception of space.[1] The child knows, for example, that something is moving towards it without being able to carry out judgements or inferences, without being able to represent to itself its spatial situation in the world, or understand such a representation. It knows these relationships because they are the "outside" of all of its perceptual acts. There is no need to represent to itself its spatial situation, because its seeing is always already spatially situated. It is because every perceptual act is inseparable from this view from the outside, that others are able to make contact with our inside. This outside, which I discover every time I try to confine myself exclusively to what it for me, is numerically the same outside that others encounter as they become aware of me perceiving, are able to see what I am looking at, and are assured that they share the mountain with me, without inferences or judgments from visible signs.

As we have seen, the argument against the retort argument presupposes that it is at least in principle possible to have a direct contact with someone else's seeing. Consequently, we cannot assume that this account of the incarnate consciousness that is perceptible to others is false, and that this impression of such a contact can be explained psychologically. What evidence is there that the gaze I encounter, which is riveted to the cable car, *is* her "becoming conscious of the cable car", her reaching in perception the cable car itself? What evidence is there that from the inside and from the outside are inseparable, that this "seeing from a point of view" is irreducible?

Let us once again reflect on the results of Rock's experiments. As we have seen, what plays a role in the recognition of the patterns is not the orientation of the retinal images, but the orientation of the patterns in the world. For as long as we attempt to make sense of Rock's results by trying to account for the behaviour of his subjects from the inside, i.e., by trying to imagine what anyone would experience who looks from an inverted position, the results remain incomprehensible. If we imagine that the subject is confronted with an upright image, plus information from kinaesthetic and other sources that he is inverted, it is difficult to understand why the image is not immediately recognized. Since the inverted image seen from an inverted position would be identical to the upright image which was seen from an upright position, Rock's subject should at least have recognised it immediately, and only afterwards have concluded from his knowledge that he is inverted, that it must be an inversion of the one seen before. Rock's way of testing establishes conclusively that there was no such recognition. Subjects were not asked to recognise patterns with the same orientation, they were asked to identify the patterns themselves.

While the results of the experiments remain unintelligible to anyone trying to imagine the perceptual images confronted by the subject, they are intelligible to anyone who attempts to share the world with the subject, who looks at the patterns *with him*, as opposed to looking *at him*, or as opposed to trying to understand or constitute in thought the process through which the subject is able to recognise images from an inverted position. If I take up the position of a fellow perceiver, the patterns on the screen emerge as *being for us*, even though he looks from between his legs. From the perspective of a fellow perceiver, the subject's eyes are powers of making contact with things and not openings for light.[12] I am not surprised that it is the orientation of the patterns in the world which affect his recognition, and not the orientation of his retinal images, because from the position of a fellow perceiver, I am open to his "seeing of the patterns themselves". As in my experience with Paula, I have no impression of the other being confined to private images, whether upright or inverted. If we maintain the dichotomy of the natural attitude, and if we insist that his perception is hidden from me, and that it involves an inference from representations, we will have to find another way of making sense of Rock's results, and we will still have to explain how I could have these "misleading" impressions.

## THE ART OF PHILOSOPHY

It is in my contact with this gaze that I am given the assurance that philosophy does not have to choose between an account of perception from the inside and an account from the outside, that we are dealing with one indivisible act.[13] Nevertheless this gaze which reaches a world for us, is not something that I could perceive directly, represent to myself, or even grasp in thought. It cannot be perceived because it is not an object in the real world, for in the real world there is no place for consciousness or interiority.[14] But as we have seen, neither is this gaze reducible to one of my subjective impressions.

The ultimate reality in terms of which we have to understand perception can be neither the external world which we can only grasp in thought, nor the subjective world of private experience. It would have to be that world within which the gaze can have its place. The ultimate subject, the position from which the reflecting philosopher must account for perception, will itself have to be that of the incarnate subject who shares the world with others, a subject who makes direct contact with a reality and with a gaze neither of which he could constitute or grasp in thought, a subject who can do nothing other than "unveil" an intelligibility already there in the world and in the grip of the other's gaze.

*University of Zululand*

## NOTES

[1] "What will always make of the philosophy of reflection not only a temptation but a route that must be followed is that it is true in what it denies. ..." (Merleau-Ponty, 1968: 32) "This movement of reflection will always at first sight be convincing: in a sense it is imperative, it is the truth itself, and one does not see how philosophy could dispense with it. The question is whether it has brought philosophy to the harbor, whether the universe of thought to which it leads is really an order that suffices and puts an end to every question." (Merleau-Ponty 1968: 31)

[2] "The passage to intersubjectivity is contradictory only with regard to an insufficient reduction, Husserl was right to say. But a sufficient reduction leads beyond the alleged transcendental 'immanence', it leads to the absolute spirit understood as Weltlichkeit, to Geist as Ineinander of the spontaneities, itself founded on the aesthesiological Ineinander and on the sphere of life as sphere of Einfühlung and intercorporeity. ..." (Merleau-Ponty, 1968: 172)

[3] "We subtract just as little from the plenitude of the world's Being, from the totality of all realities, as we do from the plenary geometric Being of a square when we deny (what in this case indeed can plainly be taken for granted) that it is round. It is not that the real sensory world is "recast" or denied, but that an absurd interpretation of the same, which indeed contradicts its own mentally clarified meaning, is set aside. It springs from making the world absolute in a philosophical sense, which is wholly foreign to the way in which we naturally look out upon the world." (Husserl, 1931: 169)

[4] Merleau-Ponty argues that if we are obliged to rely only on signs of transcendence, and that if we are genuinely incapable of experiencing the transcendence of the real, as is assumed in Descartes' argument, then notions like solipsism and the solipsists' claim that there is no real world, or that we could never be sure that there is a real world, could have no meaning. "We are not so much thinking here of the age-old argument from dreams, delirium, or illusions, inviting us to consider whether what we see is not 'false'. For to do so the argument makes use of that faith in the world it seems to be unsettling. ... The argument therefore postulates the world in general, the true in itself: this is secretly invoked in order to disqualify our perceptions and cast them pell-mell back into our 'interior life' along with our dreams, in spite of all observable differences, for the sole reason that our dreams were, at the time as convincing as them – forgetting that the 'falsity' of dreams cannot be extended to perceptions since it appears only relative to perceptions and that if we are able to speak of falsity, we do have to have experiences of truth." (Merleau-Ponty 1968: 5)

[5] "Insomuch that a man born blind, and afterwards made to see, would not, at first sight, think the things he saw to be without his mind, or at any distance from him." (Berkeley 133)

[6] This argument resembles McIntyre's argument against Putnam's "brain in a vat" argument. All sentences in vat English are false. They actually refer to images produced electronically in the brain but they are taken to refer to a situation in the world. So if I am actually a brain in a vat, then I speak vat English and the sentence, "I am a brain in a vat" actually refers to images that I have of myself as being a brain in a vat. The sentence would be true if I really did have those images, if I experienced myself as a brain in a vat. But I don't. Being in vat English, the sentence cannot say anything about whether I am actually a brain in a vat or not. "It is clear ... that the falsity of 'I am a brain in a vat' derives from its status as a sentence in vat English, and its consequence reference to brain images and vat images. The sentence is false because the brain is not an image. Therefore, the falsity of 'I am a brain in a vat' does not entail that I am not a brain in a vat." (McIntyre, 1984: 61)

[7] "When I think of Paul, I do not think of a flow of private sensations indirectly related to mine through the medium of interposed signs, but of someone who has a living experience of the same world as mine. ..." (McIntyre, 1984:61)

[8] We leave aside the complicating fact that retinal images are always inverted relative to the object which produced them.

[9] "... changing a form's orientation from its normally upright environmental position (with retinal orientation remaining normal) makes for greater difficulty in recognition than changing its orientation on the retina (with environmental orientation remaining normal)." (Rock, 1957: 493) It is thus easier to recognise the shape of an object even when looking at that object from an inverted position, resulting in an inverted image, than it is to recognise the shape of an inverted object, even when looking from an inverted position, resulting in an upright, i.e., normal retinal image.

[10] "Inside and outside are inseparable. The world is wholly inside and I am wholly outside myself." (Merleau-Ponty, 1962: 407) "There is inside and outside turning about one another." (Merleau-Ponty, 1968: 264)

[11] "... He is anterior to equality and inequality, he is the same man seen from further away." (Merleau-Ponty, 1962: 261) "As soon as I see, it is necessary that the vision (as is so well indicated by the double meaning of the word) be doubled with a complementary vision or with another vision: myself seen from without, such as another would see me, installed in the midst of the visible, occupied in considering it from a certain spot." (Merleau-Ponty, 1968: 134)

[12] "My eye for me is a certain power of making contact with things, and not a screen on which they are projected. The relation of my eye to the object is not given to me in the form of a geometrical projection of the object in the eye, but as it were a hold taken by my eye upon the object, indistinct in marginal vision, but closer and more definite when I focus upon the object." (Merleau-Ponty, 1962: 279)

[13] "We do not have to choose between a philosophy that installs itself in the world itself or in the other and a philosophy which installs itself 'in us', between a philosophy that takes our experience 'from within' and a philosophy ... that would judge it from without. ..." (Merleau-Ponty, 1968: 160)

[14] "How significance and intentionality could come to dwell in molecular edifices or masses of cells is a thing which can never be made comprehensible, and here Cartesianism is right." (Merleau-Ponty, 1962: 351)

## WORKS CITED

Berkeley, G. 1960. *A New Theory of Vision*. London: J. M. Dent and Sons Ltd.

Butterworth, G. E. and Cochran, E. 1980. "Toward a mechanism of joint visual attention in human infancy," *International Journal of Behavioral Development* 3, pp. 253–262.

Merleau-Ponty, M. 1962. *Phenomenology of Perception*. Tr. C. Smith. London: Routledge and Kegan Simone Ltd.

Merleau-Ponty, M. 1964. *The Primacy of Perception*. Ed. J. M. Edie. Evanston: Northwestern UP.

Merleau-Ponty, M. 1964(b). *Signs*. Tr. R. C. McCleary. Evanston, III: Northwestern University Press.

Merleau-Ponty, M. 1973. *The Prose of the World*. Trans. J. O'Neill. London: Heinemann Educational Books.

Merleau-Ponty, M. 1968. *The Visible and the Invisible*. Trans. A. Lingis. Evanston, III: Northwestern University Press.

Rock, I. 1957. "The Effect of Retinal and Phenomenal Orientation on the Perception of Form", *The American Journal of Psychology*. LXX, 4: 493–511.

Wait, E. C. 1995. "A Phenomenological Rejection of the Empiricist Argument from Illusions", *The South African Journal of Philosophy* 14: 3 (May), pp. 83–89.

Wait, E. C. 1997. "Dissipating Illusions", *Human Studies* 20: 2 (April), pp. 221–242.

Wait, E. C. 1998. "A Phenomenological Counter to Berkeley's Water Experiment", *The South African Journal of Philosophy*, 17: 2 (May), pp. 104–111.

SITANSU RAY

# THE CULMINATION OF REALITY: MAN IN THE UNIVERSE

## I. INTRODUCTION: THE TAGORE-EINSTEIN DIALOGUES; OUTER REALITY AND THE INNER CONSCIOUSNESS OF MAN: CAUSALITY AND CHANCE; A GIST AND CRITIQUE

Rabindranath Tagore and Albert Einstein met each other several times. Two of their meetings, one in July and the other in August of 1930 in Germany, provide us with scholastic dialogues between them that are very relevant to the pursuit of reality.

The first of the said encounters was held on July 14, 1930, at Einstein's residence in Kaputh, a short distance from Berlin. The topic of their discussion was mainly the nature of reality and its relationship to man. It is clear from the start of their dialogue that neither Einstein nor Tagore believed in any kind of Divinity isolated from the world. But, regarding reality and truth, while Tagore conceived of them as reflections of human consciousness, Einstein conceived of them as being independent of humanity. Regarding beauty, they again thought almost alike, seeing beauty as a sense of value ascribed by man to truth.

The Tagorean contention is that the infinite personality of man comprehends and subsumes the universe, and that is why the truth of the universe is human truth. This is not just poetic imagination. As solid matter is composed of protons and electrons with gaps among them, likewise humanity is composed of individuals, and these have the interconnection of human relationship that gives solidarity to the human world. The central thought behind Tagore's literature, song, religion etc. is that the entire universe is a human universe.

Einstein reacted by saying that there must be two concepts about the nature of the universe:

One, the world seen as a unity dependent on humanity.

Two, the world seen, from the physicist's point of view, as a reality independent of its human aspect.

Tagore repudiated the second conception by asserting that there can be no other conception except the human conception, for the scientific view itself is that of a human scientist. Tagore admitted the impersonal nature of science, which is not confined by individual limitations; yet that impersonal height, he



*A.-T. Tymieniecka (ed.), Analecta Husserliana LXXVI*, 19–46.

avowed, is achieved through human wisdom. The individual gives way to what Tagore called the eternal man, the supreme man, the universal being, the universal mind, etc., not in any magical sense but in the spirit of the impersonal human world of truths. Here is Tagore's sense of religion too. We achieve some standard of reason through our emotions and activities, our mistakes and blunders, our accumulated experiences, all through our illumined consciousness. So, in the Tagorean realization the domains of science, reason, religion, truth and beauty are not distant, one from the other. Pointing to the example of beauty, Tagore said that if there were no longer any human beings, the Belvedere Apollo would no longer be beautiful. Truth and beauty are akin to each other in the sense that both are realized by virtue of man's harmony with the universe or cosmos.

Einstein agreed with this conception of beauty, but not with this conception of truth. The Pythagorean theorem in geometry posits something independent of the existence of man. Reality is independent of man. Truth is relative to this reality. The negation of independent reality engenders negation of the existence of truth.

Tagore argued that the scientific truth which is to be reached through the process of logic is but the human organ of thought. The Brahman or the absolute truth cannot be conceived by the individual mind or described by words but can be realized only by a complete merging into infinity. Brahman seems not to belong to science, which deals with appearances only, with what appears to be true to the human mind, with *maya* or illusion.

Einstein commented for the sake of Tagore's argument that the said illusion must not be that of the individual only but must be one of humanity as a whole. Hence, the dialogue became complicated. Tagore said that in science we eliminate personal limitations and reach comprehension of the truth of what is called the universal mind. Einstein posed the problem of whether truth is independent of our consciousness.

Tagore now stated the whole thing in another way by suggesting that truth lies in the rational harmony between the subjective and objective aspects of reality, both of them belonging to super-personal man.

Einstein said that even in our everyday life we ascribe a reality independent of man to the objects we use. We thus connect our sense experiences in a reasonable way. For instance, the table remains in the house even when nobody is in the house. Of the same instance Tagore said that the table as a solid object is an appearance, that what the human mind perceives as a table would not exist if that mind were naught. The ultimate physical reality of the table is nothing but "a multitude of separate revolving centres of electric forces"[2] belonging to the human mind. In the apprehension of truth there is an eternal conflict between the universal human mind and the same mind when confined in the individual. This perpetual process of reconciliation is being carried on in our ethics too. Any truth absolutely unrelated to humanity must be absolutely non-existent. This is Tagore's assertion.

Tagore clarified one aspect of the concept of the universal mind, saying that the sequence of things happens to it not in space, but only in time, like a sequence of notes in music. The concept of reality for such a mind is akin to musical reality, in which Pythagorean geometry can have no meaning. Tagore drew yet another interesting analogy. The so-called objective reality of paper is eaten up by the worm, but the truth and reality of literature is invaluable to the world of the human mind or the universal mind.

The Tagorean concept of religion centres on the realization of the universal human spirit in man's own individual being. This was the subject matter of Tagore's Hibbert Lectures of 1930, entitled *The Religion of Man.*

The next encounter between Tagore and Einstein, as documented, took place on August 19, 1930, in Berlin.[3] The subject matter of this encounter was a different facet of reality associated with causality and chance.

In the realm of infinitesimal atoms chance has its play. So the drama of existence cannot be absolutely predestined in character. This very mathematical conjecture attracted Tagore's attention. Tagore and a Dr. Mendel had discussed the matter. When Tagore raised the topic to Einstein, the contention of Einstein's reply was that the facts that make science tend towards this view do not say good-bye to causality.

Both Tagore and Einstein observed and realized that the idea of causality is not to be found in the elements. Some other force builds, with various contingencies, this organized universe. The order of the universe is to be understood on a higher plane. In the minute elements the order is not perceptible. The order is there where the larger elements combine and guide existence.

Tagore then found a duality in the depths of existence, a contradiction between free impulse and directive will, working upon the existence and evolving the orderly scheme of things. Modern physics, according to Einstein, would not say that these are contradictory. A cloud appears to be such from a distance, but seen closely it shows itself to be a disordered assemblage of drops of water. Tagore then referred to the trans-disciplinary parallelism in human psychology: "Our passions and desires are unruly, but our character subdues them into a harmonious whole".[4]

Now, the question is whether similar things occur or not in the physical world. Sometimes some elements seem to be rebellious, having individual impulse. But, there is a principle in the overall physical world which dominates them and puts them into orderly organization. While Einstein conversed on how it is that the elements can never be without statistical order, Tagore pronounced, without altogether contradicting the scientist, that the drama of existence is an ongoing harmony of chance and determinism, which makes existence eternally new and living. Tagore reaffirmed the point by citing the example of the psycho-ethical problem in human affairs – the problem of freedom and determinism. There is in human affairs an element of elasticity – some freedom within a small range – which is there for the expression of our personality.

Quite relevantly, Tagore instantly jumped to the parallel reference of the musical system of India, which is not so rigidly fixed as that of Western music. The composers give a certain definite outline, a system of melody and rhythmic arrangement, and, within a certain limit the player or singer, i.e., the performer, can improvise upon it. The performer must be one with the law of that particular melody, i.e., a *raga* or *ragini*. Then only he can give extempore, spontaneous and free expression of his musical feeling within the prescribed regulations. The regulations are akin to causality; and improvisations on the spur of the moment are akin to chance events.

The composer's genius is reflected in the creation of a foundation along with the age-old tradition of the infra-structure and super-structure of melodies. But the performer inserts his own skill and artistry in the creation of variations with tonal flourish and ornamentation.

Similarly, if we do not cut ourselves adrift from the central law of existence, we can enjoy a satisfactory span of freedom within the limits of all constraints. If in our conduct we can follow the law of goodness, we can exercise optimum liberty of self-expression. In our music too, especially in the Hindustani classical music, there is a duality of freedom and prescribed order.

Einstein was further informed by Tagore of our *Kirtan* style, in which not only the melody but the words also are free to a certain extent. The *Kirtan* singer is at liberty to add his own words by introducing parenthetical comments, extempore lyrico-tonal phrases (*akhar* in *kirtan* terminology), not in the original song.

Regarding time, rhythm, and meter, Tagore told Einstein, "In European music you have a comparative liberty about time, but not about melody. But in India we have freedom of melody with no freedom of time".[5] This means that we have to maintain the full rotation of *trital*, *ektal*, *chautal* or *dhamar* (or whatever rhythmic structure a particular composition may have) throughout the performance. Furthermore, we cannot deviate from the chosen tempo of a rhythm. The tempo and rhythm of Western music may rise and fall according to fluctuations in intensity of feeling.

Einstein further learnt from Tagore that some styles of Indian music are sung with syllables having no meaning, e.g., the *telena* or what is called *tarana* nowadays. The *alapa* is sung sometimes with meaningless syllables and sometimes with prolonged vowels (mainly *a*), with no words at all. The voice is used just like a musical instrument at that time. Only tonal beauty and not articulation matters then. Thus, Tagore explained that Indian vocal music can be free from any fixed semantic content so far as the *alapa* and *telena* are concerned.

The main difference between Indian and Western music is that while Indian music is basically melodic in nature, Western classical music is based on counterpoint and harmony. Einstein said, "It seems that your melody is much richer in structure than ours".[6] The fact is that the contrapuntal and harmonic structure of Western classical music is more or less predetermined and pre-composed. There is some sort of mathematical exactness in the art of tonal score, harmonic setting, and array. That is why there is no freedom in Western classical scores. Einstein said, "Sometimes the harmony swallows up the melody altogether".[7] Despite absence of harmony, Indian music is richer in melody. We enjoy melodic freedom in Indian classical music. But we cannot deviate from the mathematical exactness of our *tala* or rhythmic form. The main performer and the percussionist may create intra-beat sub-divisions, but the total number of beats must remain the same.

Einstein's Theory of Relativity introduced the concept of time as the fourth dimension. To non-scientists the fourth dimension serves as a metaphor for some intangible abstraction or some unforeseen chance-event leading us to a realm of novel experience, towards liberation from the convention of spatial measures. Tagore conceived of an existence where "time rings as it does in music" and "the future is merely a prolonged present".[8] Musical time leads us to the infinite.

To Tagore both science and the arts are expressions of our spiritual nature, which is above our biological requirements and possessed of an ultimate value. Regarding the difference of his outlook from that of Einstein, Tagore wrote afterwards, "I could readily see that Einstein believed my universe was limited by human conception, and he was convinced that there was some truth which was independent of human mind".[9] This sort of independence proceeds to "transcendental materialism"[10] which reaches the frontier of metaphysics, to which is attributed utter detachment from the entangling world of self. Einstein held fast to the extra-human aspect of truth, while in Tagore's poetic realization the realm of truth must be in human consciousness.

To come back to music again, though the art of music is surely based on strict acoustical science, both Tagore and Einstein faced the problem that it is very difficult to analyse the effect of Indian and Western music on our minds. The tonal and rhythmic structures along with all their components can be analysed, but "what deeply affects the hearer is beyond himself".[11] Einstein added, "The same uncertainty will always be there about everything fundamental in our experience, in our reaction to art, whether in Europe or in Asia. Even the red flower I see before me on your table may not be the same to you and me".[12]

Lastly Tagore concluded, "And yet there is always going on the process of reconciliation between them, the individual taste conforming to the universal standard".[13]

The casual dialogues on serious topics ended with hyper-scholastic notes relevant to every branch of science and arts. Some region will remain beyond our explanation. Yet human endeavour in all branches of creativity and knowledge has always proceeded from the individual to the universal and come back to the individual in newer forms. This process of reciprocity is the foundation of man's cultural realm.

So we see that the sidelights of the dialogues on reality including causality and chance, illuminate the ethico-psychological or rather psycho-ethical polarities of determinism and free impulse, and the Indian musician's analogous freedom of creativity within the range of a determined set of rules.

Very recently a few physicists like Professor Prigogine and cosmologists like Sir Roger Penrose and Manilal Bhaumik have come to realize in Tagorean terms that man's consciousness cannot be separated from the space-time and mass-energy entities of reality.

## II. *VISVAPARICHAY*: OBSERVATION OF THE COSMIC WORLD; TAGORE'S WORK ON SCIENCE

Though meant for popular reading (*Lokasiksha Granthamala*, Visva-Bharati), Tagore's *Visvaparichay*[14] reflects his scientific bent of personality and his acquaintance with the updated scientific work of the east and west.

Quite objectively *Visvaparichay* deals with a span ranging from the mystery of infinitesimal particles to the amazing vastness of stardom, including the solar system, and last but not least the wonder of the evolution of life, mind and human wisdom on the earth.

Tagore dedicated the book to no less a person than the celebrated scientist Professor Satyendranath Basu. Bibhutibhushan Sen, then professor of mathematics of Krishnanagar Government College, had checked Tagore's manuscript with necessary scrutiny and corrections.

We see in *Visvaparichay* that Tagore depicts the reality of the terrestrial world as objectively as possible without any kind of fictional fancifulness. Yet he firmly states that the omnicomprehensiveness of the Absolute Reality (*Bhuma*) may be possible only in human consciousness, not in the outward space-time entity of mass.

Life may be a property of matter, it may be a purely geological fact of the transitory biosphere of the earth. Individual human life may be too short to attain a full epistemological grasp of reality in all its span and depth. Yet, it cannot be denied that the human soul is the final outcome of the universe up to its present stage of evolution. Despite all risk of falsifiability of our mental work, our soul is of a novel kind, part and parcel of the whole creation. If the whole can be realized at all, it may be possible only from doing our soulful meditation[15] in the right way. Aryan sages have done so.

The source of life among inanimate things lies in a requisite supply of radiance, mainly from the sun. Our *Gayatri Mantra* admits that all the heavenly bodies and even the stream of our consciousness and intellect have been emerging from the omnipotent source of radiance.

So, through scientific, aesthetic and spiritual pursuits, man can realize the reality that man is not just man, he is in every moment man-in-the-Universe, not as a passive component, but as an active participant.

## III. TAGORE'S CONGREGATIONAL ADDRESSES: THE QUEST FOR ETHICO-SPIRITUAL REALITY

These lectures, more than one hundred and fifty in number, are never like priestly sermons, but reveal in lucid style the ethico-spiritual reality of the creation. Tagore never imposed uncritical credulity on his listeners, rather he made them realize the precious gems of love, beauty, truth and infinity – all at par with reality.

The very first lecture, "Uttisthata Jagrata" (Rise, Wake Up), is on awakening the human soul, on bringing out human spirit from the usual matter-of-fact chores and lifting it to the higher level of the reality of consciousness.

The lecture "Samsay" (Doubt) asserts that doubt is much better than insensitivity and inactivity, resulting from unjudicious belief. Pangs arising out of doubt evoke our mind and spirit. This is true in both an ordinary and academic sense. Doubtfulness is the starting point, the stepping stone of philosophy. So, it can lead us to the realization of reality. But an incapacity for getting stirred up, ignorance of ignorance, being inactively confined to just animality, passive submission to our loss of activation, absence of reaction are miserable hindrances to realization. Contrarily, the pains of doubt and the weary search for a ray of light in the midst of the darkness of ignorance are far better than passivity of mind. Apart from this, love is the precious wealth of the human heart. Love removes inactivity and inspires mind towards the greater goal of adjustment and union. By virtue of love, a human soul can merge into the all-pervading cosmic soul. We may be bonded in our day-to-day livelihood, but we are emancipated in the realm of love.

"Atmar Drishti" (Vision of the Soul), "Pap" (Sin), "Duhkha" (Sorrow), "Tyag" (Sacrifice), "Prem" (Love), "Birodher Samanjasya" (Harmony between Contradictions) and "Ki Chai" (What Do I Want) – all these lectures were consecutively delivered in the last week of Agrahayan in 1315. The purport of these lectures points to the fact that only objects of sense-perceptions are given to us, but the reality has to be visualized through our soul. In the long run self-identity does not remain secluded but becomes identified with all for the sake of the destined communion with all souls called *Bhuma*. Sin is caused from self-confinement, which can be broken by love. Joy, combined with pains, can lead to proper self-realization. Creation is never perfect. We are bound to face and accept some sorts of oddities and unwelcome phenomena. Both bliss and curse, gains and losses form the core of our life. To cope with them, human life develops the centripetal and centrifugal forces of our psyche. Balance or harmony is developed through proportionate acceptance and rejection. As our digestive system assimilates into our organism just the necessary vitamins from our food intake and rejects improper elements through the act of evacuation, so also our life must assimilate the essential values and eliminate the oddities.

Cultivation of disinterestedness as opposed to infatuation can lead our soul to liberation from bondage. Tagore refers to the *Gita* wherefrom we learn that union through work (*karmayoga*) is possible through disinterested performance of work. But, while the *Gita* insists on severe dutifulness, Tagore glorifies love. Where wisdom and reasoning fail, it is love which wins. Love can unite the opposites. Sacrifice (*tyag*) and gain become identical in the case of love.

Tagore's month-long series of congregational addresses delivered during Paus 2–29, 1315 covers a wide range of ethico-spiritual discourse. "Prarthana" (Prayer) refers to Maitreyi's ideal, who chose to accompany her husband Yajnavalkya to *Vanaprastha* (a resort in a forest colony for higher thinking) instead of enjoying material wealth at home. If the husband-wife relationship is confined to self-seeking material comfort and sensual pleasures alone, it will divert them away from the blissful Prakriti-Purusha relationship. It will be fatal for us, like cutting down a tree to collect its fruit and flowers. Real love like that of Maitreyi is adorned with *Sri* (beauty), *Hri* (modesty) and *Dhi* (intellect and wisdom).

The lectures "Dekha" (Visual Perception) and "Shona" (Auditory Perception), relating to two basic senses of ours are unique expositions of the immediate and in-depth functioning of our senses of vision and audition. (See Section VI of this paper.)

The lecture entitled "Manush" (Man), given in the context of and during the Paus-festival of Santiniketan, is more closely related to the phenomenology of man's life. Man has his own consciousness, his own independent ego, by means of which he thinks himself separated from the whole creation even while being a part and parcel of it. By virtue of his creative will he is continuously striving for his own assertion in creation. But, his fulfillment lies in setting harmony between differences, unity among variety.

Two bi-polar lectures, "Din" (Day) and "Ratri" (Night), deal with dialectics of light and darkness and their effects on human consciousness. Subsequently there comes the dialectics of our conscious state and sleeping state, of active mood and resignation or repose. Our embodied soul takes rest in the Supreme Soul, as it were, during sleep.

The lecture "Ichchha" (Will) is the exposition of a wonderful facet of the reality of human life. Will achieves relation not by achieving freedom, which makes one estranged from others, but by merging with the wills of others. In the lecture "Basana-Ichchha-Mangal" (Desire, Will, Goodness) psycho-ethical reality is realized. While desires are directed outwards, will tends towards the inner self, evoking our *Rajasik* (royal) traits. *Rajasik* traits indulge in luxury and lavishness. It is *Mangal* (goodness) only which can safeguard a man from the strains of desire and will. "Tintola", denoting the three levels of our mind, deals first with our unrestrained natural traits, secondly, with our balanced religious traits, and finally with the tranquil spiritual traits of those of us who can reach that level.

Truth resides in orderly, joyful and beauteous aspects of reality; and that is the theme of the lecture "Saundarya" (Beauty). The reality of the Absolute Brahman is not only truthful but also joyful in His manifestations as exposed in the *Vedic* phrase *Anandarupamasmritam yadbibhati* ("All the forms of creation are His joyful expositions").

The lecture "Pran" (Life) explains the identity of life and the Absolute Brahman, who is expressing Himself like music emerging from the sitar. The body of the sitar is just matter. The player gives it life by virtue of music. Similarly, Brahman pours life into flora and fauna. Human life is more precious since man can realize Him. It is true that man and only man sings and makes music. Though Satan comes to jeopardize every good thing, man survives with all his divine qualities. After all man is *Amriter Putra* (the son of the Deathless). We are the dwellers of this heavenly earth and we are not terminated by death. Man has no other way but to know Him, whose light is coming to us piercing the darkness of death.

## IV. THE LITERARY CONCEPT OF REALITY

In a long letter to Amiya Chakravarti,[17] Tagore pronounced that science is there for knowing the objective aspects of creation and, contrarily, literary reality or rather poetic reality is closely associated with human affection. Literary creativity and enjoyment belong to the subjective world of man, a newer kind of reality akin to truth and beauty. That means that literary reality may belong to pure imagination, which may not have factual or eventual validity. Facts may take their place in a newspaper report. But mere facts devoid of artistic value cannot belong to literary content. Either the value-essence of select facts, surcharged with emotional and imaginative ethos, or pure imagination acceptable to the *Sahridaya* (a person with a sensitive heart) or *Rasika* (one who enjoys sentiments) form the core reality of literature.

What is awkward and ugly may be turned into literary beauty. Examples may be drawn from both Eastern and Western literature.

Enjoyment does not necessarily mean enjoyment of cosy and happy episodes. Tragic pain, which is unwelcome in real life, forms moving scenes in the theatre. Tragedy occupies the spectator or the reader with more emotional involvement. Intense tragic pain illumines our spirit. In the Tagorean conception, *Ananda* (blissful joy) incorporates both joy and pain, while just pleasure is shallow in nature and quite far from real *Ananda*.

The very first essay of *Sahityer Pathe* is "Bastab" (The Real),[18] which highlights in many a way the Tagorean concept of literary reality. Actually, Tagore had faced superfluous criticism from some corners that he is extra-romantic and that his literature is devoid of reality. Thus, the essay "Bastab" is a focused encounter too, aiming at denial of those jealous charges and establishing the highest idea of reality in literature.

Reality as manifested in literature is directly related to *Rasa* or aesthetic enjoyment. (*Rasa* is not identical with the set nine-*rasa* theory of the Sanskrit rhetoricians.) *Rasa* rests for and responds only to *Rasika* and not to the mass. Mass popularity, philanthropy, social reform cannot be the aim of literature.

Kalidasa did not write for the peasant community of Ujjayini. The reality of literature tends to be eternal and universal. It cannot be confined to contemporary national traits. The seat of Saraswati is on the lotus, a metaphor of timeless beauty. She cannot be seated on the subject-matter of contemporary and popular taste, confined to the naked sex-instinct, on the one hand, and the arrogance of poverty on the other. The ready-made curry power[19] of a perverted sense of reality exploits these two things in order to achieve cheap popularity.

## V. "AMI": MAN IN THE UNIVERSE

The poem "Ami"[20] (I) was written by Tagore on May 29, 1936 and was included in his poetical work *Shyamali*. *Shyamali* is the feminine form of *Shyamal*, which means green. Tagore used the word *Shyamali* as a proper noun to name not only this poetical work but also the mud house he lived in from time to time in his old age. "Shyamali"[21] is also the title of the concluding poem of the collection. The poem was composed on August 6, 1936 and is most affectionately associated with the earthen house called *Shyamali*.

"Ami" is the Bengali synonym of the Sanskrit *Aham*, the Hindi *Mai*, the German *Ich* and the English "I". Any corresponding synonym in any language denoting the first person singular subject form of the personal pronoun may be taken. Transcending its usual day to day usage that very *Ami* is the principal instrument of epistemology, the main laboratory of existentialism, the boundless span of phenomenology, the only agent of scientific objectivity and obviously the sole repository of aesthetic reality. The poem "Ami" is one of the best poetical expositions of the reality of the human mind. Conversely, the reality of the human mind finds its best integrated yet concise political manifestation in the poem "Ami".

Quite metaphorically, viz., using the rhetorical expression of transferred epithet, Tagore affirms his sense of reality that an emerald itself is not green, a ruby itself cannot be red, but, it is the encolouration of man's consciousness which tinges a piece of emerald as green, a piece of ruby as red. It is not the sun alone, but it is man's visual perception, which illumines the sky. The rose is beautiful only because man perceives it as beautiful.

If the whole idea be accused of being a stern philosophical doctrine and quite unbecoming of poetry, Tagore's answer is that this is the essence of reality and that is why it is poetry. Philosophy and poetry are not opposed to each other. They are complementary and very often identical. Tagore is proud of realizing this reality. But his pride should not be taken as self-centred fancifulness. He is proud on behalf of each and every man, or rather, on behalf of the whole of mankind.

Man is not a passive component of creation. He himself is the nominative of all sorts of creativity including the hues of precious stones, the lustre of the sky, and the beauty of the rose. That is why the poem records his self-confident assertion:

> Amari chetanar range panna holo sabuj,
> Chuni uthlo ranga hoye.
> Ami chokh mellum akashe,
> Jwale uthlo alo
> Pube paschime
> Golaper dike cheye ballum "sundar",
> Sundar holo se.[22]

This very beginning of the free-verse-style poem declares his observations on the reality of man's positive self-esteem.

In keeping with nihilistic fashion the ontologist may say that there is no emerald, no ruby, no light, no rose, and even no "I" and no "you". On the other hand, the Infinite Omnipresent has always been endeavouring to know himself through the finite reality of man's self. That very reality of man's entity of apparent finitude is called "Ami", the replica of the Infinite and Absolute Brahman as it were. In the depth of *Ami*, i.e. my self, light and shade get balanced, forms evolve enlivened with feelings. All the so-called theoretical nullifications turn into vital affirmations by virtue of an illusive magic, as it were, by virtue of lines and colours combined with delight and pain.

Man has the brush and colour in his artistic hands. He composes, draws and paints the world anew. Moreover, he has, combined with his sense of beauty, the most precious wealth of his or her heart, and that is love.

Suppose a total annihilation takes place by means of a universal cataclysm, natural or man-made. Suppose there is no trace of human civilization or human being any more. Matter and energy will remain in some form or other in the solar system; but there will be no feeling, no sense of beauty, no love. There will be nobody to say, "You are beautiful, I love you". If that stage should really come, the Creator will have to sit again for one more age-long *Sadhana* for the renewal of man, because it is only man who will utter "You are beautiful, I love you". The illuminating, dynamic and loving nature of man's self makes him greater even than his Creator, the Master of all matter, cosmic energy, and life.

There is yet another of Tagore's poems that bears the title "Ami",[20] one composed by him on February 11, 1931. This poem is included in the poetical work entitled *Parishesh* (The End). The poem reflects man's self-interpretation in existence. The gist is as follows.

The inner "I" within my self is the vital source of all my words, movements, music, and all manner of creativity. The firsthand understanding of my own self may suggest that I am the be-all and end-all of my own self. But in reality that "I" extends beyond my self. That is why I know myself better when I recollect the memory of my beloved, when I identify myself with the heroes of history and the mythology I read. Very often that inner "I" is over-shadowed by my constant daily chores. But, in reality the realm of my inner self transcends my matter-of-fact living. My inner "I" is omnipresent in the past, present and future. "I" can move everywhere at any moment. My apparent "I" perceives that inner "I".

There is still one more poem with the same title "Ami".[24] It is in the addenda to the collection of Tagore's poems titled *Shesh Saptak* (i.e., The Concluding Octave), published in 1935. The central idea of the poem is the delight of the poet's realization of his own existence. "Achhi achhi ei ye ami achhi". This means: how fine it is that I am present, whether in the midst of my natural and social surroundings or in the midst of the vast extra-terrestrial world.

In various prose writings also, this basic philosophy of the human self is the key tone sounded. One such is the essay "Amar Jagat"[25] (My World) in the book *Sanchay* (Accumulated Wealth). Here Tagore wants to establish the truth that the world is not merely the play of radioactivity, the world is after all mine. This is the basic claim of what may be designated as the essential core of reality, emanating as my consciousness. This emanating reality is the generating ground of science, philosophy, literature, music, the arts and all the realms of intellectual and aesthetic creativity.

## VI. "BIPUL VISVA-GANER BANYA" (THE DELUGE OF MAGNANIMOUS MUSIC OF THE UNIVERSE)

The uniquely metaphorical phrase "bipul visva-ganer banya"[26] occurs in Tagore's congregational address "Shona" (Auditory Perception).[27] The deluge of magnanimous orchestration of the universe is so much a vital metaphor that he can affirm in full faith: "This is neither a poetic utterance nor a rhetorical phrase, throughout space and time a continuous orchestration is being reverberated in grand fullness"[28] (translated). Tagore's enjoyment of life and creation as he expressed it in musical and aesthetic terms throughout his creativity is the subject matter of this discourse. The orchestration of the universe is reflected in the orchestration of the arts, which corresponds to the synaesthetic interrelationship among or within sound, image, colour, fragrance, movement, gesture, rhythm, word, passion, love and various other facets and attributes of creations as comprehended in human life.

As obviously seen, by virtue of transferred epithet or rather transcended epithet, the expansive sense of orchestration here transcends the factual orchestral performance of a conservatory or a symphonic group.

It would not be out of place to mention that "Shona" is preceded by its significant and complementary counterpart "Dekha"[29] (Visual Perception), the preceding congregational address by Tagore. In "Dekha", Tagore's wonder in perceiving the overspreading light and ever-evolving forms is revealed. "Ruper jharna"[30] (the Spring of Beauty) arising from and ultimately descending into "ananta rupasagar"[31] (the infinite ocean of beauty) is the theme of this address.

But the realm of beauty is not mute. It is sonorous. In emotional ecstasy Tagore records his realization:

> As this devastating deluge of magnanimous music of the universe rushes towards our soul, our single sense-organ is insufficient to receive it; we have to open up all our senses. We receive this musical flow through our eyes, ears, sense of touch and by all means of our body and spirit. We, as it were, see this grand harmonious concert, listen to it, touch it, smell it and even taste it.[32] (Translated)

Tagore mentions his own song "Baje baje ramya vina"[33] at the outset of his congregational address "Shona". In this three-stanza song, the very first stanza tells of the delightful *vina*-music throughout the natural surroundings around us, such as in the beautiful lotus, in the moonlit night, in the dark cloudy sky, in the fragrance of flowers. The second stanza tells of the orchestral aspects of the sun and the stars, the rivers and the oceans, the rhythmic emergence and extinction of life, the rhythmic flow of aeons. The third stanza depicts the delightful embellishments of the blue sky, the beauty of dawn and dusk, of the colourful earthly dust, of the devoted souls of the rich and the poor. Orchestration along with dancing and the embellishment of everything in the universe is surcharged with passionate love (*prem*) as it were (*preme preme baje/nache/saje*).

All the branches of Tagore's creative writings including his songs and dramas abound in musical terminology and orchestral metaphors. These metaphors are vibrating, as it were, with tonal and rhythmic components.

Towards the aim of a holistic realization, let me divide the theme into the following sub-topics and cite only a few select examples accordingly.

### *The Significance of* Omkara: *Orchestration on the Macro Level of Space-Time*

Tagore, well versed in chanting the Vedic hymns, realizes the significant *O . . . m* on a steady and prolonged tone before starting and after completing a hymn. This *Omkara*, the sonic contemplation of the ever-evolving Universe, is the music of the boundless infinitude, as Tagore says.[34] *Om* synchronizes *Rik* and *Sama*, word and tune, truth and life. *Om* is the tonal affirmation of vibrating life within the mass and energy of the universe.

Tagore values *Om* for its musico-spiritual effect rather than for any mystery or mysticism within it. He says that when a piece of music ends, it does not create a nihilistic darkness of silence; rather it is merged into and unified with *Om*, the vibrating spirit of the space-time entity of the Universe.[35] *Om*, right down from the aeon of the Aryan sages, is the tonal metaphor of the whole gamut of cosmic feeling. In a specific portion of the book *Chhanda*[36] (Rhythm), Tagore writes that the tonal effect of the hymns rather than their semantics leads our spirit to the infinite *Brahman*.

Let us take into account the observation of other writers. *The Pelican History of Music* records: "Thus it is that the *Ragas*, like the *mantras* or sacred formulas, are regarded as aspects of and hence as approaches to *Sabdabrahman*, that is, the Absolute (Brahman) conceived as sound".[37] *The Oxford History of Music* says about *Om*:

"In the advanced culture of India the syllable *Om* (arrow) (which is pitched very high in the udgita song) is the nail which pierces the whole world and holds it together".[38] The continuous tonic drone of the *tanpura* or *tambura* is absolutely akin to *Om*. The celebrated aesthetician Ananda Kentish Coomaraswamy says:

> We have here the sound of the tambura which is heard before the song, during the song, and continues after it. That is the timeless Absolute, which as it was in the beginning, is now and ever shall be. On the other hand there is the song itself which is the variety of Nature, emerging from its source and returning at the close of its cycle.[39]

### *Man in the Universe*

Physical grasp of time-space vastness including innumerable luminaries is impossible. The advance of modern space-technology has led man only up to the moon, the nearest heavenly body of ours. Man is just a tiny creature within the vastness of time-space. Human knowledge and civilization, which have made man much greater than other creatures, are quite negligible and transitory in the unbounded arena of time-space.

Yet it may be taken for granted that human consciousness is the only vital sense organ of the universe. Otherwise, the universe is nothing more than an inanimate play of mass and energy following the unconscious physical law of radioactivity and quantum mechanics. So, apparently negligible man is great indeed, not as a discrete being, but as a unique phenomenon in the course of the evolution of the universe. The universe can know itself through man's self only. Herein lies Tagore's winning point in the Tagore-Einstein dialogues.[40] Man is unique in the sense that he can at least conceive of the wonder of the creation; he can appreciate and create beauty; mankind is the vital repository of wisdom and love.

Now, musical tone is the sonic paradigm of the universe including man's realization of love and wonder. All other arts and crafts supplement it in advancing the wholeness of man's creativity. Hence, orchestration of the arts brings forth the macro level of time-space eternity into man's realization. The temple tops suggest the unending height of spatial dimensions; sculptures create a tangible model of the cosmos; pictures and paintings represent visual charms as comprehended by the artist; literature and the theatre create a parallel illusion of man's involvement in his world. Music, being the etherial wealth of man's tonal creativity, is far above earthly representation. Music is the *Nada-Brahman*, the infinite Brahman conceived in a tonal entity. Seen from the opposite perspective, Tagore declares that through music, i.e., our music, creation expresses itself. "Prakasha-piyasi dharitri bane bane/sudhaye phirilo sur khunje pabe kabe".[41] (At the very first dawn of creation, music was there for the expression of the creation.)

### *The Colourful Sky Corresponding to the Imagery of Tuneful Orchestration*

Tagore often reacts to the rising and setting sun, and his reaction is musical. Besides his numerous songs of dawn and dusk he describes his musical reactions in various portions of his works. Let us take the very first poem of the poetical work *Patraput*. The venue is the mountain-top of Darjeeling. The time is that of sunset. All the jubilant excursionists are silently enjoying the colour of the western sky along with the terrain below. The *esraj* with them for their pastime music is laid silent on the ground. All of a sudden the poet turns back and as by chance it happens to be a full moon day, the rising full moon on the eastern horizon comes to his sight with all amazement. The unforeseen commingling of the golden west and the silvery east is depicted as a marvellous musical imagery in the concluding portion of the poem.

The artiste plays *alapa* (melodic improvisation) on his *vina* every day. On that fine day, when there is no bustle in the whole environment, a chance-phenomenon occurs – that is the harmonizing orchestration of the tunes of the gold string and the silver string.

In the next moment the sublime orchestra descends into ultimate silence. The player's instrument bursts out into melting ecstasy, as it were.

Being the unexpected human witness of this phenomenon, I exclaim: "Wonderful!"[42]

(Approximate translation)

Examples are many. To be selective enough, let me now mention the colourful operatic play of the sky and the sea as described by Tagore in the *Japanyatri* (The Voyage to Japan, a travelogue or travel-diary). He enjoys the abundance of the moving forms and colours of the clouds of the sky as well as those of the waves of the rolling sea especially during the sunrise and the sunset. The whole experience is imaged as a dynamic music and dance show.[43]

While in Nator as the Maharaja's guest, Tagore takes an afternoon stroll along a road in open terrain and afterwards writes of his experience and imagination in an intimate letter to his niece Indira Debi:

Over the utter silence all around, what a deep, calm and heartening tune resounds everywhere from the earth to the realm of the stars! This is exactly what is actually happening. The cosmic vibration which stimulates our eyes is light and that which strikes our ear is sound. If we will try with concentrated mind, we can translate the enormous harmony of all the light and colour of the universe into a grand symphonic orchestra.[44]

(Approximate translation)

This is the core reality behind what is called synaesthesia, i.e., the interrelation of the senses, in modern psychological researches.

### *Orchestration on the Micro Level in the Human Body-Mind*

The mystery behind profuse use of musical terms in the cases of psychic actions and reactions lies in the otherwise inexplicable relational aspects behind transferred epithets. I mean that musical epithets are transferred in the description of body-mind phenomena. Far from being arbitrary, these uses are quite natural to a sensitive mind. We cannot but associate our emotional faculties with the serene joy of enjoying music. Let me furnish examples.

The hero of the short story *Aparichita* (The Stranger Woman) identifies the appearance and voice of Kalyani with music. Her gentle utterance "garita jayga ache"[45] (There is room in this compartment of the train) resonates continuously in the hero's mind like the refrain of a song, and the rhythmic sound of the running train is felt by him like the accompaniment of *mridanga* (percussion), made of iron as it were. Next morning, in a junction station, he sees her and feels that here is music having taken the visionary form of her charm.

Now, how can music be a paradigm of the speech and beauty of Kalyani? What is the common factor between the sweet passion for a woman and that for a musical tune? Therein lies the mystery of the intricate phenomena of human life, mind and consciousness – all seated in the microcosm of the cells of the human body.

The imagination of tactile sensation creates a unique imagery of orchestration. The following is an approximate translation of just a few lines of the poem "Manas Sundari" (i.e., The Handsome Lady-Love of Imagination) from the poetical work *Sonar Tari* (The Golden Boat).

O my beloved, I listen to the inner mystery of yours while in close embrace with you. Like fingering, your heartbeats will strike the strings of my heart and raise the waves of sound of music.... I myself shall vibrate with the tuneful ripples.[46]

Microcosmic orchestration spreads throughout the macrocosmic span in the following lines from "Jyotsna Ratre" (In the Moonlit Night) a selection from the poetic work *Chitra*:

Shiver my person with the memory of your embrace. Let my nerves and veins resonate with the orchestration of the infinite. Let my heart burst into ecstatic joy and spread over space like the flowing of a tune.[47]

(Approximate translation)

Tactile, visionary and auditory imageries are juxtaposed or rather fused together. Very often our organic body is compared to a *vina*. Besides the main strings there are so many strings for resonance and overtones in a *vina*. Similarly our body consists of a very complex yet sensitive nervous system. Our sensations are epithetically transferred to orchestral tunes. In parallel manner our heartbeats and pulsations are juxtaposed with orchestral rhythms. Affectionate or passionate tactile sensation generates musical sensation. In his love song "Amare karo tomara vina", Tagore says: "Do make me your *vina*, take hold of me and let my strings resonate with your lovely fingers"[48] (translation).

In the poem "Dhwani" (Sound), from the poetical work *Akash Pradip* (Sky-Lamp), Tagore recollects his childhood sensitivity to what is called the soundscape: "I was born with my psychic strings acutely tuned. Sounds produced all around had their vibrations and resonances encircling my delicate nervous system"[49] (approximate translation).

In two letters written to his niece Indira Debi, Tagore records his inner musical response and reaction:

I am like a vital piano with a number of strings with mechanistic intricacies inside its darkness. I don't know who plays it and when. It is very difficult to understand fully why it is played. I understand only what is being played – whether the tunes are of pleasure or pain, whether the tones are sharp or flat, whether the pieces are rhythmic or non-rhythmic. I know further the lower and the upper ranges of my octaves. No, do I even know that properly? I am furthermore puzzled as to whether I am a sympathetic grand piano or a cottage piano[50] (translation).

I was tuning a new song of mine. The tune is not very new. It is like a *Kirtan* style *Bhairabi*. Yet, while singing with rhythm, the whole blood circulation of my body is engrossed in music as it were. The whole of my body and mind is being vibrated like a musical instrument. The waves of that music spread out from my body and mind towards the whole outer world. Thus an orchestral harmony is taking place between my soul and the Universe. When a *vina* is being played, its strings look blurred; likewise through a musical tune the whole world seems misty and vibrating as it were[51] (translation).

Through the above examples we find that the microcosm and the macrocosm were woven together by virtue of human consciousness, perpetually creating harmonized orchestration, as it were, in every significant phenomenon.

### *The Imagery of Nataraja in Tagore's Songs and Verses*

Nataraja is the mythical king-God of music and dance. The universe is conceived as a grand theatre hall and Nataraja, its conductor throughout ageless infinity. On this theme Tagore composes his famous song "Nrityera tale tale, Nataraj, ghuchao, ghuchao ghuchao sakala bandha he"[52] (Remove All My Fetters through the Rhythms of the Unbounded Span of Tunes of Your Dance Music). The gist of the remaining portion is as follows:

The cosmic ripples are created by your dance steps. Saraswati's (Goddess of dancing and music) shoreless lake is full of those ripples. The lotus of Saraswati's lake is emitting divine fragrance.

Tomar nritya amita bitta bharuk chitta mama. (Let the immeasurable wealth of your dance fill the span of my psyche – this prayer is the refrain of the song coming at the end of each stanza.)

Your liberated embodiment is revealed in your dance, all of your illusions are reflected in your cosmic dance. The whole universe as well as the infinitesimal atoms are under the spell of your dance-movements.

In the swings of your tuneful rhythmic dance, bondage and freedom become balanced through ages and aeons. I am amazed at your infinity.

Rebellious atoms get beautifully regulated under the spell of your dance. The sun, the moon and other luminaries are orchestrated along with the luminous anklets, jingling around your feet.

Cosmic consciousness is due to the life-force of your dance. Your ecstatic joy brings forth cycles of pain and pleasure into a balance through infinite spans of time by virtue of tune and rhythm.

Through your evolutionary dance, I (the human being) have come into the realm of your orchestral whirl. O the hermit, O the beautiful, O the pacifier, O the awful! With your ageless orchestration and dance you carry on the rotation of life and death. I have come down to thee. Let the immeasurable wealth of your dance be poured into my psyche[53] (a sketchy translation).

Thus we find that Tagore's concept of Nataraja is quite far from that of conventional Shiva. Nataraja is the cosmic embodiment from the core sources of astronomy, cosmology, physical science, religion, music and dance and man's psychic realm too.

There are various other songs too depicting the imagery of the other aspects of Nataraja such as "Pralay nachan nachle yakhan"[34] (When You Perform the Dance of Annihilation, Your Matted Lock is Loosened Causing the Release of the Musical Stream of *Jahnabi*), "Kaler mandira ye sadai baje"[55] (The Cymbals of the Aeon Are Being Played For Ever).

Tagore's play *Nataraja – Riturangashala*[56] (Nataraja and the Pageant of Seasons) depicts the colourful ethos of the cycle of six seasons. In its preface Tagore declares that man's liberation lies in realizing that the evolution of creation in the outer space and the evocation of *rasa* in man's inner firmament – are both caused by the dance of Nataraja.

### *The Imagery of the Bucolic Flute Player God*

Tagore elevates the bucolic flute player god's imagery from the conventional Krishna cult mythology towards the vast astronomical expanses. The boundless space is visualized as the pastoral ground of Vrindavana. The sun and all the stars and planets are as it were the cattle. All of them are grazing since Rakhal, the bucolic God, is playing his flute somewhere above the space-time vastness. This is the metaphorical theme of Tagore's song "Ei to tomar alokdhenu"[57] (These Are Your Luminous Cattle). In the last portion of the song the whole metaphor is transferred to the human mental ground, wherein our hopes and desires are grazing astray. The God of the universe is also "mor jibaner Rakhal"[58] (the mentor of my life). It is prayed that with His flute music He will tend to our desires and longings in His Shelter at our day-ending.

Krishna has another identity too. With the tune of His flute He attracts His lover Radha and Radha's *sakhis* (confidante damsels). In Tagorean aesthetics

Krishna represents the eternal masculine, Radha the eternal feminine, and the tune of Krishna's flute becomes the symbol of intense love-attraction between man and woman. The love song "Sakhi, oi bujhi banshi baje; banamajhe ki manomajhe"[59] (O Sakhi: Listen to the flute music, I wonder whether it sounds in the grove or in my mind) as if in the voice of Radha signifies the romantic pangs of love associated with flute music. Another such a song may be designated as "Call of the Flute", that being "Mari lo mari; amay banshite dakechha ke"[60] (I Feel Helpless, I Am Called by the Flute).

*Unfolding the Metaphors*

Nataraja and Lord Krishna may exclusively belong to Hindu culture alone, but the idea of the music of the spheres is very much common to both Indian and Western cultures.

Just two decades ago newer research was undertaken and carried on under the auspices of UNESCO in the field of soundscape. All the sonic properties of the universe were examined with acute seismographic minuteness and lastly it was inferred that: " … the universe is held together by the harmonies of some precise acoustic design, serene and mathematical".[61]

Man's perception rises from some properties of the environment up to the level of silent spheres. The luminaries have been associated with the realm of orchestration from the time of Pythagoras.

Some sort of explanation has been provided by Tagore in his *Shabda-Tattwa* (Principles of Phonetics), *Pancha-Bhoota* (Five Basic Elements) and in various other works.

In *Shabda-Tattwa* Tagore refers to the mystery of the usage of adjectives and adverbs of sound (including onomatopoetic words) even when there is no sound at all.[62] As for example, the word "loud" is an adjective of sound, but we very often speak of a "loud colour". Actually all our five senses are subtly interlinked with one another in our psychic region. This very interrelationship has been termed synaesthesis by the modern psychologists. We speak of the "tone of a picture" to indicate its balance of lines, colours, and light-and-shade. To express physical pain we speak of "cutting pain", "gnawing pain", "tearing pain" or "bursting pain" – all bearing adjectives attributed to sound. Conversely, sound is also sometimes attributed to adjectives associated with our sense of vision. As for example, – good music that illuminates the whole atmosphere along with our soul. Thus light and sound have a mutual agreement with each other. Gustation also joins hands when we speak of a "sweet tune." We have already seen that tactile sensation is expressed in musical terms. We very often speak of a "soft colour" or a "soft tune". Olfactory sense does not stand apart from this since in a Tagore song we come across the expression "sangitasaurabh",[63] i.e., musical fragrance.

In *Pancha Bhoota* too Tagore explains that the emotional faculty of the human mind is keenly sensitive to all the objects of light, colour, sound, etc. There is a latent but powerful effect of the vibrations of light, heat and sound on the sympathetic vibrations of our nervous system. Simultaneously, the effect of speed is also there. The sun, the moon and all the stars and planets are circling in their respective orbits. The grand phenomena of their movement and light gave birth to the idea of the music of the luminaries among ancient Greek thinkers. In just such a teleological but lucid narration Tagore conjectures his inference of the "Jyotishkamandalir Sangit"[64] (music of the luminaries).

In *Alochana* (a collection of various discussions) Tagore quotes from an English poet:

> There's not the smallest orb which thou beholdest
> But in his motion like an angel sings.[65]

and justifies the poet's intuition as being on par with scientific and philosophical truth. Actually simile and metaphor are based on some obvious truth relevant to human consciousness. Music is not kept in a closed-door milieu; it is inseparably associated with other aspects of being.

Tagore sings in his *Puja* (devotional) song "Preme prane gane gandhe":[66]

> "Overflowing the earth and the universe
> your nectar is being showered in love,
> life, music, fragrance, light and joy."
> (Approximate translation)

The orchestration of the universe is a sense of musical value ascribed to the creation. The unfolding of the metaphor generates more and more metaphors. The span of our understanding extends and the depth of our realization plunges into more and more affections. In another *Puja* song Tagore sings:

> *Tumi naba naba rupe eso prane*
> *Eso gandhe barane eso gane?*[67]
> (Do come into my life in newer and newer forms.
> Do come in fragrance, in hues and in tunes.)

Just look at another *Puja* number, "Jagate ananda-yajne"[68] wherein we come across the heart-filled contentment of the composer as he sings:

My human life is gratified as I have been invited into the pomp of the ceremony of the universe. My vision roams to my Heart's content in the realm of the beauteous and my listening is absorbed in the depth of tunefulness.

You have entrusted me with playing the flute and I sing songs of the pains and pleasures of life. Before I depart, I offer my thankful ovation to you.

(Rough translation)

### *Some More Songs on the Music of the Universe*

It is very, very interesting to notice that quite a number of consecutive songs in the devotional and love categories are songs on music, both human and cosmic. The very first song of the theme of nature is a grand song on the orchestration of the natural world around us. The first seven songs of *Vichitra* (Sundry Delicate Feelings) are on dancing. Apart from these, there are yet other various songs on music and dance strewn through all the categories of *Gitabitan*. This is because music and dance are at the root of our spiritual and erotic life, our close relationship with nature, our inmost feelings and thoughts. Thus music and dancing become the bridge between the human life-world and the endless universe surrounding our life and being.

In many of Tagore's devotional songs, mainly in those songs which take up the theme of music, the Creator is conceived as a Singer, a Music Master, a *Vina*-player or a flute-player. In the very second song the Creator is addressed as *Surer-Guru*[69] (the Tune Preceptor). The flow of *Mandakini*,[70] the morning star and the *Kanakchampa*[71] are initiated in His tune. We music-loving human beings beg initiation into tunes from Him. The third song "Tomar surer dhara"[72] develops the image of the creations as a tonal stream beside which the poet has his resort and with which he wants to be in tune always. The fourth song images the Absolute as a Great Singer and the poet as an amazed listener. Just notice the extended metaphors in this song. "Surer alo" (the light of tune) spreads throughout the universe, "surer haoya" (the breeze of tune) blows throughout the sky and "surer suradhuni"[73] (the divine river of tune) flows through the rocks. The sixth song employs a unique metaphor, that is, the blaze of tune (surer agun)[74] illuminating everything including the human heart. It is interesting to note that Rabindranath was inspired to compose this song after seeing a painting of Saraswati with a *vina*, painted by Asit Kumar Haldar.[75] This reminds us of medieval Indian *Ragamala* paintings corresponding to *ragas* and *raginis* (classical melodies) and their meditational imageries as described by the court poets then.

*Agni-vina bajao tumi keman kore*[76] (How Do You Play the Luminous *Vina*), song number 158 in the *Puja* category brings forth the imagery of the reverberating sky full of orchestrated luminaries.

Let us also observe the song *Prabhu, tomar vina yemni baje andhar majhe omni phote tara.*[77]

O my Master! when you play your *vina* in the dark, the stars are lit up. Let the same *vina* be played in my life too. The darkness of my heart will be eradicated with tonal illumination and your bliss will glorify my life forever.

(Concise English prose.)

The very first song in the category of nature may be specially marked as a song of cosmic orchestration. This song is "Visva vinarabe visvajana mohichhe"[78] (All the Human Beings of the World Are Enchanted with the Cosmic Orchestration). The earth, the water, the forest and the sky – all are always orchestrated beautifully, combined with expressive orchestral accompaniment. The spring, the rains and the autumn bring forth tonal and orchestral variations.

Song number eight in the nature category is very relevant to our present topic. Heavenly bodies including the sun and the stars abound in the sky. The world is full of life. It is a great wonder that I am here amongst and with all. This very wonder evokes music in me. My blood-circulation responds to the cosmic rhythm of the limitless time-tide. With this wonder, music awakens in my soul. My footsteps tread the grassy terrain. I get a start from the soothing fragrance of a bloom. Creation is profuse with blissful gifts. Music emerges from my being before these wonders. My life and senses are open to all such gifts. My quest for the unknown is caused by the known finite things in my surroundings. This is the wonder which generates my music. The song "Akash bhara surya tara, visva-bhara pran"[79] is the lyrico-tonal expression of the poet's consciousness reacting to the wonders of creation.

"Eso eso basanta dharatale"[80] (Invocation of the Spring Season), the concluding song of the dance-drama *Chitrangada*, depicts all beauty *par excellence*. In clusters of fine poetic-cum-tonal clauses one will find here newer tonal flows (*naba tan*), newer songs (*naba gan*), new life, hilarious fragrant breeze, in-depth cosmic consciousness, jubilant swinging with exultant rhythms, flowery groves for amorous play and all such surplus wealth overflowing with the fullness of life and creation. Furthermore, the spring is welcomed for bringing forth overflowing nectar during noctural trysts, resounding with music. The spring is manifest in the glow of the dawn, in moonlit night, in the cosy gardenhouse, in the still water of the lake as well as in the rolling sea just like the downpours of thunderstorms. The spring overcasts the cities, the countryside and forests. We receive the spring in our chores, words and minds. Spring is reciprocated in jingling anklets on rhythmic feet and in the renderings of sonorous voices. Nature responds in young shoots, leaves and blooms. The spring is welcome in youthful vigour, which conquers ageing.

On the whole, mankind and the whole of creation join together in the orchestration of the universe.

The remote past is brought forth in the living present in many a song. One such song is "Oi ase oi ati bhairava harashe"[81] (Yon Come the Rains with Tremendous Joy). At the onset of the rains the poet recollects ancient imagery of beautiful women in natural costumes and cosmetics. They are engaged in grand orchestral bands including instruments like the *mridanga*, the *muraja* and the *murali*.

Torrential rain and storm remind the poet of the magnanimous orchestration of nature and of the rain-songs of the poet-composers of bygone ages.

In the course of lectures delivered in America, Tagore said in his lecture entitled "The World of Personality":[82]

Our individual minds are the strings which catch the rhythmic vibrations of this universal mind and respond in music of space and time. The quality and number and pitch of our mind-strings differ and their tuning has not yet come to its perfection, but their law is the law of the universal mind, which is the instrument of finitude upon which the Eternal Player plays his dance music of creation.[83]

A musical personality like Tagore's does not believe in utter silence after death. Even that inevitable event is musicalized.

Addressing his affectionate intimates, the poet says: "Grant me leave from the web of light and shade of all that is familiar in life. In the nameless solitude let me tune up myself with the multi-stringed orchestration of the infinite."[84]

(Approximate translation)

Man and his music cannot be absolute and autonomous. Man derives his music from the orchestration of the universe, and finally he has to merge his music with the same infinite orchestration of the universe again.

## VII. CONCLUSION

Reality must be Absolute and One. We can endeavour to approach it from various perspectives only. The Tagorean concept of Reality is pluralistic in manifestation, though Monistic at the End.

In the Tagorean sense, appearance and Reality are not necessarily opposed to each other. All significant appearances fall within the realm of Reality. Appearances emerge from Reality since Reality appears by virtue of appearances only.

Man-in-the-Universe is the most precious aspect and culmination of the evolution of the Universe since it is Man who endeavours to know and realize Reality.

Despite possibilities of mal-observation, illusion, delusion and falsification, man's yearning for Reality has created all the arts and sciences.

*Visva-Bharati University*

## NOTES

[1] Rabindranath Tagore, *The Religion of Man* (London: George Allen and Unwin Ltd., 1931, Fifth Impression 1958). Appendix II 'Note on the Nature of Reality', pp. 222–225.

[2] *Ibid.*, p. 225.

[3] Rabindranath Tagore, *Sangit-Chinta* (a posthumous anthology of Tagore's thoughts on music) (Visva-Bharati, 1932 B. S.), pp. 342–347. It was earlier published in *Asia*, March 1931 (New York, ed. R. J. Walsh). The issue is not available here. However a typescript of the same is kept in the Rabindra Bhavana archives (Visva-Bharati University, Santiniketan) under the call no. T928 F3 TT. An edited version of the conversation was reprinted in *Asia*, March 1937 (pp. 151–152).

[4] *Ibid.*, p. 342.

[5] *Ibid.*, p. 345.

[6] *Ibid.*, p. 346.

[7] *Ibid.*, p. 345.

[8] From the typescript in the Rabindra Bhavana archives (T928 F3 TT), Visva-Bharati University, Santiniketan, *op. cit.*

[9] *Ibid.*

[10] *Ibid.*

[11] Rabindranath Tagore, *Sangit-Chinta, op. cit.*, p. 347.

[12] *Ibid.*, , p. 347.

[13] *Ibid.*, p. 347.

[14] Rabindranath Tagore, *Visvaparichay* (1st edition, 1344 Bengali era), Lokasiksha Granthamala, Visva-Bharati, 1401.

[15] Rabindranath Tagore, "Meditation," *Personality* (1917), (Macmillan India Ltd. 1980 edition), pp. 151–166.

[16] Rabindranath Tagore, *Santiniketan, Rabindra Rachanavali* (Tagore's Works), Birthday Centenary Edition (Govt. of West Bengal, 1961), vol. 12, pp. 97–509. Henceforth *Rabindra-Rachanavali* is abbreviated as *R. R.* Along with the place name Santiniketan, *Santiniketan* is the title of the anthology of Tagore's congregational lectures delivered in the span between Agrahayan 17, 1315 and Paus 7, 1321 (1908–1914) at the temple of Santiniketan.

[17] Tagore wrote various letters to Amiya Chakravarti. Here the letter, dated 8 Ashvin 1343, compiled as an introduction to *Sahityer Pathe* is our concern. *R. R.* 14, pp. 291–293.

[18] *R. R.*., 14, pp. 295–300.
[19] "Sahitye Nabatwa" (The Newness [or Originality] of Literature), *Sahityer Pathe*, *ibid.*, p. 334.
[20] "*Ami*", *Shyamali*, *R. R.* 3, pp. 392–393.
[21] "*Shyamali*", *ibid.*, pp. 435–436.
[22] *Ibid.*, p. 392.
[23] "*Ami*", *Parishesh*, *R. R.* 2, pp. 880–881.
[24] "*Ami*", *Shesh Saptak*, *Sanyojan* (addenda), *R. R.* 3, pp. 235–237.
[25] "*Amar Jagat*", *Sanchay*, *R. R.* 12, pp. 559–565.
[26] "*Shona*", *Santiniketan*, *R. R.* 12, p. 127.
[27] *Ibid.*, pp. 126–128.
[28] *Ibid.*, p. 127.
[29] "*Dekha*", *Santiniketan*, *ibid.*, pp. 124–126.
[30] *Ibid.*, p. 126.
[31] *Ibid.*, p. 126.
[32] "*Shona*", *op. cit.*, p. 127.
[33] Rabindranath Tagore, *Gitabitan* (Collection of Songs), Part I, song no. 321, "*Puja*" (the devotional category), Visva-Bharati, 1970 edition, p. 135.
[34] "*Om*", *Santiniketan*, *RR 12*, pp. 256–258.
[35] *Sahityer Pathe* (Towards the Path of Togetherness, i.e., literature and the arts), *RR* 14, p. 310.
[36] *Chhanda*, *RR* 14, p. 268.
[37] *The Pelican History of Music* (1960), Vol. I, p. 36.
[38] *The Oxford History of Music*, Vol. I (London: Oxford University Press, 1957, reprint 1960), pp. 196.
[39] Ananda Kentish Coomaraswamy, *The Dance of Shiva* (New York: The Noonday Press, 1957), pp. 95–96.
[40] Sitansu Ray, "The Tagore-Einstein Conversation: Reality and Human World, Causality and Chance", *Analecta Husserliana* Vol XLVII, ed. A.-T. Tymieniecka (Dordrecht: Kluwer Academic Publishers, 1995), pp. 59–65.
[41] From "*Bhumika*", the introductory lyric of Tagore's *Gitabitan*, *op. cit.*, p. 1.
[42] *Patraput*, *RR* 3, p. 350.
[43] *Japanyatri*, *RR* 10, pp. 492–494.
[44] *Chhinnapatravali* (A Cluster of Scattered Letters), *RR* 11, p. 82.
[45] *Galpaguchchha* (A Cluster of Stories), *RR* 7, p. 699.
[46] "Manas Sundari", *Sonar Tari*, *RR* 1, p. 390.
[47] "Jyotsna Ratre", *Chitra*, *RR* 1, p. 469.
[48] *Gitabitan* (Collection of Songs) *RR* 4, p. 218.
[49] *RR* 3, p. 636.
[50] *Chhinnapatravali* (Collection of Scattered Letters), Letter No. 119. *RR* 11, p. 133.
[51] *Ibid.*, letter no. 148, p. 165.
[52] *Gitabitan*, *RR* 4, *op. cit.*, pp. 417–418.
[53] *Ibid.*, pp. 417–418.
[54] *Ibid.*, pp. 418.
[55] *Ibid.*, pp. 418–419.
[56] *Nataraja: Riturangashala*, *RR* 5, pp. 619–663.
[57] *Gitabitan*, *op. cit.*, p. 159.
[58] *Ibid.*, p. 159.
[59] *Ibid.*, p. 253.

[60] *Ibid.*, p. 228.

[61] Prof. R. Murray Schafer, "The Music of the Environment", *Cultures*, (UNESCO, Paris), Vol. 1, No. 4, p. 17.

[62] *Shabda Tattwa*, *RR* 14, pp. 38–39.

[63] "Bhumika" (Preface), *Gitabitan*, *RR* 4, *op. cit.*, p. 1.

[64] *Pancha Bhoota*, *RR* 14, pp. 673–675.

[65] Quoted by Tagore, *Alochana*, *RR* 14, p. 594.

[66] *Gitabitan*, *RR* 4, p. 102.

[67] *Ibid.*, P. 58.

[68] "Puja", song No. 317, *Ibid.*, p. 102.

[69] "Puja", song No. 2, *Ibid.*, p. 3.

[70] The Ganges of Heaven as believed by Hindus.

[71] A kind of sweet scented golden flower.

[72] "Puja" No. 3, *Gitabitan*, *op. cit.*, pp. 3–4. The whole creation is conceived of as an overflowing sonorous stream.

[73] "Puja" No. 4, *ibid.*, p. 4. *Suradhuni* is another synonym of *Mandakini*, i.e. the Ganges of Heaven.

[74] "Puja" No. 6, *ibid.*, p. 4.

[75] Santidev Ghosh, *Rabindra Sangit* (Calcutta: Visva-Bharati, 1365 Bengali Era), p. 297.

[76] "Puja", No. 158, *ibid.*, p. 55.

[77] "Puja", No. 35, *ibid.*, p. 14.

[78] "Prakriti" (Nature), No. 1, *ibid.*, pp. 329–330.

[79] "Prakriti", No. 8, *ibid.*, pp. 331–332.

[80] The dance-drama *Chitrangada*, *ibid.*, pp. 551–552.

[81] "Prakriti" No. 27, *ibid.*, pp. 337–338.

[82] Rabindranath Tagore, *Personality* (London: Macmillan, 1917), Indian reprint 1985, pp. 41–76.

[83] *Ibid.*, pp. 54–55.

[84] Rabindranath Tagore, Poem No. 43 (written as a letter to Amiya Chakravarty); *Shesh Saptak* (The Extreme Octave), *RR* 3, p. 218.

BEATA SZYMAŃSKA

# AN EXPERIENCE OF PURE CONSCIOUSNESS IN ZEN BUDDHISM

It is generally accepted that the phenomenological method as well as the concepts derived from it are particularly suitable for the analysis of some philosophical systems of ancient India, China and Japan. This approach has resulted in a number of interesting studies, more or less directly related to phenomenology. This happens especially in Japan; the Kyoto School became particularly interested in the phenomenological movement, and its most prominent exponents studied it in Germany. In the West, too, some scholars pointed out the analogies between phenomenological approach and the teaching of ancient Eastern masters.

In this short study I intend to concentrate on but one very interesting aspect of this analogy. I will analyze the applicability of some phenomenological categories to Chinese Taoism and Chan Buddhism. The concept of "cleansing the mind" and the related metaphor of mind as a mirror, concepts which appeared in those philosophies, will also be included and discussed.

The starting point of these reflections is the conviction common to both Taoism and Zen Buddhism that in order to "recognize" reality and its relationship to the mind which is recognizing it, one must renounce all previous beliefs and notions. The idea of *epoché*, bracketing or "suspension" of judgement about existence is already implied in the legendary Buddha's silence when confronted with a question of the existence of God and of an individual soul. In my opinion, however, one can fully appreciate the phenomenological and eidetic reduction only in relation to Chinese and Japanese Zen Buddhism. It may also apply to the early school of Taoism (which preceded them) because in Taoism suspension of judgment also takes the form of silence.

One must then accept a certain interpretation of "being silent in relation to something." Such a postulate is contained both in the writings of Zen Masters and in the beginning of one of the most important Taoist treatises, *Daodejing*, whose first verse states that: "The Tao that can be told of is not the eternal Tao."[1]

The assertion that the essence of being cannot be expressed in discursive language does not necessarily mean that it is inaccessible to knowledge. Even though language can only point to the existence of reality that cannot be described in words, individual experience reveals it. The existence of the



*A.-T. Tymieniecka (ed.), Analecta Husserliana LXXVI*, 47–56.

great Taoistic treatises of philosophical heritage of Soto Master Dogen, in particular, his work *Shobogenzo* (a text which forms the basis of my arguments), as well as contemporary works of both Suzukis, demonstrates this approach.

"In the first place, Zen proposes its solution by directly appealing to facts of personal experience and not to book-knowledge.... By personal experience it is meant to get at the fact at first hand and not through any intermediary, whatever this may be," states Daisetz Teitaro Suzuki.[2]

But it is only in Zen that we can talk about the identity of a pure experience and enlightenment, which is expressed by the term *samadhi* **perception**.

We have many examples of sudden insights into truth induced by sensory experience. The best-known example is that of the Japanese poet Basho – who attained his enlightenment on hearing the splash of the water caused by a frog jumping into an old pond.

Such an experience may be stimulated by the sight of a flower, a reflection of the moon in water or even a sharp pain evoked by the master's stroke. I have used the term "stimulated" even though it is not entirely accurate. The perception of the stimulus becomes everything that is in consciousness. There is nothing else, the awareness of the self disappeared and the perception becomes an enlightenment. Such an experience, though potentially available to any man, cannot occur if the mind is contaminated by previous assumptions. (I have in mind any assumptions, not necessarily mistaken assumptions as commonly believed.) At most – following Descartes – we can claim that all previous knowledge should be considered doubtful due to methodological considerations. Thus what is suspended or bracketed is the very question concerning the existential modus of the outer world – the question of whether it exists independently of perceptions, or if it is possible that the world is a kind of illusion (although our mind cannot be said to have created it). Moreover, all judgements based on everyday experience, evaluations, emotions, carnal experiences should be eliminated.

The consideration shows that we can speak about certain *epoché*, although the very analogy applies only to a limited extent. First and foremost, the most important fact is that this exclusion is not an intellectual act of *epoché* of our judgements. It is rather a result of a peculiar spiritual training, which either gradually or suddenly brings the mind to a state of emptiness. One cannot, however, point to a definable technique of emptying the mind. (Here one can discern the differences between the Southern and Northern versions of Chan Buddhism. These schools were divided over the issues concerning either a sudden or gradual attainment of this state of purity of the mind.) It is possible to find (in different texts) analogous descriptions of stages of liberating the mind not only from all presuppositions but also from emotions and body experiences.

We find such a metaphorical description in an ancient Taoist treatise *Zhuangzi*, written by Zhuang Zhou. It is often recalled by followers of Zen Buddhism. The following fragments describe the gradual steps of emptying the mind in the process of "forgetting" or "mind fasting." Here we find the story about the master craftsman making musical instruments. He tells how he has attained the full understanding of the essence of the bells he creates:

"When your servant had undertaken to make the bell-stand I ... felt it necessary to fast in order to compose my mind. After fasting for three days, I did not presume to think of any congratulation, reward, rank or emolument.... After fasting five days, I did not presume to think of condemnation or commendation which it might display. And the end of the seven days, I had forgotten all about myself: my four limbs and my whole person."[3]

This fragment illustrates well a process of "freeing the mind" of emotions, body experiences and, ultimately, of self-awareness. But the descriptions of this process are not always identical. Taoist texts do not actually provide exact instructions on how to attain this stage of "forgetting." Only Chan Buddhism or, more exactly, its Japanese version described various sets of techniques, such as *zazen* and *koan* leading to "cleansing of the mind." It is the task of the Master to teach the student the appropriate methods.

One of the oldest texts of Chan Buddhism is the collection of teachings of Huineng the Sixth Patriarch, entitled the *Platform Scripture of the Sixth Patriarch* which contains the following description:

"I suddenly realized the original nature of True Thusness. For this reason I propagate this doctrine so that it will prevail among later generations and seekers of the Way will be able to achieve perfect wisdom through sudden enlightenment, each to see his own mind, and to become suddenly enlightened through his own original nature."[4]

The teachings of the masters, which are so important in Zen Buddhism, are merely offered to point out the ways to "those who are not able to enlighten themselves" thus to attain the desired state of mind. Hence teaching has one purpose only: practicing. "You have now found yourself, from the very beginning nothing has been kept away from you.... In Zen there is nothing to explain, nothing to teach, that will add to your knowledge. Unless it grows out of yourself, no knowledge is really of value to you, a borrowed plumage never grows."[5]

The most important issue from the epistemological point of view is the "emptying of mind," leading in essence to the primary state of a *tabula rasa* (this state is often described as the *original mind* or *original nature*).

There are numerous works devoted to the notion of "primordial mind" but the issue is too complicated to be fully presented here. Undoubtedly it is an *original mind* which is pure, freed from all presuppositions. "Man's nature is originally pure. It is by false thoughts that True Thusness is obscured."[6]

The process of its "contamination" has been described even before Buddhism. It is important to note, however, that we deal here with natural processes, with the ordinary mental activity of – perpetually working – consciousness. "Switching off" the latter's functions seems – from the viewpoint of ordinary, everyday experience – to be something unusual and difficult to obtain.

Pure mind is often described in negative terms denoting the state of its being empty, the state of "disappearance." The saying of Nyojo, a Buddhist monk: "Za-zen is to cast off body and mind," is often quoted. It is Dogen who referred to this maxim of his old master and many commentators regard it as a key notion of the Zen Buddhist. David Shaner in his inspiring article about Dogen states that "*casting off*," interpreted phenomenologically, is parallel to the neutralization of thetic positings characteristic of first order body-mind awareness. Dogen is emphatic in his emphasis that casting off is not a denunciation (negative positing) of body and mind. One must cast aside all thetic positings, for example, "accepting good" and "rejecting evil."[7]

The terms describing pure mind often use negative prefixes. For example, Chinese Buddhism uses the prefix *wu* meaning "non" or "there is not" or "nothing." We find expressions like *wu-xin, wu-nien*, denoting *no-mind* or *no-thinking*. *Xin* is an equivalent of the Cartesian *cogito*, encompassing all acts of consciousness such as judging or evaluating, as well as emotions. Moreover *xin* includes what may be described approximately as the "unconscious."

Taken in its literal sense *wu-xin* denotes the disappearance of all conscious elements (functions). That is why there is another term in Zen: *emptiness*.

I believe, however, that all these terms are used metaphorically, and authors of ancient texts, especially Dogen, warned against the literal interpretation of "empty mind" as nothingness. Emptiness is not nothingness. Dogen wrote: "The Fifth Patriarch has said: 'Because the Buddha-nature is empty, we say there is no-Buddha nature.' As is clearly expressed here, emptiness is different from non-being."[8]

The complete understanding of this "no-mind," "no-Buddha," is identified with the illumination – *Satori*. "When we understand the true meaning of *no-Buddha-nature* we become a Buddha."[9] Becoming a Buddha though, does not mean the end of knowing (e.g., empirical knowledge) as it occurs in everyday experience. The flow of data comes in all the time. Under the influence of *satori*, however, the content of consciousness is transformed and the character of cognitive acts changes. Such a stance has been expressed by the saying that "before the attainment of enlightenment the grass was green while the sky was blue." At the moment of consciousness' transformation, both the grass and the sky lose their inherent qualities, but when the illumination occurs, "grass is green again and the sky blue." It is worth mentioning here the statement ending Husserl's *Cartesian Meditations*. "Il faut d'abord perdre le monde par l'époché, pour le retrouver ensuite dans une prise de conscience universelle de soi-même."[10]

A mind "cleansed" like that returns to things and perceives them according to their real nature, in their true essentiality. An individual object preserves in the act of perception its own "uniqueness," which is in Buddhism denoted by the Sanskrit term *tathata* – "suchness." The perception by a "clear mind" offers at the same time insight into the essence of the object as a universal.

We do not have original Zen texts analyzing in detail this problem, but I believe that we can refer to frequent descriptions of creativity (writing poetry, painting) which illustrate the transcending of individual features in order to achieve insight into the essentiality ("suchness").

An artist who is trying to paint a cat is supposed to "catch" not only actual similitude to the object but its essence, its *catness*, which pertains to every individual cat. To achieve this result an artist must be gifted, but – at the same time – owing to a long-term concentration he must purify his mind and concentrate on the depicted object as he tries to grasp its essentiality. "When we paint a spring landscape we must not only paint willows, or red and green plums and peaches, we must paint spring itself."[11]

Let us concentrate on the subject of an artwork. In reflections on Japanese aesthetics various authors often quote a well-known haiku poem by Basho.

From the pine tree
Learn of the pine tree
And from the bamboo
Of bamboo.

Here is Keiji Nishitani's comment: "He does not simply mean that we should 'observe the pine tree carefully.' Still less does he mean for us to 'study the pine tree scientifically.' He means for us to enter into the mode of being where the pine tree is pine tree itself, and the bamboo is the bamboo. He calls on us to betake ourselves to the dimension where things become manifest in their suchness, to attune ourselves to the selfness of pine tree and the selfness of bamboo."[12]

This quotation illustrates that such emptiness of mind is a readiness to accept the revealed *eidos* of things.

So far I have been discussing various efforts, specific methods and ways leading to the cleansing of the mind. It is important, however, to note that the end of the process means also an end to efforts, an end of all acts. The mind must become only a mirror in which reality is reflected.

The metaphor of the mind as mirror is an essential element in the Chan treatise *Platform Scripture of the Sixth Patriarch*. It was the motif of the two famous verses (*gathas*) written in reply to the call of the Fifth Patriarch. Those verses constitute a kind of dialogue dealing with – among other things – the nature of mind. The important thing was the statement (formulated by Huineng) that the mirror-mind is – from the very beginning – clean.

The issue of the nature of mind and the question of under what conditions it becomes clear, i.e., is able to perceive the essence of things, was a subject of dispute between later Southern and Northern Chan schools. In Dogen's treatise *Shobogenzo* the mirror is the subject of a separate chapter, *Kokyo* – "An Old Mirror." Dogen, the follower of Huineng, shared the conviction that "sudden enlightenment" could be attainable by all men, under many conditions. This results in a dilemma, namely, how to reconcile the two different kinds of enlightenment.[13] If there is a clean mind and original enlightenment (*hongaku*) and acquired enlightenment (*shikaku*), why does the initially clean mind require such intensive work and effort in order to attain this state? Why are we not enlightened from the very beginning? As a matter of fact enlightenment is not knowledge coming from outside. It is given to everyone, although it is not always revealed in all of us. Dogen sought the answers to this question in all of his reflections (e.g., *Shobogenzo*). In my opinion his implicit conclusion is that the clear mirror in which there is nothing, not even the experience of one's own "I" must be regarded as a metaphor. We are faced with the Husserlian question then: What else remains when one "suspends" the whole world including ourselves and all our *cogitare*? The elimination of self-awareness seems rather to imply a "shifting" of intention. In the words of Nishitani, "By pulling away from our ordinary self-centered mode of being (where, in our attempts to grasp the self, we get caught in its grasp) and by taking hold of things where things have a hold on themselves, so do we revert to the 'middle' of things themselves."[14]

Enlightenment in the sense of understanding must be a revelation of some kind of primary cognition, or at least some disposition towards it, some "universal" *a priori* which makes possible "grasping" the sense of something.

This fact justifies the work of searching for one's original nature. In Zen – let us remember – there is no knowledge which comes (in the act of illumination) from outside. Neither is there a kind of reasoning leading up to it. One must – as Husserl says in the Fifth Meditation – "expliciter le sens que ce monde a pour nous tous, antérieurement à toute philosophie."[15]

It is true that the experiencing of a "primordial, original nature" precedes all experience (it is a primary given), but attainment of conscious insight into the "I" becomes possible only after the cleansing of a person's consciousness. The process of purification is described as extended in time, a process of a gradual getting rid of all impurities, but the insight into the pure nature is momentary. It is an eidetic recognition, grasping of the essence of things. "If you want to understand the true meaning of Buddha-nature, you should see its momentary manifestation. When the right time comes, the Buddha-nature manifests itself."[16]

This demonstrates the fact that the division between subject and object is only a result of everyday experience and language. A special feature of the concept of Zen seems to be that the aim of all our activities is the attainment of a state of total passivity. The mind is a mirror immobile and static. Things are reflected in it, they appear and disappear on their own, while the mirror remains the same: immobile. In such a process of passive perception there is no distinguished object. Things appear without a background, always on the first plane of consciousness and each thing disappears as the next one takes its place, because attention is "switched off" and the mind does not concentrate on any reflection. It is not clear, however, if this stream applies only to perceptions or includes internal representations. I believe that the primary object of our experience is not self-awareness but the subject of perception. (It does not matter if and how the ontological question will be answered later.) The mind does not perceive itself, as expressed in the metaphor of the eye "that can see, but cannot see itself," and there is no act in which mind becomes its own *noema* ("the impossibility of grasping the mind with the mind"). But the recognition of one's own essence is attainable through acceptance of a totally passive perception making possible the reception of an ontological identity of subject and the perceived object. This does not mean, however, that things are creations of our mind or illusions. "The whole world is not produced as a result of illusion, for the whole world is clearly a manifestation of the Truth."[17]

Our attention is always focused on the external, on what is a *noema* whose mode of existence is not defined since it has been bracketed. As a matter of fact one may go as far as saying that the most characteristic feature of Zen is that it does not go beyond the phenomenological reduction. This means that the question concerning the existence of the world is never answered. In other words, the ontological status of the world is neither affirmed nor denied.

At the same time, as pointed out by Nishitani, Zen does not claim that there is some sphere of reality, a "thing in itself" which by its very nature is not knowable. It is possible to presume, however, that such a reality exists. It can be recognized but not described. Every attempt to resolve this question breaks away from a basic postulate of Zen. At the same time the postulate of eidetic reduction remains valuable.

It is worth noting, however, that the mirror metaphor does not answer the problem; it attributes existence to the external world even though this reality has only a phenomenal character. If a reality is "reflected" in the mind, then the recognition of this fact is identical with the thetic act: it is the recognition of something which exists, which is reflected and whose essence is recognized.

The very reflection of things, however, is not the ultimate goal of knowledge. The goal is always reaching one's own nature. This does not contradict the previously mentioned conviction that the mind does not perceive or experience itself. The object of knowing is always an intentional one – irrespective of whether it is an individual object or a set of its counterparts because the eidetic insight reveals the essence of Whole which is man and all of reality. In this sense the loss of experience of the Self is not equal to a loss of awareness. It is rather a qualitative transformation best rendered by the Upanishad's formula: *tat tvam asi.*

In Zen Buddhism the formula "everything has the nature of Buddha" is accepted, while enlightenment is often summed up by the statement "You are Buddha."

It goes without saying that this does not resolve the question of the mode of existence. For neither the notion of Buddha nor "Buddha-nature" (*buddhata*) are identical. As far as an individual undergoing a transformation of mind is concerned, that question makes little difference because it is a theoretical problem. For scholars who study the problem of enlightenment, however, it becomes the crucial issue. True enough, this problem is treated metaphorically.

For instance, in the *Platform Scripture of the Sixth Patriarch* it is difficult to find any concrete solution. Dogen "set aside" the suspension of the question of the ontological status of the reality with which we form a whole. His answers are various or perhaps only formulated in different ways. Some of these answers refer to Mahayana Buddhism, especially "venerable Nagarjuna" whom Dogen likes to quote (e.g., the theory of emptiness, *śunjata*). The recognition of the fact that what is reflected is emptiness leads us to the conclusion that the mind itself is an emptiness in the literal sense of the term. But emptiness is not nothingness.

Another answer is: *All is Buddha*. In this case mind as a mirror also reveals its own Buddha-nature. For such philosophers as Dogen this is no contradiction, since both emptiness and Buddha are only linguistic creations which cannot be transcended.

Intuitive insight is an experience which cannot be described, but a Master, as mentioned before, can transmit the knowledge of how to attain the possibility of this kind of knowing.

Zen masters are convinced that the attainment of such an insight is something simple and within the range of abilities of every man. Truth cannot be gained through discursive understanding. Moreover we do not need any theoretical background. In the words of a contemporary Soto Zen Master, Shunryu Suzuki: "Big mind is something to express, not something to figure out. Big mind is something you have, not something to seek for."[18]

*Jagiellonian University*

NOTES

[1] Laozi, "Daodejing," in Wing-Tsit Chan, *A Source Book in Chinese Philosophy* (Princeton: 1973), p. 139.

[2] Daisetz Teitaro Suzuki, *Essays in Zen Buddhism*, First Series (New York: 1961), pp. 18–19.

[3] "Zhuang Zhou, Zhuangzi," James Legge, *The Sacred Books of China, The Text of Taoism*, Part II (New York: 1962), p. 22.

[4] Huineng, "Platform Scripture of the Sixth Patriarch," in Wing-Tsit Chan, *A Source Book in Chinese Philosophy* (Princeton: 1973), p. 439.

[5] Suzuki, op. cit., p. 245.

[6] Huineng, op. cit., p. 435.

[7] David E. Shaner, "The Bodymind Experience in Dogen *Shobogenzo*: A Phenomenological Perspective," in *Philosophy East and West*, 35: 1, p. 26.

[8] Dogen, *The Shobo-genzo*, transl. Yuho Yokoi (Tokyo: Sankibo Buddhist Bookstore, 1986), p. 29.

[9] Ibid., p. 31.

[10] Edmund Husserl, *Méditations Cartésiennes* (Paris: 1931).

[11] Dogen Zenji, *The Eye and Treasury of the True Law. A Complete English Translation of Dogen Zenji's Shobogenzo*, by K. Nishiyama and J. Stevens, Part II (Tokyo: Nakayama Shobo, 1975), p. 150.

[12] Keiji Nishitani, *Religion and Nothingness*, trans. J. Van Bragt (Berkeley, Los Angeles, London: 1983), p. 128.

[13] Shaner, op. cit.

[14] Nishitani, op. cit., p. 140.

[15] Husserl, op. cit., p. 129.
[16] Dogen Zenji, *The Shobo-genzo*, transl. Yuho Yoko op. cit., p. 22.
[17] Ibid., p. 24.
[18] Shunryu Suzuki, *Zen Mind, Beginner's Mind* (Tokyo: 1988), p. 90.

MARTA KUDELSKA

# AN ATTEMPT AT A PHENOMENOLOGICAL DESCRIPTION OF THE SELF-KNOWLEDGE PROCESS IN THE CHANDOGYA UPANISHAD

We have been trying to take advantage of the phenomenological method in order to explicate the classical Hindu text and answer the inspiring question of whether the notion of atman-brahman might have any implication for phenomenology. As is widely known, atman-brahman is not only described as a conscious being, but it is said to be a pure consciousness. The latter – treated as pure subjectivity – is an authentic, autotelic being – brahman. "Oncoming" layers or circles of consciousness are equivalent to subsequent stages of reduction. These are the counterparts of particular states, modi of consciousness (awareness). The noemas seem to be real, coherent and active on a given temporal stage until one investigates their nature, until "the systematic, essentialistic eidetic analysis is achieved." As will be stressed (in this text), the subsequent stages of reduction concern the very object which turns out to be a pure subject – atman. Thus all stages of the "appearing" of the world must be constituted by a conscious subject. The latter intentionally aims at something transcendent, "going" deeper into the very layer of constitution itself. It seems that all acts are anchored into something deeper; one may say that consciousness looks for its "interior," makes circular movements which eventually lead it to the level where a subject intentionally aims at itself.

The text we have chosen for the analysis is 2,500 years old. It belongs to the classical group of the "sacred" books. My fascination, as well as methodological fear, consisted in a deep belief that the Husserlian approach to an analysis of our human reality is deeply rooted in a very long tradition. Thus the following article is an ardent attempt to verify the Husserlian postulates: his universal method of describing reality.

While studying the first volume of Husserl's *Ideas* I have been constantly encountering fragments which might be an exact commentary on the "Upanishad." Due to space and time limitations I have to make scant references to the discovered similarities and intellectual (spiritual) affinities found in both texts. However, let me begin by pointing out certain undoubted differences. They concern initial suppositions of methods. Husserl reaches the (very) concept of consciousness treated as a phenomenological residue without a final limit (point of "destination") while in Chandogya it is Indra (following the teachings of Prajapati) who reaches this primordial point.



*A.-T. Tymieniecka (ed.), Analecta Husserliana LXXVI*, 57–66.

Prajapati knows everything about the reality which has actually been defined by him. This is the ultimate reality. Husserl – as we all know – advocates an analysis, an investigation based on no presuppositions. In Chandogya such a presupposition is openly and clearly accepted. The final goal (the point of destination) is univocally defined at the very beginning and steers all cognitive processes. In view of this, one may raise the question of whether there are any similarities in so far as the application of the phenomenological method is concerned. Setting aside all differences (as I am deeply convinced), it is worth analyzing the methods themselves, so similar in so many respects. I think that for Husserl (who regarded phenomenology as "first philosophy") such evidence of a pre-phenomenological (in a historical sense) approach would be an inspiration for constructing the fundamental science.

Let us now turn to the conversation in the final parts of Chandogya Upanishad (VIIIth Book). God – the master of all creation – Prajapati describes all reality in terms of its being the source of the whole universe and the goal of all our striving (purposes and actions included).

"The self (atman) that is free from evils, free from old age and death, free from sorrow, free from hunger and thirst; the self whose desires and intentions are real – that is the self that you should try to discover, that is the self that you should seek to perceive. When someone discovers that self and perceives it, he obtains all the worlds, and all his desires are fulfilled. So said Prajapati."[1]

The above words were heard by all – both gods and demons striving for power over the world. They sent their representatives to Prajapati to obtain some knowledge by which to get the upper hand over their enemies. Thus Indra (the representative of gods) and Virochana (the representative of demons) had been staying at the master's abode for thirty-two years waiting for this knowledge to be imparted to them. After that period Prajapati received that first explanation. That possibly satisfied Virochana, but Indra after second thoughts asked Prajapati for more exact and more accurate explanations. After the second turn of thirty-two years, the situation was exactly the same. So Indra decided to stay for some period of time again, and after his subsequent return he needed only five years to receive deeper knowledge. The latter (obtained after 101 years) may not have differed from the knowledge he had already possessed. Thus it was the method itself which really mattered.

What conclusions could be drawn from this initial definition of reality given by Prajapati? How can this be the meaning of human endeavours? We deal here with the description of atman as pure consciousness. As we know from the Upanishads, atman-brahman is not only described as a conscious entity but is nothing other than pure consciousness itself. On the other hand, consciousness – as pure subjectivity, atman – is also authentic being, brahman. So it seems to be the basis of that which "gets to know" (which is in the process of knowing) as well as the basis of that which is known. We shall soon see how this reality makes appearances, reveals itself through the subsequent layers of consciousness.

The first layer of application (concerning that which brahman-atman is) was presented by Prajapati after his first thirty-two years of learning.

"Prajapati then told them: This person that one sees here in the eye – that is the self (atman) that is the immortal; that is the one free from fear; that is brahman."[2]

Here we find a classical play on words. The word atman denotes the fundamental microcosmic element, but it also functions as the demonstrative pronoun: me-myself. The latter is often used to mean (the totality of a knowing subject, including the body) itself. The term for person – purusha – is also ambiguous. It denotes a man (a male in current usage) as well as the fundamental cosmogonic entity. As the result of a self-division of purusha, its self-sacrifice, the whole perceptible world has emerged. Referring to ambiguous (unequivocal) terms, Prajapati explains that atman (sought after by so many) is nothing but themselves. Whether they are beautifully dressed or not, the reflection they throw is themselves: it is the world itself. Thus, the reflection in question must be the world itself – which is pleasant or unpleasant for the perceivers.

> The whole *spatio-temporal world*, to which man the human Ego claims to belong as subordinate, singular realities, is *according to its own meaning mere intentional Being*, a Being, therefore, which has the merely secondary, relative sense of a Being *for* a consciousness. It is a Being which consciousness in its own experiences posits, and is, in principle, intuitable and determinable only as the element common to the [harmoniously] motivated appearance-manifolds, but *over and beyond* this, is just nothing at all.[3]

This explanation satisfies Virochana – the representative of the demons. But Indra – on his way home – sees many incongruities in such a description of the world. How is it possible that such a real, genuine, authentic reality as atman-brahman (independent, pure, constant consciousness) must be dependent upon external factors? Thus he returns to Prajapati again.

Indra explains that the description of Prajapati does not satisfy him as it does not correspond with that in which atman-brahman is treated as an autotelic reality. The definition says that (ontologically autotelic) reality, which is something constant, is something that is "inflected," defined in an immanent way, so that all terms used to define it must pertain to it in an effective way. That which we perceive (including ourselves and the *milieu* we live in), i.e., the world, may under certain circumstances be something from the state of some other circumstances. It all depends on incidental accidental factors such as: garments, "dressing" or different points of view. This is not atman, pure consciousness, but a specific counterpart of consciousness – noema.

Such is the attitude on the part of Virochana. For him the appearance (presence) of consciousness equates with autonomous consciousness. Not seeking to clarify the issue of the reality of this noema, he regards the latter as the only real entity (being). Indra's approach is different, however. On the one hand, he does not resolve the problem of the reality of this particular form of consciousness (he "brackets," suspends this question); on the other hand, he asks about the conditions of its appearance. This situation is approached by two different persons. On the one hand, we have a psychological attitude in which (and this is the natural stance) the direct perception of the knowing subject turns directly to sensations as states of sensation; on the other hand, we have a phenomenological stance. The first leads – although not necessarily – to the inherently possible adoption of the phenomenological stance on the part of the subject. The latter (during the act of reflection) excludes that which the first transforms into being and turns to that which is absolute – pure consciousness.[4]

A peculiar type of apprehending or experience, a peculiar type of "apperception," completes what is brought about by this so-called "linking-on," this realization of consciousness. Whatever this apperception may consist of, whatever special type of manifestation it may demand, this much is quite obvious, that consciousness itself in this appreciative interweaving, in this psychological relation to the corporeal, forfeits nothing of its own essential nature and can assimilate nothing that is foreign to its own essence, which would indeed be absurd. Corporeal Being is in principle a Being that appears, declaring itself through sensory perspectives. The consciousness that is naturally apperceived, the stream of experiences, given as human and animal at once, in close empirical connexion with corporeality, does not itself become of course through this apperception something that appears perspectively.[5]

Indra saw that the first explanation of Prajapati concerned only the manner of appearing of autotelic being, not himself, so he returned to obtain further knowledge. After thirty-two subsequent years of learning, he received the following answer:

"The one who goes happily about in a dream – that is the self; that is the immortal; that is free from fear; that is brahman."

Indra then left, his heart content. But even before he had reached the gods, he saw this danger: "It is true that this self does not become blind when this body becomes blind, or lame when the body becomes lame. This self is clearly unaffected by the faults of the body – it is not killed when this body is slain or rendered lame when this body becomes lame. Nevertheless, people do in a way kill it and chase after it; it does in a way experience unpleasant things; and in a way it even cries. I see nothing worthwhile in this."[6]

Searching for the experience of atman-brahman Indra maintains a gaze constantly turned towards the sphere of consciousness. Moreover, he analyzes what it inherently contains. Without bracketing the level of judgements, he essentially analyzes consciousness. Such an analysis requires an understanding of consciousness as such, particularly the extent to which – out of the essence of consciousness – the natural reality is given. This second stage in Chandogya is, as it were, a modification of the first one. The sensations in one's dreams come from our consciousness, but they are the results of our natural attitude and are treated – like the aforementioned sensations – identically. In the natural stance we deal with immanent and transcendent acts. This distinction (in the Chandogya text) concerns the first two stages – but the first is the most important one. It will soon turn out (by way of definition) that what is transcendent will ultimately become immanent. Again, we attain the experience of autotelic being. As might be expected, then, Indra returns after another sequence of thirty-two years and receives the following explanation:

"When one is fast asleep, totally collected and serene, and sees no dreams – that is the self; that is the immortal; that is the one free from fear; that is brahman."

Indra then left, his heart content. But even before he had reached the gods, he saw this danger: "But this self as just explained, you see, does not perceive itself fully as, I am this; it does not even know any of this being here. It has become completely annihilated. I see nothing worthwhile in this."[7]

Looking for the experience of atman-brahman, Indra plumbs ever move deeply the layers of the presence of autotelic being. One can discern many affinities between the phenomenological method and the lessons imparted by Prajapati. This becomes the method for human self-knowledge. We find this motif distinctively in the Upanishads, as those texts (for the first time in the Hindu tradition) point to the fact that atman is "myself." Thus knowing atman means knowing myself – my deepest nature (essence). "Throwing away" in gradual stages the world's appearance (which means getting rid of the layers of the consciousness of appearance, emergence) – we reach the interior, immanent space. The Upanishads define it as the utmost inner space in our heart, owing to which all is (becomes) apparent. Consciousness exists in its own *milieu* constituting itself. According to the Upanishads consciousness of atman-brahman is the basic element of both modi of intentionality. In a way according with Husserl is insight, it becomes an autotelic world of its own, imbued with its own indelible qualities and features. It is the most important feature: that of consciousness. For it is intentionally that which must direct itself toward anything transcendent (including itself). If we assume this in our analysis, it will stand to reason that the third state (the only evidence of which is its prior being) will appear to Indra as emptiness. (For all natural features disappear in it.)

What does this mean? What might emptiness mean in this context? It is not the *śunyata* of Madhyamaka Buddhism – as that is an ultimate and positive being. Indra does not "stay" at this stage and soon leaves it, having found no good in it. This third state (deep sleep) leaves only a memory of it. Waking up one can only say that it has happened, that the sleep was good or not – but no sensations are formed. Resorting to our terminology, we can say that here consciousness does not constitute anything, does not function in the three temporal dimensions. After the moment of waking up, there is only an experience of the past.

In the analyzed Upanishad we find a distinct evaluation of the particular layers by which consciousness makes its appearance. The first three (having been "entered into" by Indra in so far as its essence is concerned) seem to him to be good. The deepest level is by its nature the everlastingly present, moreover, perennial and non-determined, "now." The third level intentionally directed towards the memory (remembrance) of its noema is deprived of something – it is emptied of everlasting presence ("now"). "In stating this we have not wished to say that the recollective consciousness has no independent right of its own, but just that this is not one of 'seeing'."[8] It is a kind of retentional modification – it is not a fundamental, presentifying consciousness.[9]

Atman does exist at all the stages. It is sometimes "overshadowed" by its appearances. The same holds good with the "I" ("ego"), which seems to be present all the while, which seems to be "necessary."[10] The third stage – although freed from all raw impurities seems to Indra to be empty – but that has no consequence regarding its absolute value. Indra is looking for that state in which all qualities of authentic being would immanently pertain to it, and these are not to be the correlates of consciousness – first appearing (emerging) as things, then as sensations and memories.[11]

To attain this fundamental experience, Indra goes a fourth time to Prajapati. This time the training lasts only five years. Indra then receives the following instruction:

This body, Maghavan, is mortal; it is in the grip of death. So, it is the abode of this immortal and non-bodily self. One who has a body is in the grip of joy and sorrow, and there is no freedom from joy and sorrow for one who has a body. Joy and sorrow, however, do not affect one who has no body.
The wind is without a body, and so are the rain-cloud, lightning, and thunder. These are without bodies. Now, as these, after they rise up from the space up above and reach the highest light, emerge in their own true appearance, in the very same way, this deeply serene one, after he rises up from this body and reaches the highest light, emerges in his own true appearance. He is the highest person. He roams about there, laughing, playing, and enjoying himself with women, carriages, or relatives, without remembering the appendage that is this body. The lifebreath is yoked to this body, as a draught animal to a cart.[12]

Thus the fourth state of consciousness is presented in a positive way. It is stable and certain: unchangeable in its essence, with immanent qualities of immortality, eternity, luminous energy, truth and bliss.[13] Searching for reality (owing to which Indra might attain all the world and fulfill all his desires) goes to the very source of it. This whole process (being at the same time the act of our self-knowledge) shows that by embracing fundamental consciousness we learn about the subsequent correlates that emerge as different layers of reality. Learning about the structure of the world, we may master the mechanism of it and reach the Ultimate.

*At all stages in which we deal with an aware cogito, it*, "belongs to the essence of the stream of experience of a wakeful Self that the continuously prolonged chain of *cogitationes* is constantly enveloped in a medium of dormant actuality (*Inaktualität*), which is ever prepared to pass off into the wakeful mode (*Aktualität*), as conversely the wakeful into the dormant."[14]

All those four worlds – which have been "waded" by Indra (overlap) are closely linked with one another. Loading for source Indra – treats each emergent, external, level as (determined) on a deep level. "the intentionalities in noesis and noemat rest on one another *in descending levels*, or rather *dovetail into one another* in a peculiar way"[10]... "over against the empirical experience, and *as the assumption on which it depends for its very meaning*, stands the *absolute* experience."[11] Resorting to the language of the Upanishad one may state that it is only pure consciousness which really exists. Its appearances, correlates cannot be said not to be of consciousness but their meaning is determined by an autotelic independent being. Thus they do and do not exist! We cannot pass any statement in so far as their metaphysical value is concerned. They are described in regard to the very process of knowing them. After the reduction there only remains pure consciousness as the matrix of intentional acts while the real world disappears from our view "leaving us" with the world-sense and the thirty sense.

At the beginning of this essay we have noticed the importance of a search for the reality of reality. This is both atman, the deepest, most fundamental element of microcosms and brahman, the most fundamental element of macrocosms. Although we think that there is a relation of identity the text was to show how it gradually developed. It stands to reason that Indra looking for the reality as such is actually looking for the real (authentic) source of it that is brahman. Indra is doing it in order to master all the world. This takes place in the process of self-knowing through the revelation of the genuine nature (essence) of atman. When finally we reach the fourth stage – we experience the moment of the identity of atman–brahman. Thus we recognize the nature of this primordial (original) consciousness seeing how all the worlds come from it in their appearances. For those who cherish (keep to the natural stance) it looks like an act of creation as well as (staying in one's consciousness). What Indra tries to reject (turn down) is the very intentionality of consciousness – e.g. the fact that it is always directed towards something transcendent. "It is intentionality which characterizes *consciousness* in the pregnant sense of the term, and justifies us in describing the whole stream of experience as at once a stream of consciousness and unity of *one* consciousness."

Knowing brahman equates with knowing atman – that is oneself. The identity of atman–brahman determines the conditions of veracity and one's ability to know something. Atman–brahman is described in Upanishads as a reality, consciousness and bliss *sat-cit-anada* (as it is shown in the final part of the text in question) "with the general understanding of the essence of the reason, of the reason stretched to its widest to cover *all varieties of the positing act*, including the axiological and the practical, the general elucidation of the essential correlations which unite the *idea of true (wahrhaft) Being* with the ideas of truth, reason and consciousness must *eo ipso* be secured. (...) Since the thesis of reason should be an original one, it must have its rational ground in the *primordial givenness* of that which is in a full sense determined: The X is not only meant in its full determinacy, but therein primordially given."

What we have tried to do is the phenomenological analysis of the object atman. Having presented what should pertain to its/his nature (qualities, terms), nature we have found the peculiarity of the analyzed object. Atman is both (at the same time) consciousness and reason. Knowing and a pure subjectivity. The pure "I" appears on many levels of consciousness. It seems to be the determining factor of the world-appearance so it will be a kind of the world–matrix. If atman is brahman then the principle of objectivity is identified with the principle of objectivity. We see here observed like in phenomenology the principle of identity of knowing and truth.

*Uniwersytet Jagielloński Kraków*

NOTES

1 Upanishads, transl. P. Olivelle, Oxford University Press, 1996, VIII.7.1.
2 *Ibid.* VIII.7.4.
3 Edmund Husserl, *Ideas*, transl. W. R. Boyce Gibson, London, 1931, p.153
4 c.t. 165.
5 *Ibid.* p.165.
6 Upanishads, *op.cit.*, VIII 10. 1–2
7 *Ibid.* VIII. 11.1
8 Edmund Husserl, *op.cit.* p.379
9 c.t. 392
10 c.t. 172
11 c.t. 220
12 Upanishads *op.cit.*VIII.12.1–3

(*Left to right*) Louis Houthakker, Robert Sweeney and Jorge García-Gómez

JORGE GARCÍA-GÓMEZ

# REFLECTIONS ON JOSÉ ORTEGA Y GASSET'S NOTION OF *ÀLÉTHEIA*

One may distinguish between several different, albeit interconnected, senses of the word *verdad*, or truth, in Ortega's work, namely:

1. *Adaequatio*, correspondence, or veracity;
2. Coherence or compatibility (or even logicality, the name he assigned to it when it was carried to the limit with the greatest possible rigor);
3. Self-coincidence, authenticity, or the lived agreement between one's thought and experience, on the one hand, and the beliefs at the core of one's life, on the other;[1] and
4. *Àlétheia*, unconcealment, or dis-covery.

Here the concept of the truth as self-coincidence, despite its great importance, will be left out of consideration, except incidentally or when it is unavoidable. A proper study of it will be left for another occasion. By contrast, I have dealt elsewhere[2] with the ideas of *adaequatio* and coherence *in extenso*, a fact allowing me to take the liberty of dealing with them here in summary fashion. To that end, let me avail myself of John Holloway's succinct formulation of their principled nexus and say that truth as such must certainly meet at least two requirements: one, to "'make sense' (that is, satisfy a criterion of internal coherence)"; another, to abide by "a criterion of 'correspondence with reality'...."[3] Ortega could not have failed to examine these two conceptions of the truth (and the theses in which they are advanced), based as they are on a logico-metaphysical tradition of long standing, be it that one interprets it in the light of realism or in that of idealism. Yet his analysis of the two ideas proves insufficient, whether one regards it as a function of his own doctrine or in view of the "nature of things," for they can be shown not only to be interdependent, but also to constitute the source of difficulties.[4] Moreover, it seems that any attempt to think them through leads one to a consideration of the other two dimensions or aspects of the truth, namely, self-coincidence and *àlétheia* (and their relationship). In this paper, I would like to engage only in a preliminary examination of the latter, to which task I proceed forthwith.

## TRUTH AS *MÉT'HODOS*

It is possible to approach the reality of the truth from a linguistic-historical standpoint.[5] In so doing, one would be well advised, following Marías, to



A.-T. Tymieniecka (ed.), *Analecta Husserliana LXXVI*, 67–97.

begin by placing it in its proper cultural context, according to which *three* different words would correspond to it in the Western tradition, namely, *àlétheia*, *veritas*, and *emunah*. Such a decision is of assistance in determining the senses given expression by those locutions taken from Greek, Latin, and Hebrew, respectively; that is to say, the senses that are "primarily ascribed to them." However, as Marías hastens to add, the restriction he introduces by means of "primarily" is not accidental, "because the belief would be illusory according to which the other significations would not be *marginally* lived in each one of the [three] cases."[6] And that would seem to be the case, I would say, not only because of the nature of the occasions by virtue of which such significations have come to form part of the given historical context, occasions that constitute, after all, a whole fraught with contingency, but also by reason of the fact that the significations in question are "decisive and essential dimensions of the notion of the truth...," as Marías himself points out by way of justification.[7]

Be that as it may, one could argue that the word *àlétheia*[8] in particular seems to correspond to a complexus of meaning consisting of three dimensions, to wit:

1. A "privative character... [since] it is the *un*-concealed ([as is suggested] if one thinks of *lanthánein* [= to be un-concealed])."[9] In my opinion, this aspect could be characterized primarily as the "objective" or noematic dimension of *àlétheia*, although these terms must be understood here in a non-idealist and pre-reflective manner.
2. "... [T]hat which is *not* forgotten ([as is clear] if one avails oneself of [words like] *léthos* or *léthe* [= forgetfulness])."[10] This aspect could be primarily characterized as the "subjective" or noetic dimension of *àlétheia* (the use of these qualifications being subject, again, to the same reservations as in [1]).

   Marías brings these two interconnected significations together when he says that "in both cases something *passive* is involved,"[11] by which assertion one certainly cannot do complete justice to the depths of experience to which such words correspond. This was clearly suggested by Ortega himself when he asserted that the classical Greek term (*àlétheia*) "originally meant the same as the word *àpokálupsis* later, that is, dis-covery, re-velation, or rather, un-veiling, re-moving a veil or cover."[12] This points in the direction of the fact that the truth is not something already there, ready-made and complete, but rather something that demands a process in which the truth would be verified or constituted, since, as Ortega remarked, "... [h]e who wants to teach us a truth should place us in a position to *dis-cover* it ourselves."[13] Or to put it bluntly: an active encounter (*àpó* = from or away from) appears to be the means necessary to deal with what exists, insofar as its self-manifestation or appearance is concerned.

3. In conformity with this, one would arrive at a third component of the meaning of the word *àlétheia*, one which would – at once – presuppose and surpass the prior two. Ortega had said as much simply by employing the locution *àpokálupsis* as a synomym of *àlétheia*, since the former serves precisely to refer to the "*dis-covery*, or the result of an action consisting in *dis*-covering (*apò*) what has been covered up or lies concealed (*kaluptein*)."[14]

Now then, if this is the case, one would also have to speak of the givenness of the essential possibility of the opposite or counterpart of the truth, that is to say, of "falsehood ... [which would be], by contrast, the [event of] *covering-up*."[15] By analogy, it would also follow that falsehood presupposes, as well, the characters of concealedness and forgetfulness, a conclusion that would point, if I may put it thus, to *falsification* as the active concealment and forgetfulness of reality, a possible compromise effected too by human beings, however many the various ways of covering-up may correspondingly be.[16] The paradoxical aspect of this may be given expression – and even underscored – by referring to *security* as a necessary component of *àlétheia*, since it would then seem legitimate to wonder whether or not one could find its match as a dimension of the concealment and forgetfulness brought about by the covering-up that falsification is. In other words, can a human being come to rest by living on the grounds of the fruits of falsification? In my opinion, this must be the case, even if one does not take into account the phenomena pertaining to the "relativization" of being, or to its removal from the place and function of last recourse and foundation at the level of appearance, for, if one did, the situation would be even more serious.[17]

It is not reasonable, however, to leave these matters at the level of principle, if one's objective is to gain access to the truth in all its complexity and concreteness. To this end, one must, first of all, emphasize the *manner* in which the truth emerges in our lives. According to Ortega, truth originarily presents itself "before" us as the *non-mediated absence of itself* or, to put it in positive terms, as "a certain thought [coming to] meet an *intellectual need previously* sensed by us.... The truth is, therefore, that which *stills* our *intellect's disquiet*...."[18] Consequently, what is involved here is no *mere* gap or lack one would experience, whether on the subjective or the objective plane. Such a gap or lack would be something that one could fill at will, or even *ad libitum*. One may clearly come to this finding if only one insists on the fact that, in order adequately to think of these questions, one must move exclusively on the basis of that which necessarily pertains to the lived experience of the truth, for, otherwise, one would be led to an understanding of the said disquiet as a simple mental "state" which, as such, would belong to the realm of the contingent and the perishable. Moreover, one would also come to deal with the possible noematic correlate of the disquiet – i.e., the truth as the fruit of veri-fication – as something optional, and with one's freedom to pursue it as a "condition" motivated by simple curiosity and even sheer arbitrariness.[19]

This, however, is not what one finds at all in the actual experience of the truth. In order to make things clear, let me formulate the question differently. The "subjective" source of the truth – its spiritual *negative*, so to speak – is *not* a *mere desire*, and *neither* is its "objective" or noematic correlate something *already* known to us, even if only as contained, in principle, in one's present capacity to determine its intrinsic possibility.[20] Now, this cannot be otherwise, since, as Ortega pointed out, a "desire does not exist unless, previously, the thing desired existed, whether in reality or – at least – in the imagination."[21] Accordingly, here I am not thinking of some vague *desiderium veritatis* (which would eventually result in the anticipation of something, be it by way of calculation or even of an inclination bordering on frivolity), but rather of a more fundamental pre-determination that would function as the *living equivalent* of the real and genuine truth. Now, if I am not mistaken, this is the same as affirming that one is dealing at this juncture with a *subjective need*, such that the correlate thereof would not have to exist already, or as a matter of course; instead, one would be face to face with a requirement and call to *do* and *make* something, though, to be sure, not in the sense of a mere possibility that is up to the mind to fashion and bring about,[22] but in that of something I *feel* as that which I *should* feel, even if it eventuates, as it certainly does, on a deeper plane than the one corresponding to one's usual moral experience. In order to do justice to the sense of the truth at this level, one must therefore distinguish between *mere* feeling (even if its character is necessitating, say, because of pathological or some other reasons) and a *genuine* subjective need, just as one must correlatively distinguish between the objective satisfaction of a desire (even one whose character is necessitating) and the necessitating and possible "object" of a desire, for, as Ortega told us, a "...genuine need exists without that which could satisfy it having to pre-exist it, *even if it is only in the imagination*."[23]

But, why is it that one has to engage in the search[24] for the truth? Why is it that the truth is not just found there, within reach, ready, so to speak, to be picked at pleasure? To be sure, the reason would not always be one's lack of knowledge. Often enough, it is just the opposite, for, as Ortega pointed out, "...if there were no *facts*, there would be no *problem* or *enigma*, there would be nothing concealed one would have to bring out of concealment or un-cover."[25] In other words, the difficult aspect of the genuine truth lies, first of all, with one's *separation* from it. And as Ortega himself indicated, the separation in question is frequently based on our *familiarity* with the facts, which, by dint of excess, may become an *obstacle* to our coming to *know* them or gaining access to the truth, i.e., to what things essentially are.[26] What is the truth about, after all? Undoubtedly, it is *not* about the facts or data of experience, but, rather, about the *reality* of things, which is precisely what is *veiled* because of the multiplicity of facts about them with which we are already acquainted (or with which we can become acquainted, by one or another of our anticipatory ways).[27] For, as Ortega hastened to add, "facts *cover up reality*,"[28] and this is so – as he made it clear elsewhere – because the facts constitute "everything that '*has to do*' with ... [the thing in question] but is not the thing itself...."[29] The facts *conceal* the *reality* of things, or at least are essentially capable of doing so,[30] since, by virtue of their boundless "emergence," they do not place us without mediation at the truth, but only "in the *chaos* and *confusion*"[31] typical of our "spontaneous living."[32] At best, our manifold familiarity with facts – whether actual or possible – could become the *occasion* for *inquiry*, inasmuch as it is possible for them – given a genuine subjective need in us – to spur or "instance" us to decipher them.[33] The facts *may*, in that sense, point us in the direction of the truth, but the charting of the path and its *viability* are not assured thereby (and much less is our success in the venture), since the traversing of the distance on our part is founded upon, and arises out of, the *living nexus* between our possible goal (i.e., the truth) and our present "consciousness" of chaos (i.e., our confusion or befuddlement).[34]

In nearly a summary fashion, Ortega identified for us, as follows, the milestones that would be found along the path toward the truth:

1. One initiates the passage at the level of perception, the forms of which would apparently have to correspond to the nature of the things experienced, if one is to be successful in the endeavor, though usually sensory perception provides the point of departure. Now then, this means, if I have understood Ortega correctly, that the facts do not play the role of

*mere* data; they also act – and do so essentially – as *motors* and *occasions* of "inquiry,"[35] since they present themselves to someone as a *function* of his or her genuine subjective need of *rendering an account* of what he or she is perceiving (or has perceived). As such, the need in question and the correlative "knowledge" that may or may not be obtained in light of the need are already dimensions constitutive of the "substance" of everyday life, at least in an inchoate form; they do not require, or automatically prompt, the creation of specialized forms of behavior (e.g., science, philosophy, or theology), which depend for their initiation and practice on *further motivations of a very particular nature.*

In one's many-sided everyday experience, one already finds the various nexuses one must rely upon in order to surpass it by striving to determine what is abiding about things (and, in principle, about oneself). Moreover, one does not come to realize whether the ways of perceiving and understanding typical of everyday life have reached a crisis point, unless one is immersed therein as a matter of course. And one certainly is. Now, the moment we are living at such a juncture, one has good reason to attempt to go beyond ordinary rational procedures, consisting as they do "in orient[ing ourselves] ... *from given to consequence*...,"[36] for being placed in such a situation is tantamount to being incapable of living and continuing to live in our already established ways.

Consequently, it is in light of the pockets of absurdity, with which one may be faced in everyday life, that the possible existence of a path different from the *usual* begins to take shape within that life itself and for its own reasons, that is to say, for the sake of continuing to live in terms of the realization of one's life projects. What I have in mind, principally, is the activity of theorizing, which, whatever its form, involves a radical intellectual style of *lógon didónai*. The order of proceeding of the latter is indeed the reverse of the customary one, motivated as it is by one's having taken cognizance, *at the level of everyday life*, that the normal ways and their products are insufficient not only to live, but also to continue living with a modicum of sense, once the crisis point has been reached. Under such circumstances, if one is to survive meaningfully, one must become aware of the fact that the *poverty* of things – and of one's life, insofar as it consists in dealing with the things of the world – is *not* coextensive with the degree of one's *un*-familiarity with their consequences, possible or actual. When one comes to this awareness, one is moving, or has already moved, from the practical to the theoretical order or plane of experience (or, at least, one is being faced with the need to consider – or even invent – theory as a *real* condition of possibility for the fulfillment of one's life). This is so because, precisely during that period, one's life becomes co-terminous with the realization that things are, *a radice*, "latent ontological traged[ies] ...."[37] To put it equivalently: one then comes to understand that the *poverty* of things (and, therefore, one's own, insofar as things and one's having to become acquainted with them are essential components of life) exceeds their "*misery of fulfillment*."[38] Such a state of deprivation cannot be overcome by the mere renewal of an insistent and systematic concern for consequences. The reason for that seems to lie in the fact that things – and the self, by implication – are "in want of *arkhé*, or principle,"[39] for the perceptual nexuses of life are without foundation. Thus, things at that point are felt to be *enigmatic*, because, to a significant degree, they cease to be mere instruments or tools for living and become *problems* to be solved, if one's life is to continue to be endowed with sense. But, whether or not one succeeds in discharging the tasks involved in their resolution, the fact is that the situation in question comes to pass when we "see" that the things of the world, essential as they are to living, nonetheless fail to ground themselves, and their foundation is nowhere to be found and remains concealed.[40]

2. Now then, this is hardly sufficient, because, to put it in Marías' words, "... [w]e must ask ourselves *how* this unveiling of reality *is possible*, how it is possible to go beyond the surface to reach the latent and make it patent, how it is possible to ... bring it to the surface, to place it in the light and thus to *make* it true...."[41] To meet with success in doing this, one has to learn to see the aspects and behaviors of things as so many "phenomena" or appearances in the radical sense, that is to say, as arising from their proper *arkhé*. This serves clearly to indicate, it seems to me, what the "nature" of the search for the truth is, consisting, as it would, in seeking the very principle (or principles) of things. And this means, if one wishes to arrive at firm and concrete conclusions, that one must proceed carefully from argument to argument, a procedure subject not only to the requirement of advancing in a logically valid manner, but also to the condition that at least its ultimate premises give expression to essential intuitions. This is suggested by Ortega's contention that, in order to set out "to un-cover [*àletheúein*] reality ... we must, for the moment, *withdraw* from the facts surrounding us, and remain alone with our minds."[42] To accomplish this, we have to radically avail ourselves of our imaginations, something that may only be done cogently under the

control of the doubt.[43] According to Rodríguez Huéscar, the significance of this recourse is twofold:

On the one hand, the facts, when they are regarded in light of one's genuine subjective need (and, I would add, only then, if fitting),[44] become available as "'*instancies*'[45] *of interpretation,*" and the truth – that is, the process of veri-fication in which the truth *qua* product is constituted – "is tantamount to interpreting them."[46]

On the other hand, the *withdrawal* that Ortega points to does not imply that, by such a move, we would "lose touch altogether with reality,"[47] but the diametrically opposite view, for our minds would endeavor thereby to be in the most *intimate* terms possible with things, precisely because they had become problematic to us. Consequently, our endeavor to remain exclusively by ourselves, "except for the reality of things," is tantamount to removing everything that may lie "between them and us,"[48] that is to say, to bringing about a situation in which neither everyday experience nor any idea in fashion (however dominant or firm it may appear to be) nor the tissue of beliefs rendering both possible is allowed to serve as a bridge between things and us. This means, if I am not mistaken, that to achieve the *highest degree of solitude* one must come to be in the *greatest proximity* of the things themselves, or to keep them *company most closely*. To envisage it correctly, this outcome has to be taken as the opposite of straying in the midst of things (a predicament in which we seem to find ourselves in the wake of empiricism, positivism, naturalism, and other currents of thought issuing from the decline of realism), or of the abandonment of things (which appears to be one of the fruits of idealism). "Truth" is just the name assigned by Ortega to such a result,[49] a name the application of which must be well-delimited, if one is to do him and reality justice, for it does not primordially refer to any state or condition (even if it satisfies certain decisive requirements), or fundamentally to any mental events (which as such would correspond to the process of veri-fication), or even to the judicative product that the latter may lead one to. Rather, according to Ortega, it signifies, above all, a movement away from oneself and in the direction of things, one that would certainly require on one's part an attempt to form a precise concept of the nature of the thing in question. It is by means of such a concept, as the fruit of the rigorous employment of one's imagination, that one would come to establish a dwelling place in the close vicinity of the thing under scrutiny. That concept corresponds to the formula resulting from the adequate comprehension of the reality of the thing. Assuming one can and does successfully take that last step, one would then become assured of the real possibility of having surpassed the radical crisis point, that is to say, of living and acting, beyond it, with a modicum of sense in the only context where that can occur, namely, in the world and by means of the things of the world.

Ortega did not rest his case, however, at the level of such generalities. On the contrary, he took one further step in order to make it clear that rigorous imagining or thinking, if it is to become the royal road to the truth, had to go through the following two stages, which, though successive, implicate one another:

First, and "at our own risk, [we are] to imagine … a reality,"[50] i.e., precisely the *reality of those things that have become problematic for us*. But this is equivalent to *endowing* them with an *abiding nature* of which they are devoid, to begin with, at the level of *appearances*, or that they were *deprived* of on that plane when they became problematic for us, be it as the result of our metaphysical or scientific reflection, or because the ideative system already established on the basis of the belief regimen in force (*status quo ante*) gave way, at least significantly, by having reached a point of historical crisis. This seems to correspond to the formation of a privileged, hypothetical interpretation Ortega called the "idea of reality."[51]

Second, "by *keeping* to the solitude of our imagining, we [may] *discover* which aspect, which visible patterns of organization, in brief, which *facts* such an imaginary reality would produce"[52] for us, were we actually to live in a world that is, presently, only given by way of anticipation.

These two stages of thinking – when they are conjoined and also conceived in a sufficiently broad fashion – constitute the notion and sphere of what can be called, following Ortega, truth as *coherence, compatibility*, or *logicality*.[53] Consequently, the fruit of the first stage (i.e., the "idea of reality") must be regarded, so to speak, as being both the source of and the means for testing the products forming part of the domain of the true in that sense.

3. Then, still engaged in the activity of thinking understood as the exact employment of fantasy, and *by means of it*, we proceed to "leave behind the solitude of our imaginations … and *compare* the facts that the reality *imagined* by us would produce with the *actual* [or perceived] facts surrounding us."[54] But this is precisely what has been referred to as the phenomenon of the truth *qua* correspondence or *adaequatio*, and which Ortega also called veracity.[55]

At least for the moment, this is the way in which thinking brings the circle of its trajectory to a close. As was pointed out earlier,[56] it began as *ànábasis* or withdrawal away from things by making the perceptual world (or the thing or event in the pertinent region thereof) its point of departure. The withdrawal in question amounts to going, so to speak, against the current, and yet, paradoxically, it may eventually result in a return to the point of origin. Let me insist on this, for the formula I just employed serves to give expression to one of the essential requirements to which life is subject. Since "we are given over to ..."[57] the immediate, as Marías admirably puts it, when, to such ends, we engage in the disciplined employment of the imagination, what we do is only to avail ourselves "of a device, of an instrument,"[58] by which to constitute a new region of *irreality* (or inner world) that would allow us to endeavor to establish a settlement in the midst of the circum-stance, which had turned inhospitable to the extent that it had become enigmatic. This is what Ortega exactly described as "the tactical turn which we have to take in order to cope with the immediate,"[59] and which Marías, on his part, clarifies by asserting that such are the means that permit us to exercise our dominion over things, and to possess them,[60] in the wake of their having previously proved inaccessible to us by having become unintelligible *for our purposes*. In brief, the viability and continuation of life (as a totality of sense in the making) depends on the (relative) success of the work we do in the world on the basis of our imagining, insofar as our life can be effectively fashioned by us only by our coming radically to understand things. This notion leads us, however, to a new difficulty (or, to be precise, to the other side of the difficulty already under examination), for it is possible to ask whether thinking can truly return to that from which it originally departed, inasmuch as it had presented itself as something enigmatic and now presents itself to the extent that the inner world we have been shaping casts its light on it. In other words, the circumstance has been transformed in view of that inner realm and has thus become a world, i.e., a whole organized as a hierarchy of planes.[61]

Suppose we find ourselves ready to take a step that would meet the conditions stipulated by Ortega.[62] At best, the possible facts that were then portrayed in our imagination – i.e., those given shape on the grounds of the "idea of reality" as a pure product of our fantasy – and those given in perception would dovetail or, at least, would prove to be compatible with them.[63] But, if this is so, we could then assert that "we have deciphered the hieroglyph," that is to say, that "we have un-covered *reality*," which, as such, *always lies beneath or behind the appearances*, because "the facts *covered* [it] *up* and *kept* [it] *secret*."[64] Or to put it equivalently: we would have then succeeded in formulating the origin and justification of the facts, by establishing that they are what they are because they are consistent with the *reality* of the thing in question, whether logically or by way of adequate conformity. Now then, were we to translate such notions into classical terms (which, as we now can appreciate, are both well grounded and derivative in character), we would have to argue as follows:

A. On an "objective" plane, we would have gone from the level of *appearance* to that of *essence*. We would no longer rest content, therefore, with understanding things as a function of the use we put them to, or could, in light of that which is taken for granted (i.e., the beliefs socially in force in the given period)[65] and the practical patterns derivable therefrom, on the one hand, and our motivations,[66] on the other. This way of proceeding would ultimately result in the formation of various series of means which, on such a basis, would be ordered to action.[67] By contrast, we would be able to ground the intelligibility of things in enduring structures, which, while hidden from our practical gaze, would nonetheless "serve to render an account of the manifest behavior"[68] of things. In other words, we would no longer be involved just in an effort to "know … for the sake of action,"[69] but rather in a search for the *being* of things, with which they would be endowed only as correlates of the inquiry determining it, since being is, according to Ortega, its end and product.[70] At the level of everyday experience, where we count or rely on things, "things therefore *are-not*; they are *no-thing*,"[71] because on that plane they are not given as "enduring structures" resting in themselves, "but only as [changing] actions and responses with regard to [the successive objectives and plans of] my life"[72] and in respect of whatever happens therein and befalls me. One may then assert that *being*, in that sense, "is … not what I live and have, but, rather, *that which is not and is lacking* in everyday experience. Being is the *radical* name for the insufficiency and poverty of life…,"[73] as well as a remedy thereto, for being is precisely that which things are in need of under specific circum-stances, if they are to come to *be themselves*. This point becomes clear only when we find ourselves called upon to trace things to their ultimate roots, if we are to go on living in the world meaningfully, and discover that, for such a purpose, that which is constituted on the grounds of what is taken for granted is hardly enough. Now, this is the case not only insofar as the process of rendering *genuine reality* manifest takes place before someone who is in need of it, but also to the extent that what is rendered manifest to him or her becomes manifest, paradoxically, both *from itself* and as the *dynamic correlate* of one's radical inquiry. In this sense, one must assert that *being* is that which – in one's life and by virtue of its neediness and motivations – is constituted as principle (*arkhé*) and living source (*phúsis*) of the appearances the sense of which has become problematic.[74] Consequently, it is not only legitimate, but also necessary to affirm both that *being* or *essence* is the true reality of things and that the truth is the "true *reality* (*alethés òn*)…."[75]

B. Now then, this very conclusion allows us to complete the picture of living, if we move to a "subjective" plane that is nothing but the correlate of the "objective" one we have just considered, since "it is precisely thinking which struggles to remedy the constitutive opacity of life."[76] In virtue of the fact that "being, the thing *itself*" (or the reality thereof) "is essentially that which lies concealed, covered up…,"[77] as Ortega himself pointed out, we must assert not only that being corresponds to a *certain* intellectual operation, as I have already indicated, but also that the emergence of being requires it. The operation in question is no other than the one that "leads us to find … [being] under its concealments…," and it amounts to the special style of mental performance before which the thing is given in a manner that is special too, namely, by "*lying bare* before us, and this is its 'truth.'"[78] Now, as has already become evident, an operation of that sort is anything but simple, since, being as it is a "process of dis-covery," it consists, in Rodríguez Huéscar's words, in enduring life

> between two enigmas: the enigma serving as *one's point of departure* (i.e., that of the "facts" or interpretations [that have become problematic]) and the essential enigma *one* [would] *arrive at* (i.e., that of genuine reality, or reality freed from interpretations).[79]

Or to give it expression in equivalent noetic terms: *veri-fication* is the process of "illumination"[80] (or *àletheúein*) that is provoked in our lives by the conjunction of problematicity and genuine subjective need, is developed by the *imaginative* and uncertain progression toward *coherence*, and is at its climax when thought is brought to adequate *correspondence* with the concealed *reality* of things.

This outline of thinking, as an *ironic* enterprise in which one attempts to do justice to things in depth, would remain incomplete, however, if one did not come to a formulation that would be more appropriate to the *terminus* of the imaginative process involved. Let me attempt to come to it by approaching the said terminus insofar as it is contained in the process in question as its implicit, albeit only possible, end; that is to say, to the extent that, in the progression of thought, one aims at constituting "... the *lógos*, that is to say, the "sense" [that is embodied in a concept and serves to] establish *connections* and gradually 'binds one thing to another and everything to us, in a firm essential structure'...."[81]

To discharge that task, I do not believe it would be idle to appeal to texts that, in spite of belonging to different authors, are nevertheless intimately linked, inasmuch as they give expression to two *genuine* efforts to come to terms intellectually with the *same thing*. Let us take a look at them:

A. First of all, let me consider for a moment a passage in which Rodríguez Huéscar endeavored to characterize Ortega's metaphysical thinking as both a constructive and a destructive enterprise. To that end, he argued that "... the result of the *view* to which I have been led by eliminating the interpretations of others cannot be but another interpretation of reality that would have its foundation in that view, namely, *my own interpretation*."[82]

Accordingly, one could say that Ortega's metaphysical thinking is in the nature of a "*lógos* that would render something manifest." This formula may be regarded as accurate on the condition that it be taken to refer only to the disciplined way of imagining that is "limited to *interpreting* the reality [kept] in view"[83] as one strives to carry it to its ultimate consequences. In other words, it is a question of holding a *view* of what things *themselves* are, of an act of intellection by means of which one would bring the *reality* of things to *e-vidence*. The "*lógos* that would render something manifest" would therefore correspond to "'saying' *how* I see ... [the reality kept in view], *how* I have seen it.... The essence of interpretation" – and metaphysical thinking is necessarily *interpretive* or constitutes itself as the *lógos* of *e-vidence* – "consists, in effect, in rendering a '*being*'*-as* manifest."[84]

The correct understanding of this point requires, it seems to me, at least two observations:

First, the fruit of the *lógos* of *e-vidence* (or the *correlate* of thinking in form, i.e., of radical or metaphysical thinking) cannot be regarded, as a matter of course, as an element of the endowment of everyday life. On the contrary, it corresponds to a function that is activated exclusively when the need arises, that is to say, only when a certain binomial pattern is constituted by the co-givenness – in active reciprocity – of one's circumstance (to the extent that it has become enigmatic)[85] and the personal dimension of one's experience (insofar as one lives therein under the spur of a genuine subjective need to overcome the emerging unintelligibility of appearance). In his splendid remarks on a passage of Ortega's *Meditaciones del Quijote* concerning clarity as an intrinsic dimension of life,[86] Marías made precisely that point when he said:

To be a man is to illuminate, clarify, cast light on things, and thus to uncover them, reveal them or to unveil them, to make them patent and place them on view, to make them *true*. Ortega said before that truths, once known, acquire a utilitarian *crust*, and no longer interest us as *truths*, but only as useful formulae. This is literally the case: they cease to be true as such, in the sense of *alétheia*, since they are left "covered" with a *crust* ... and lose their quality of patency and clarity.[87]

The word "man" does not ultimately refer us, therefore, to one's everyday experience, but, rather, to the *tense manner of living* called "authenticity" or "self-coincidence," for, as Ortega put it, "[c]larity is not life, but it is the fullness of life."[88] Authenticity or self-coincidence, then, as the acme of life, may be presented, as a function of this discussion, as one's "dying" (*desvivirse*) to make truth manifest and justify its condition and presence in life.[89]

Second, the notion of "'being'-as" may lead us to an interpretive error consisting in placing the radicalness of Ortega's metaphysical position at a level not its own. This is the reason why Rodríguez Huéscar hastened to record, after the passage quoted above,[90] the following reservation:

"'*Being'-as*" thus becomes, for Ortega, a fundamental "ontological category" which includes, within itself, as one among many, the very *interpretation* of reality proper to traditional philosophy, namely, the one that takes reality *as being*.[91]

Mark this point well. Accordingly, it would be more accurate to say "*living-as*"[92] than "*being-as*." This realization allowed Rodríguez Huéscar legitimately to underscore, as a fact, that "this [notion] has little to do with the Heideggerean account of *adaequatio* as the 'as' ... of the thing 'represented' in the proposition..., because Heidegger is moving within [the scope of] the interpretation of reality *as being*...."[93] In saying this, Rodríguez Huéscar did not mean to reject or scoff at Heidegger's interpretation of reality; on the contrary, he intended thereby only to place it where it belongs, namely, on a plane that would have its niche in life, when conceived, as in Ortega, as the primordial reality at the level of appearance. But that is tantamount to finding its proper justification, according to which Heidegger's inquiry into being would constitute a derivative enterprise "encompassed" and "grounded" by living reason.

B. For his part, Marías remarked that "... the radical sort of human security" – that is, the one achieved, as was pointed out above,[94] by way of a *tactical turn* rendered possible by the formation of concepts in the disciplined exercise of the imagination – amounts to "... the *clarity* in which the meaning, or *lógos*, of things *appears* or manifests itself, is made *patent*...."[95] This simple but far-reaching assertion permits us to forestall a possible *subjectivistic* misunderstanding of the interpretive stance adopted by Rodríguez Huéscar.[96] In fact, it has nothing to do with it, as is clear from the following: *lógos* as a noetic process, or as thinking, is not only necessarily other than *lógos* as the noematic product of that process, or as the sense (or reality) of things, inasmuch as the former is the essential correlate of the latter, but it is, also and above all, an action in which one goes in search of the sense or reality of things, a movement that is, therefore, always characterized as the *pursuit of the truth*, not as the truth itself that would be its counterpart, were the verification to meet with success. Now then, *e-vidential* thinking, or thinking in form, by contrast with the constructive sort or with the one unfolding by way of one kind of calculation or another,[97] would achieve fulfillment in the experience of correspondence or *adaequatio*, which may eventuate when "we *compare* what we think or utter of reality with"[98] the things themselves, as given in perception, and of which we are motivatedly thinking. Thinking is undoubtedly a search for security; indeed, it amounts to security when it meets with success, at least in view of the delimited objectives it is intent on confronting on the given occasion. In other words, thinking is, ideally speaking, the human action yielding the "interpretation which permits us to know what to hold to in relation to the real; that is, with respect to what *there is* and with which we have *to have it out*, by accounting for it...,"[99] in order to live and go on living meaningfully. That interpretation is what Ortega called the "idea of being" and characterized "as a plan for this 'holding to' with respect to things."[100]

Let me insist on this point for the purposes of accuracy. Even though security is a state or condition someone may come to and, therefore, is a "subjective" determination (i.e., ultimately belonging to my life, be it at the global level or on the particular plane of the functioning of an objectivating self in a given problematic nexus), that does not imply, however, that security amounts to being a further specification of "subjectivity." On the contrary, security is the state or condition of life which is grounded in "that interpretation which permits us to know what to hold to in relation to the real," that is, with regard to (the reality of) the things with which we fashion our lives. But, to use Marías's words, "the idea of *being* as a plan for this 'holding to' with respect to things" is the *product* of thinking, i.e., the sense or *lógos* by means of which "one thing [is bound] to another and everything to us, in a firm essential structure."[101]

### *ÉROS* AND *ÀLÉTHEIA*

In the work from which this assertion by Ortega is taken, he speaks to us of philosophy as "the general science of love,"[102] which, in his view, "represents the greatest impulse toward an ... [all-embracing *connection*] within the intellectual sphere...."[103] Connection, or connectedness, and love are seen, then, as intimately related, but only in a special way, which is conveyed, if I am not mistaken, by Ortega's notion of *salvation*, the play of which, as an exercise of reason, amounts to this: "given a fact – a man, a book, a picture, a landscape, an error, a sorrow – to carry it by the shortest route to its full significance."[104] And this essentially means, as Marías put it, that "... [l]ove, connectedness, [and] reason are the three elements that converge in the constitution of philosophy."[105]

Now, in order to do justice to both Ortega's doctrine and the reality of things, one must come to determine the sense in which love plays such a fundamental role in philosophy. Even though this is not the place to deal with such matters fully, I must nonetheless take into consideration, as very significant, the fact that Ortega, to that end, availed himself of the concept of *éros*. In particular, he referred to a passage of the *Symposium* in which Plato said of *éros* that it intervenes *hóste tò pân autò autòi sundedesthai*,[106] which reads – in Jowett's translation, for example – " [so that] in him all is bound together...,"[107] but that Ortega beautifully and insightfully translated by means of the Spanish equivalent of the clause, "so that everything in the universe might... [live in connection]."[108]

The accuracy of Ortega's version lies, paradoxically, in its imprecision,[109] which, I believe, was not indeliberate. Let me attempt to show that this is so. The clause in question acquires its full significance, of course, as a function of the meaning of *éros*, its referent. Ortega chose the Spanish word *amor*, which, as "love," its English counterpart, has a much broader sense[110] than that of *éros*, though, to be sure, it may include it. In keeping with that choice, Ortega contended that (a) "love... binds us to things...,"[111] and that (b) the "beloved object," or the thing insofar as it is the correlate of one's love, "is, for the moment, indispensable. That is to say, we cannot live without it...."[112] Up to this point, the erotic dimension of love would carry the day, for, as Plato said, "I would have you consider whether 'necessarily' is not rather the word. The inference that he who desires something is in want of something, and that he who desires nothing is in want of nothing, is in my judgement... absolutely and necessarily true."[113] This, it seems to me, presents us with the basic sense of Ortega's assertion to the effect that "love... binds to things." However, he did not rest his case with that proposition, for he added the decisive point that "...[t]here is... in love an extension of the individuality [of the lover] which absorbs *other* things into it, which unites them to us."[114] In affirming that, Ortega certainly placed himself on the plane of *éros* (for, as Plato contended, in loving we seek to *possess* the beauty – or the goodness – of the beloved object, of which we are bereft and in need),[115] but, in so doing, Ortega moved to another plane, at once holding on to *éros* and transcending it, inasmuch as this "union and interpenetration enables us to… [*delve* into] the properties of the beloved object."[116] It is by means of this remark that we may begin to appreciate the paradoxical accuracy of employing an imprecise term like *amor* for *éros*.

To be sure, because of the erotic dimension of love, we are eager, even *anxious*, to make the beauty or goodness of the beloved object ours and to keep it to ourselves. In fact, it is by virtue of it that we abide in the "consciousness" of our essential poverty, for, however tangentially or implicitly, we remain aware, even at the moment of success, that we are still bereft of something important we are in need of. As Plato put it, the lover "desires that what he has at present may be preserved to him in the future, which is equivalent to saying that he desires something which is non-existent to him, and which as yet he has not got."[117] However, it is not enough to insist on this. We must also focus on the condition that specifically renders the beloved objectively lovable to us. Ortega took a decisive step in this direction when he pointed out that "… [delving into] the properties of the beloved object"[118] allows us to "observe that the beloved object is, in turn, part of something else that it requires and to which it is bound…."[119] In other words, it is the eye of love which opens our lives not only to what we need and is embodied in the beloved object, but also to what the latter requires; and it is capable of doing so by placing it where it necessarily belongs, namely, in the *context of its connectedness*. Accordingly, when we love, that which is "indispensable for the beloved object,"[120] namely, that which "it requires and to which it is bound,… also becomes indispensable to us …".[121] The expansion of the lover and that of the beloved object are thus found to be essentially correlative dimensions of love. In other words, we can truly be said to love if we desire both what we require and what the beloved requires. The lover, in order to be active as such, must transcend the erotic dimension of love; and he or she does so only by being of service to the beloved object. This action aims at complementing *éros*, a goal that may be achieved if the lover opens up to a different dimension of the same love, namely, the one traditionally known as *benevolence*.[122]

It is worth noting that Leibniz, for one, would have agreed with the doctrine of love just presented here *in nuce*, at least insofar as love is fulfilled in the interaction and reciprocity of persons acting as such. This can be appreciated by means of the following. First of all, Leibniz asserted the proposition that "to love or to cherish is to find pleasure in the happiness of another, or what amounts to the same thing, to accept the happiness of another as one's own...,"[123] a point admirably condensed by Pieper in the formula, "shared joy."[124] Then, Leibniz proceeded to derive from it a corollary of major importance, since it provides a way of resolving the contradiction apparently involved in the passage from *éros* to *philía*, and in their conjunction in one and the same act of the lover. It serves likewise to clear up any lingering doubts about the possibility of real disinterested love, of that love "which is free from hope and fear, and from every consideration of utility...."[125] This is the corollary in question: "The happiness of those whose happiness pleases us is *obviously built into our own*, since things which please us are desired for their own sake."[126] Now then, this is not only compatible with the thesis I advanced above, namely, "the lover must transcend the erotic dimension of love... [in] complementing it [by means of] ... *benevolence*," but in fact it is also equivalent to the said thesis. This contention would be justified if one could carefully *generalize* the scope of the corollary, so as to include – as part of the extension of the concept of beloved object – not just persons, but things too, although, to effect the transition legitimately, it may very well be necessary to avail oneself of some mediating principle, perhaps one that could run like this: "The sphere of lovely (and, therefore, of lovable) things *as such* is constituted only *for* persons, and in light of their purposes and projects." If this were not so, it would not be permissible for Leibniz to say, as in fact he did, that "to *love* is to be brought to take pleasure in the perfection, good, or happiness of the beloved object."[127] As he made it clear, his mind here was no different from what he had formulated in the "Preface" already quoted, [128] and yet he did not rest his case with that. Indeed, he pushed his analysis further, and this led him to introduce a restriction in the employment of the concept of beloved object, for at that point he thought he had discovered an ambiguity in its use. As he then argued,

> ... properly speaking, we do not love anything that is incapable of pleasure or happiness, and we enjoy things of this nature without, for that reason, loving them, except by way of prosopopeia, and as if we imagined the things themselves enjoying their perfection.[129]

Undoubtedly, Leibniz was correct in detecting a difficulty here, but his attempt to resolve it, by his appeal to the use of the figure of speech known as prosopopeia or personification, is hardly satisfactory. It is not enough to identify a rhetorical practice, even if one is right on target in doing so, as Leibniz certainly was. One must go further and determine the metaphysical principle at its basis, so as to be able to legitimate the practice. Now then, that was precisely what I was suggesting above, when I pointed to the mediation seemingly called for in the process of generalization of the concept of beloved object.

Consequently, in following Ortega, it would be necessary to insist on the fact that the lover is concerned with his or her own perfection and with complementing the object of his or her love. Moreover, one would have to emphasize the fact that he or she strives after both in unison, since significant success in the former endeavor is ultimately achievable by means of the latter one. Yet, even if this could be done out of self-interest, even on a reflective, calculating basis, it does not seem to me that love can be sustained for long, except when the lover seeks his or her own perfection and the complement of the beloved object for each other's sake, since, paraphrasing Leibniz, the perfection of the beloved object is obviously built into our own, and, further, "things which please us are desired for their own sake."[130] Therefore, if one's love is in principle to have the capacity of complete fulfillment, then it must essentially involve, to use Ortega's formulation, an effort to bind "one thing to another and everything to us, in a firm essential structure."[131]

Let us try to understand this point with greater precision. Elsewhere, Marías insightfully points out that "every thing is to transcend [*salir de*] itself, if it is to achieve its perfection...."[132] This notwithstanding, he may mislead us if we are not careful, for, as a way of introducing his remark, he asserts that the "perfection of the *beloved object*... is not an 'intrinsic' constituent of it...."[133] However, the text itself, to which Marías is referring us in his commentary, in fact belies it, for in it Ortega affirmed that "... [e]ach thing is a fairy whose *inner* treasures are concealed beneath poor commonplace garments...,"[134] as if to protect its own being, while it lies in wait for the arrival of someone who, by means of the right word and deed, would cast the spell allowing it to gain access to its own interiority[135] and thus to become capable of "enjoying... [its own] perfection," to use Leibniz's phrase. To be sure, in loving, the lover is seeking his or her own perfection, and yet the latter would ultimately remain beyond its reach and resources. If this were not so, a lover could meet with success simply by the practice of self-love, and the ineradicable experience of self-transcendence – as a necessary ingredient of love – would become unintelligible. Rather, the lover's perfection can only be secured, in principle, by bringing about the perfection of the beloved object (that is to say, in Plato's terms, by increasing its goodness or beauty). This is clearly suggested by Ortega's use of words like "inner" and the opening phrase in the sentence, "within every thing, there is an *indication* of its possible plenitude."[136] For one thing is what one is seeking (by virtue of lacking and being in need of it), another the *means* to engage in the search. Most perceptively, Ortega employed the locution "virgin"[137] to refer to the beloved object, and that points to the fact, if I may put it thus, that the beloved object is *pure from the virtual reality* which is, nonetheless, its very own, the reality which is, for that reason, its truth.[138] Let me put it otherwise: the reality or truth of the beloved object is not something external to it, or *imposed on it from without*, although that may appear to be so at times, for *falsification*, the commission of *error*, and even the phenomenon of *lying* are among the possibilities to which love and its attendant "inquiry" are essentially open for diverse reasons. The reality or essence of the beloved object pertains to it precisely as its virtuality, [139] and this is the reason why it requires the mediation of the lover who – as a possible "mover" that would act on the grounds of his or her pertinent motivations – is found inscribed in the virtual makeup of the thing as its would-be "cognitive" activator (or, reciprocally, the virtualities of the object are "built into our own," to give it expression by means of Leibniz's formula).[140] In order to be able legitimately to elicit the reality of something, or to render it manifest, one must do it by departing from the thing as given in perception, even though that will prove insufficient, as is clearly suggested by Ortega's metaphorical employment of the word "fruitful" when, in reference to the beloved object, he asserted that it is "a virgin that is to be wooed or courted if it is to become fruitful."'[41] In other words, one is to un-cover (*àletheúein*) the reality of the beloved object, or render it patent, by having recourse to the beloved object as source, i.e., as the concealed, intrinsic foundation which, in the thing itself, functions as the origin of its appearances and behaviors.[142]

Finally, let me recapitulate or summarize the findings of this analysis. Love is, to begin with, *éros*, that is to say, the movement of life taking place by virtue of one's lack and need of perfection, a movement that is directed toward the other *qua* beloved object and is bodied forth by means of it. And yet, paradoxically, it has to develop and transform itself into the practice of *generosity* in regard to the beloved object. This is an essential condition to be satisfied by *éros* as a dimension of love, if *éros* is to remain itself, i.e., the action by means of which the lover seeks to achieve his or her own perfection. Yet in philosophy – as one of the sublime forms of *éros* – one does precisely come face to face with one's own sheer poverty. If however that were to be believed, one would only come to confuse the philosopher with what Plato called the *philothéamon*,[143] whom he defined in terms of mere curiosity, or the unsatisfiable "taste for every sort of knowledge...."[144] Surely, we cannot say that we are entirely dispossessed as long as we move on the surface of things, for they make available to us, at every turn, their wealth of appearance. We are nonetheless poor at the level of principle. The *arkhaí* of things remain concealed from us as long as our "eyes" are pleased with resting on their surface. This sort of poverty can only be helped by engaging in – and to the extent that one succeeds at – the *task* Ortega referred to by the word "love," which signifies the endeavor to *save* things, or to "carry ... [them] by the shortest route to ... [their] fullest significance."[145] But that demands that *we* place them, one by one, at the *center of the universe*, as it were, in order to magnify them by the realization – in the twofold sense of the term – of their inner wealth. However, the explication of the internal domain of things does not culminate in a new (excess of) appearance; rather, it is the unfolding of *their own* phenomenal *system* of virtualities brought about by placing them – imaginatively, conceptually – in the "all-embracing connection"[146] to which they belong necessarily, albeit only potentially, "so that everything in the universe might ... [live in connection],"[147] to use the English equivalent of Ortega's own version of Plato's words. But is this not, in the final analysis, the objective intended, *mutatis mutandis*, by both Plato and Ortega when they philosophized, or set out in the pursuit of the truth? Accordingly, this way of thinking *in extremis*, this strict *discipline of life* which philosophy is, can be rightfully characterized – to use the words of Plato – as the love "not of a part of wisdom, but of the whole."[148] Thus, it is evident at this point that, however mindful of appearances we may be as long as we live (and for the sake of living and continuing to live), by reason of both liking and necessity, we should not mistake philosophy for a mere taste of surface and novelty, no matter how nuanced the former or seductive the latter could be. Rather, if one judges the question in view of the goal the philosopher strives to fulfill, possess, and enjoy, we must come to identify philosophy with the "vision of the truth,"[149] which is not just at the successful "end" of life (if I may speak in such terms), but is already present *and* at work – as the intrinsic objective of the movement of life – in the very charting and traversing of the path[150] in which we find ourselves when we endeavor to uncover (*àletheúein*) the foundation of things – and this is so to the extent that such a foundation is the creative origin of the appearances by means of which the reality of things comes to be expressed at their surface, a context of phenomena that indeed spurs us toward that origin, if only a genuine subjective need thereof is operative in us, and we do not close ourselves to it. In this light, it might be both justified and opportune to restore to philosophy its "original name," to wit: *àlétheia*,[151] as the enduring name it deserves insofar as it signifies inquiry or verification,[152] essentially involving as it does the "endeavor at discovery and at deciphering enigmas to place us in contact with the naked reality itself."[153]

*Long Island University*
*Southampton, New York*

## NOTES

[1] Cf. Julián Marías, *Introducción a la Filosofía*, 9th. ed., in *Obras* (2nd. ed., Madrid: Revista de Occidente, 1962), Vol. II, p. 95 and my paper "La teoría orteguiana de las ideas y las creencias. Una dificultad interpretativa", *Humanitas* (Universidad Autónoma de Nuevo León, México), 1999, pp. 137–138.

[2] Cf. my paper, "Caminos de la reflexión. En torno a la teoría orteguiana de las ideas y las creencias", *Revista de Filosofía* (Universidad Complutense de Madrid), 3ª época, Vol. XI (1998), Nos. 19 (pp. 28ff) and 20 (pp. 122ff).

[3] J. Holloway, "A Reply to Sir Peter Medawar", in Peter B. Medawar, *The Hope of Progress: A Scientist Looks at Problems in Philosophy, Literature, and Science* (Garden City, N. Y.: Anchor Books/Doubleday, 1972), p. 43. Cf. P. B. Medawar, *ibid.*, p. 27.

[4] As already pointed out, I have taken into account these two ideas and their implications elsewhere, cf. *supra*, n. 2.

[5] Cf. J. Marías, "Comentario", in José Ortega y Gasset, *Meditaciones del Quijote* (Madrid: Ediciones de la Universidad de Puerto Rico/Revista de Occidente, 1957), pp. 296–304; Holger Helting, "*à-létheia*. Etymologien vor Heidegger im Vergleich mit einigen Phäsen der *à-létheia*-Auslegung bei Heidegger", *Heidegger Studies* (Berlin: Duncker & Humblot), Vol. XIII (1997), pp. 93ff; Rudolf Bultmann *et al.*, *àlétheia*, in *Theologisches Wörterbuch zum Neuen Testament*, ed. G. Kittel (Stuttgart: W. Kolhammer, 1933); and Ceslas Spicq, *Notes de lexicographie néo-testamentaire* (Fribourg: Éditions Universitaires, 1982), Vol. I.

[6] J. Marías, *Introducción a la Filosofía*, p. 86. (The emphasis is mine.) However important and interesting it may be, the question of showing the presence or givenness (by way of interconnectedness) of the other senses of "truth," in each one of the acceptations of the term, it cannot be adequately dealt with here. I will limit myself briefly to calling the reader's attention to what Marías says in that regard: "*Veritas* involves a straightforward reference to the act of saying, which exceeds the reference to a declarative or *apophantic* manner of saying. It is the shade of meaning conveyed by the Spanish word, *veracidad* [veracity or truthfulness]." (Cf. José Ortega y Gasset, *El hombre y la gente*, in *Obras Completas* [Madrid: Alianza Editorial/Revista de Occidente, 1983], Vol. VII, p. 145. Henceforth I shall be referring to this collection as *OC*, and Ortega's name will not be identified as the author's in subsequent citations of any of his works.) Concerning the Hebrew locution, *emunah*, Marías points out that it "contains a personal reference: it is about the truth in the sense of *trust* [emphasis added] ... it points, then, to the fulfillment [of a promise made by someone, particularly by God, a promise rendered true thereby], to something one hopes for and which *will be*" (J. Marías, *op. cit.*, pp. 86–87). Cf. Exodus 3: 14; Fray Luis de León, *De los nombres de Cristo*, i, "Faces de Dios", in *Obras Completas Castellanas de Fray Luis de León*, ed. F. García (3rd. ed., Madrid: Biblioteca de Autores Cristianos, 1959), p. 425; Hans Freiherr von Soden, "Was ist Wahrheit?" (1927); "Apuntes sobre el pensamiento, su teurgia y su demiurgia", *OC*, Vol. V. p. 536, n. 1; Xavier

Zubiri, "Sobre el problema de la filosofía", *Revista de Occidente*, 1ª época, No. 118 (April, 1933), pp. 94ff. and "Nuestra situación intelectual", *Naturaleza, Historia, Dios* (5th. ed., Madrid: Editora Nacional, 1963), p. 14, n. 1.

[7] J. Marías, *Introducción a la Filosofía*, p. 86.

[8] Cf. "Meditación preliminar", § 4, *Meditaciones del Quijote*, in *OC*, I, pp. 335f. *Vide* J. Marías, "Comentario", pp. 296ff; Martin Heidegger, *Sein und Zeit*, in Gesamtausgabe (Frankfurt am Main: V. Klostermann, 1977), Part I, Vol. II §§ 7 and 44, pp. 33f, 280ff, 290ff, and 293f.

[9] J. Marías, *Ortega. I. Circunstancia y vocación* (Madrid: Revista de Occidente, 1960), p. 472. Henceforth I shall be referring to this work as *Ortega I.*

[10] *Ibid.*

[11] *Ibid.*

[12] "Meditación preliminar", § 4, *Meditaciones del Quijote*, p. 336. English transl.: *Meditations on Quixote*, trans. E. Rugg *et al.* (New York: W. W. Norton & Co., 1961), p. 67. The transliteration from the Greek has been slightly modified. (For the term, *àpokálupsis*, see Albrecht Oepke, *kalúpto, àpokálupsis*, etc. in *Theologisches Wörterbuch zum Neuen Testament*, ed. G. Kittel, Vol. III.) Cf. Antonio Rodríguez Huéscar, *Perspectiva y verdad* (2nd. ed., Madrid: Alianza Editorial, 1985), p. 222: "When he discovered logical thought, Parmenides believed he had *un-covered* genuine reality, and he lived that thought as 're-velation' [*à-létheia*] ...." On this basis, one may distinguish between truth as re-velation and that which is re-vealed thereby, a distinction serving as the ground for conceiving of what Ortega called *belief*, in contradistinction to *idea*. (See my articles entitled, "Caminos de la reflexión", *supra*, n. 2.) Cf. *La idea de principio en Leibniz y la evolución de la teoría deductiva*, in *OC*, VIII, p. 210: "True thought is thought; it becomes true inasmuch as it ceases to be thought and turns into the presence of reality itself." (Henceforth I shall be referring to this work as *La idea de principio.*) This point is well taken, and yet one must be careful not to misunderstand it and relapse thereby into idealism, especially into the psychologistic or extremely subjectivistic form of it. To that end, one must keep in mind, as Rodriguez Huéscar prudently warned us, that "to dis-cover or re-veal is an operation one performs, while [a] *re-velation* is always [the re-velation] of a thing, of the real itself ...; it is always [a] '*transcendent* re-velation'" (*op. cit.*, pp. 229–230). Accordingly, it is clear that reality and its coming to the truth not only correspond to each other necessarily, but also exist exclusively *in absolute reciprocity*, a point Ortega himself made as follows: "Due to some curious *contamination* [emphasis added] between that which is un-covered = reality, and our act of un-covering or denuding it, we often speak about the 'naked truth', a tautology. That which is naked is reality and denuding it is the truth, inquiry, or *àlétheia*." (*Origen y epílogo de la filosofía*, in *OC*, IX, p. 386; English transl.: *The Origin of Philosophy*, trans. T. Talbot [New York: W. W. Norton & Co., 1967], p. 63.) To this one could add that, from a different point of view, "... what to Parmenides is genuinely *evident* – and not just a *belief* – is not the fact that true reality behaves in conformity with 'logical thought' under the rule of the principle of contradiction (that is what *is believed*), but, rather, *that there is* [such a thing] *as 'logical thought', which is compulsory, 'anankic', necessitative, non-contradictory*. This is still evident; it continues to be the truth [*àlétheia*], and will continue to be such as long as there is a thinking being capable 'of placing itself face to face' with the fact or reality that the given statement makes manifest [in the process of coming to the truth or *aletheúein*, of re-vealing, rendering evident, or un-covering]." (A. Rodríguez Huéscar, *op. cit.*, p. 224).

[13] "Meditación preliminar", § 4, *Meditaciones del Quijote*, p. 336. English transl.: p. 67. (The emphasis is mine. The spelling has been slightly modified.)

[14] J. Marías, *Ortega. I*, p. 472. Cf. p. 567, nn. 21 and 22. *Vide* Critias of Athens, *Súsuphos*, B 25, v. 26, in H. Diels and W. Krantz, *Die Fragmente der Vorsokratiker*, 6th. ed. (Berlin: Weidmann, 1951–1952), Vol. II, p. 388, I. 4; *apud* J. Marías, "Comentario", p. 298: "*pseudei kalúpsas tès alétheian lógoi*." Cf. *Ancilla to the Pre-Socratic Philosophers*, trans. K. Freeman (Cambridge, Mass.: Harvard University Press, 1962), Fragment 25, p. 158: "covering up the truth with a false theory [word]."

[15] J. Marías, *Introducción a la Filosofía*, p. 86.

[16] Cf. *ibid.* and A. Rodríguez Huéscar, *op. cit.*, p. 426, n. 124: "...Zubiri points out [that there are] three components to the notion of *àlétheia* or truth: being, manifestness, and security." *Vide* X. Zubiri, *Naturaleza, Historia, Dios*, p. 14, n. 1.

[17] It seems that a very complex objective-subjective dialectic is at work here. Cf. M. Heidegger, "Vom Wesen der Wahrheit", § 5ff., in *Wegmarken*, Gesamtausgabe, Part I, Vol. IX (1976), pp. 192ff.

[18] "Sobre el estudiar y el estudiante", in *OC*, IV, p. 546. The empasis is mine.

[19] *Vide* Plato, *Republic*, v, 18, 475 d–e and 480. Cf. *¿Qué es conocimiento?* (Madrid: Revista de Occidente/Alianza Editorial, 1984), Part II, pp. 67ff and 76–77.

[20] Cf. *El tema de nuestro tiempo*, in *OC*, III, Chap. 3, pp. 160–161.

[21] "Sobre el estudiar y el estudiante", p. 548.

[22] I have in mind Ortega's term, *mentefactura*. Cf. his "Rectificación de la República", in *OC*, XI, p. 416.

[23] "Sobre el estudiar y el estudiante", p. 548. Perhaps it may not be groundless to attribute the role of a lantern, so to speak, to a "genuine subjective need," in the sense in which "lantern" is taken in the Gospel parable. (Cf. Luke 15: 8; St. Augustine, *Confessions*, 10.18.27.)

[24] The search in question does not necessarily imply a *creative* endeavor on one's part, i.e., an effort in which one would be intent on increasing the "objective wealth" of knowledge at one's disposal, since the only requirements to be met by one's attempt in that direction are: (a) that one come to possess, by one's own means, a truth that is not yet one's own, and (b) that one provide a justification of that truth in terms of one's own reasons, at least an account that would be arrived at because of the motivations permitting one's access to the reality of the real. But more generally, though implicitly, the search would be rooted in one's radical self-transcendence, which is one's response to a call for justification (if one is so constituted as to require it), and which would be effected in whichever manner is appropriate in the particular case and to the person involved in the given theoretical matter. Such self-transcendence would open us up to, and be matched by, a field of transcendence, which would allow the sought-after justification to arise. Accordingly, the pair "self-transcendence/field of transcendence" would "unexpectedly lead ... *back* to the particular again ... [but] the particular can no longer have the weight of an absolute...." (Hans Urs von Balthasar, *Der Christ und die Angst* [Einsiedeln: Johannes Verlag, 1989], iii. English transl.: *The Christian and Anxiety*, trans. D. D. Martin *et al.* [San Francisco: Ignatius Press, 2000], p. 130). "Essence" is the name chosen by Ortega to refer to the fruit of such self-transcendence. (Cf. *infra*, n. 26.) Whether the field of transcendence is that of Being (as claimed by Balthasar here or by Heidegger, e.g., in *Sein und Zeit*, §§ 5–6 and 69c, pp. 364–366) remains here an unresolved problem, which can only be tackled by considering the relationship between being and life, as the latter is conceived by Ortega, namely, as the dual, dynamic unity and totality by reciprocity of self (as self-transcendence) and circum-stance (as the instancing occasion for metaphysical perplexity).

[25] *En torno a Galileo*, in *OC*, V. pp. 15–16. (The emphasis is mine.) As usual, Ortega employs the word "enigma" in its Greek-etymological sense of "riddle."

[26] Cf. *¿Qué es conocimiento?*, Part III.

[27] Cf. Plato, *Republic*, VII, 516 c–d and *Gorgias*, 501 a.
[28] *En torno a Galileo*, p. 16.
[29] "Apuntes sobre el pensamiento, su teurgia y su demiurgia", in *OC*, V, p. 525.
[30] *Ibid.*
[31] *En torno a Galileo*, p. 16. For the "cognitive" sense of the experience of "chaos" (and of the attendant "confusion" or befuddlement as a form of "consciousness"), cf. my article, "Caminos de la reflexión", *loc. cit.*, No. 19, pp. 12–21.
[32] Cf. A. Rodríguez Huéscar, *op. cit.*, p. 235.
[33] Cf. *ibid.* For the relevant concepts of "instancy" and "instancing", *vide* A. Rodríguez Huéscar, *José Ortega y Gasset's Metaphysical Innovation. A Critique and Overcoming of Idealism*, trans. J. García-Gómez (Albany, N. Y.: State University of New York Press, 1995), Part II, Chap. 4, §5B, pp. 104ff.
[34] Cf. *¿Qué es conocimiento?*, Part III, p. 100; Immanuel Kant, *Kritik der reinen Vernunft*, B 211–212, 233ff, and 275ff; and Henri Bergson, "Introduction à la métaphysique", *La pensée et le mouvant*, vi, in *Oeuvres* (Paris: Presses Universitaires de France, 1963), pp. 1397–1398.
[35] Cf. *supra*, n. 33.
[36] J. García-Gómez, "The Problematicity of Life", in *José Ortega y Gasset*, ed. N. de Marval-McNair, Proceedings of the "Espectador Universal" International Interdisciplinary Conference (New York: Greenwood Press/Hofstra University, 1987), p. 34.
[37] *Ibid.*
[38] *Ibid.*
[39] *Ibid.*
[40] Cf. *En torno a Galileo*, p. 15 and "Meditación preliminar", § 5, *Meditaciones del Quijote*, p. 373: "When [today] we look for reality we search for appearances. But the Greek took reality to be the opposite: the real is the essential, the profound and latent; not the external appearance but the *living sources* of all appearance." (English transl.: p. 124. The emphasis is mine.) Marías interprets this passage as follows: "Ortega contrasts the attitude dominant in his time, and against which he reacts, to that of the Greeks; the identification of the real with that which is accessible to the senses, which is the point of view of Positivism in all its forms.... The ultimate expression of the Orteguian sentence we have quoted is the deepest interpretation of *àlétheia* ... [which encompasses] its metaphysical significance: the truth is the true reality (*alethès òn*), that which is patent; that is, it *makes* patent [emphasis added] or manifests *what truly is...*, that which *gives life to appearance*, that from which appearance arises, the *living springs of appearance*. This expression, *living spring*, would translate admirably what the Greeks understood by *physis*, that from which springs or arises what is shown and uncovered in its appearance, and which for this reason can be interpreted as principle (*arkhé).* Recall the use of the word 'hontanar' (source) in Ortega's work" Ortega. I, § 81, p. 473. English transl.: *José Ortega y Gasset. Circumstance and Vocation*, trans. F. M. López-Morillas [Norman: The University of Oklahoma Press, 1970], pp. 445–446). Marías's understanding of the text is not only correct, but of decisive importance, since, in his commentary, he does not rest content with considering the level of appearances; rather, he regards them in a *dynamic* fashion, as is made clear by his use of the phrase, "it makes patent or manifests...."
[41] *Ibid.* (The emphasis is mine.) Marías employs here the expression "make it true" to underscore the active role played by human beings in the experience of the truth.
[42] *En torno a Galileo*, p. 16. The emphasis is mine.
[43] Cf. my articles, "Caminos de la reflexión," *loc. cit.*, No. 19, pp. 8ff and No. 20, pp. 132ff and 141ff.

[44] But if that is so, then pure feeling and even passion, no matter what their intensity may be, prove inadequate in the search for the truth; one must, therefore, clarify and justify them by means of well-articulated arguments to be developed on an intuitive basis.

[45] Cf. *supra*, n. 33.

[46] A. Rodríguez Huéscar, *Perspectiva y verdad*, p. 235. As the author hastened to add, "... one must observe that such 'facts', precisely as given in one's spontaneous living ... [or as the result of] scientific investigation, already are *interpretations* ..."; that is to say, that what they signify for us in everyday life – and, to a point, in positive science itself – is mediated by the beliefs on the basis of which we live at a given juncture of history. Cf. my article, "Caminos de la reflexión", *loc. cit.*, No. 20, pp. 114–120.

[47] A. Rodríguez Huéscar, *op. cit.*, p. 236.

[48] *El hombre y la gente*, p. 145.

[49] Cf. *ibid.*

[50] *En torno a Galileo*, p. 16.

[51] Cf. my article, "Caminos de la reflexíon", *loc. cit.*, No. 20, pp. 124ff.

[52] *En torno a Galileo*, p. 16.

[53] Cf. *La idea de principio*, p. 109 and my article, "Caminos de la reflexión", *loc. cit.*, No. 20, p. 122.

[54] *En torno a Galileo*, p. 16. The emphasis is mine.

[55] Cf. *La idea de principio*, pp. 68, 104, and 121–122; A. Rodríguez Huéscar, *Perspectiva y verdad*, pp. 204–205; and my articles, "Caminos de la reflexión", *loc. cit.*, Nos. 19 (nn. 30, 44, and 79) and 20 (pp. 122–123).

[56] Cf. *supra*, pp. 71–72.

[57] J. Marías, *Ortega. I*, § 83, p. 478.

[58] *Ibid.*

[59] *Meditaciones del Quijote*, p. 321 (English transl.: p. 44).

[60] J. Marías, *op. cit.* This does not *necessarily* involve a will-to-power one would realize by means of things, and even perhaps at their expense. To exercise true dominion over things and genuinely to possess them, which is the yield obtainable as a result of engaging in genuine thinking, consists, rather, in taking on the *power proper to things*, and this can only be accomplished by letting them be what they are in the medium of one's thought. In other words, it is a question of deliberatively striving to bring about a non-distortive co-incidence of things and thought, i.e., an endeavor that would allow thinking to come to an adequate correspondence with the reality of things, a reality that, as such, would arise from them as a response to one's care-ful inquiry.

[61] Cf. "Meditación preliminar", § 4, *Meditaciones del Quijote*, pp. 336–337 and J. Marías, "Comentario", pp. 305–306.

[62] Cf. *supra*, p. 73.

[63] In the case of compatibility, as opposed to that of logical derivation, the employment of the imagination would have to be pursued further in order to produce other members of the inner world in question (i.e., the one set up and rendered characteristic in terms of the operative or relevant "idea of reality"). The discharge of this task would be called for as a function of the theoretical difficulties that may arise in the course of thinking.

[64] *En torno de Galileo*, p. 16. The emphasis is mine.

[65] Cf. my articles, "Caminos de la reflexíon", *loc. cit.*, No. 20, pp. 114ff and "La acción y los usos intelectuales. En torno a la problemática de las ideas y las creencias en la filosofía de José Ortega y Gasset", *Torre de los Lujanes* (Real Sociedad Económica Matritense de Amigos del País), No. 34 (October, 1997), pp. 117–138.

66 For the functional distinction between the two sorts of interconnected motives at play in life at every juncture (i.e., the "because of" and the "for the sake of" motives), cf. *¿Qué es conocimiento?*, Part I, pp. 45ff. *Vide* Alfred Schutz, *Der sinnhafte Aufbau der sozialen Welt* (2nd. ed., Vienna: Springer Verlag, 1960; 1st. ed., 1932), §§ 17 (p. 95) and 18 (p. 100); English transl. *The Phenomenology of the Social World*, trans. G. Walsh *et al.* (Evanston: Northwestern University Press, 1967), pp. 88 and 91; "Projects of Action", iii, in *Collected Papers* (The Hague: Martinus Nijhoff, 1962), Vol. I (ed. M. Natanson), pp. 69ff; and *Reflections on the Problem of Relevance*, ed. R. M. Zaner (New Haven: Yale University Press, 1970), Chap. 2, § E, pp. 45ff.
67 Cf. *Meditaciones del Quijote*, p. 321; J. Marías, "Comentario", pp. 254–255; M. Heidegger, *Sein und Zeit*, § 15; and my study, "The Problematicity of Life", p. 35.
68 Cf. my study, "The Problematicity of Life", p. 35.
69 Jacques Maritain, *Creative Intuition in Art and Poetry* (New York: Pantheon Books/Bollingen Series XXXV-1, 1953), Chap. 2, § 2, p. 46.
70 Cf. "Filosofía pura. Anejo a mi folleto 'Kant'", in *OC*, IV: "The discovery that being is only endowed with sense [if understood] as a question posed by a subject ... could have only been made by someone who [like Kant] has severed the two meanings of the term 'being' [namely, the in-itself and the for-itself] ..." (p. 56). "And yet that *being* [should prove to] be a question and, because of it, [that it should prove to] be thought, did not oblige Kant at all to adopt an idealist solution.... That being [should prove to] be devoid of sense and incapable of signifying anything, if one were to consider it apart from the knowing subject – and, therefore, that thinking should play a part in the *being* of things by *positing it* – does not imply that entities, [or] things, in being or not being, become thought, as two oranges do not turn into something subjective, because their equality exists only when a subject compares them" (pp. 56–57). "... [B]eing, what is objective, etc. are meaningful [notions] only if one goes in search of them, [the search] essentially consisting in one's moving toward them. Now then, human life, or a human being *qua* living reason, is the subject in question. Radically speaking, a human being's life is [tantamount to] his or her being occupied with the things of the world, not with him- or herself" (p. 58).
71 J. García-Gómez, "The Problematicity of Life", p. 35.
72 *Ibid.*
73 *Ibid.* Cf. my paper, "José Ortega y Gasset's Categorial Analysis of Life", *Analecta Husserliana*, Vol. LVII (1998), pp. 150–158.
74 Cf. *supra*, n. 40.
75 *Ibid.*
76 J. García-Gómez, "The Problematicity of Life", p. 35. Cf. *Unas lecciones de metafísica*, in *OC*, XII, p. 72.
77 "Apuntes sobre el pensamiento, su teurgia y su demiurgia", p. 525. The emphasis is mine.
78 *Ibid.*
79 *Perspectiva y verdad*, p. 236. Cf. *supra*, p. 69 and n. 25.
80 Cf. "Apuntes sobre el pensamiento, su teurgia y su demiurgia", p. 525 and "Meditación preliminar", § 4, *Meditaciones del Quijote*, p. 335: "That pure, sudden illumination which characterizes truth accompanies the latter only at the moment of discovery" (English transl.: p. 67). *Vide* J. Mariás, "Comentario", p. 295: "The Orteguian interpretation of the truth marks its beginning with the idea of *light*, by way of a sudden *illumination*. Far from being a fleeting image or an occurrence, it constitutes the heart of the doctrine." (Cf. "Meditación preliminar: la luz como imperativo", § 12, *op. cit.*, pp. 356ff; J. Marías, "Comentario", pp. 297 and 354ff; and M. Heidegger, *Sein und Zeit*, §§ 7 and 28.) Marías comes to the conclusion that "life [cannot be] reduced to [a zone of] clarity. Clarity is *internal* to life." ("Comentario", p. 359; cf. *Meditaciones del Quijote*, p. 358.) Nonetheless, as he had remarked earlier, "in [Nicolai] Hartmann [cf. *Platos*

*Logik des Seins*, 1909, *apud* J. Marías, *loc. cit.*, pp. 302–303] two notes are missing [which are part of the idea of truth *qua àlétheia* and which are] particularly interesting. [This is in contrast with his acknowledgement of the role played by forgetfulness and remembrance in that concept. Those two notes] are found in Ortega's passage, and they are: the connection between *àlétheia* and *àpokálupsis* ... and, above all, the reference to *light*, the character of 'sudden illumination' that belongs to the truth *at the moment of discovery* [emphasis added], that is to say, insofar it is *àlétheia*, a theme that is to reappear in Heidegger" (*loc. cit.*, pp. 303–304. The transliteration of the Greek has been slightly modified).

[81] J. Marías, *Ortega, I*, p. 478. The passage in quotes is from *Meditaciones del Quijote*, p. 313 (English transl.: p. 33). Cf. *infra*, pp. 79ff.

[82] *Perspectiva y verdad*, p. 245.

[83] *Ibid.* The fundamental sense of this assertion becomes clear to us in terms of what Rodríguez Huéscar affirmed elsewhere, to wit: that, in Ortega's opinion, "evidence [etymologically understood, i.e., as deriving from *ex* (= out of) and *videns* (= seeing)], is an act of vision [*videncia*], an unmediated [act of] vision ... an intuition. Therefore, it cannot be [a matter of] feeling, because feeling is *blind* or, better yet, an-optic ('devoid of eyes')." (*Ibid.*, p. 209.) This allows us to realize that thinking, when it is *in form* – i.e., when it is an act of *philosophizing* – consists in endeavoring to overcome the mediation of beliefs (not primarily of ideas) in order justifiedly to arrive at the *reality* of the things *themselves*. As Rodríguez Huéscar pointed out, "intellectual evidence, then, is irreducible to the 'cataleptic' character of the so-called 'evidence' afforded by a belief" (*ibid.*), which by definition is "incompatible with theoretical or cognitive truth", inasmuch as a belief plays the role of something which, as a matter of course, is taken for granted (or even identified with reality itself). For the Stoic notion of "cataleptic" imagination, cf. *La idea de principio*, § 25, p. 253. See *¿Qué es Filosofía?* in *OC*, VII, pp. 350ff and my paper, "La acción y los usos intelectuales", p. 127 and n. 36.

[84] A. Rodríguez Huéscar, *Perspectiva y verdad*, p. 245.

[85] This would be so whenever it is given as lacking the character of world, be it as a totality (as is the case in metaphysical reflection, at least originally) or in some respect (as it usually happens in everyday life).

[86] Cf. *Meditaciones del Quijote*, pp. 357–358.

[87] J. Marías, *Ortega. I*, § 81, p. 483 (English transl.: p. 455). This passage refers us to *Meditaciones del Quijote*, p. 335.

[88] "Meditación preliminar", § 12, *Meditaciones del Quijote*, p. 358. English transl.: p. 99.

[89] For Ortega's notion of hero, cf. *ibid.*, p. 390 and J. Marías, "Comentario", p. 428; for Ortega's notion of truth as self-coincidence, see *En torno a Galileo*, Chap. 7. Cf. the notion of *Eigentlichkeit* or "authenticity" in M. Heidegger, *Sein und Zeit*, § 9, p. 43.

[90] Cf. *supra*, p. 77 and n. 84.

[91] A. Rodríguez Huéscar, *Perspectiva y verdad*, p. 245.

[92] *Ibid.*

[93] *Ibid.* Cf. p. 437, n. 320 and M. Heidegger, "Vom Wesen der Wahrheit", ii, pp. 182ff.

[94] Cf. *supra*, p. 74 and n. 59.

[95] J. Marías, *Ortega. I*, § 83, p. 480. (English transl.: p. 452). Please note that here the word *lógos* is used to signify the product of the process of verification, not the process itself.

[96] Cf. *supra*, p. 77.

[97] Cf. "Sensación, construcción e intuición", *OC* XII, pp. 487ff.

[98] A. Rodríguez Huéscar, *Perspectiva y verdad*, p. 208.

[99] J. Marías, *Ortega. I*, § 83, p. 480. (English transl.: p. 452).

[100] *Ibid.*
[101] *Meditaciones del Quijote*, p. 313 (English transl.: p. 33).
[102] *Ibid.*, p. 316. (English transl.: p. 38).
[103] *Ibid.* (The emphasis is mine.) As indicated, the translation has been modified to reflect the original more closely.
[104] *Ibid.*, p. 311. (English transl.: p. 31). Ortega does not seem here to be employing the term "fact" as I have done thus far, in keeping with his later practice. In my opinion, at this point it means anything that is observable or noted at the level of phenomena. If this is correct, the phrase, "its fullest significance" would refer to an *expansion* of what is given, which expansion, when carried to the limit, would find a rightful place for the phenomenon in question in the system of appearances that is the equivalent of the universe. Accordingly, the phrase, "its fullness of significance" should not then be taken as an expression for the intensification of, and the concentration upon, that which is *essentially proper to* the thing under consideration, as the later notions of "fact" and "reality" would lead us to expect. The two concepts of "fact" are not, however, incompatible (nor are they, *en revanche*, unrelated, as would be those referring to two different spheres that just merely happen to coincide at a certain juncture, be it spatio-temporal in character or otherwise), since it is possible to interpret them in such a fashion that the *reality* of the thing is seen as the *fundamental core* and *source* of the expansible system of appearances serving as its correlative manifestation. If this is the case, then one would have to add, to the dialectical interplay of the metaphors of circumference and center, the radical dimension of depth, with which they would have to be brought to synthesis.
[105] J. Marías, "Comentario", p. 224.
[106] *Sumpósion*, 202 e in *Platonis opera*, ed. J. Burnet (Oxford: Clarendon Press, 1973), II.
[107] *The Dialogues of Plato*, trans. B. Jowett (New York: Random House, 1937), I, p. 328.
[108] *Meditaciones del Quijote*, p. 313. (English transl.: p. 33). The translation has been slightly modified, so as to follow the original more closely.
[109] Cf. Aristotle, *Nicomachean Ethics*, I.iii.1.
[110] Cf., e.g., Erich Fromm, *The Art of Loving* (New York: Harper & Row/Perennial Library, 1974), ii, §§ 2–3, pp. 32ff.
[111] *Meditaciones del Quijote*, p. 312 (English transl.: p. 33).
[112] *Ibid.*
[113] Plato, *op. cit.*, 200 a 8 – 200 b 1, in *The Dialogues of Plato*, trans. B. Jowett, I, p. 325.
[114] *Meditaciones del Quijote*, p. 313 (English transl.: p. 33). The emphasis is mine.
[115] Plato, *op. cit.*, 200–201 and 204 d.
[116] *Meditaciones del Quijote*, p. 313 (English transl.: p. 33). As indicated, the translation has been modified to reflect the original more closely.
[117] Plato, *op. cit.*, 200 e in *The Dialogues of Plato*, trans. B. Jowett, p. 326.
[118] Cf. this page and n. 116.
[119] *Meditaciones del Quijote*, p. 313. (English transl.: p. 33).
[120] *Ibid.*
[121] *Ibid.*
[122] Benevolence, or *amor benevolentiae*, deserves the special name of *philía* or *amor amicitiae* when reciprocity as a fundamentally possible relationship between lover and the beloved object is intended, that is, when both are persons acting as such with regard to each other. For the distinction between *éros* (or love as *dilectio concupiscentiae* or *cupiditas*) and *philía* (or love as *dilectio amicitiae sive benevolentiae sive beneficentiae*), cf. Aristotle, *Nicomachean Ethics*, viii–ix; St. Thomas Aquinas, *Summa theologiae*, I–II, qq. 25 (a. 2) and 26 (aa. 1, 3, and 4); and A. E. Taylor, *Plato* (New York: The Dial Press, 1936), p. 233 and n. 1. In connection with *philía*,

one cannot fail to mention the notion of charity or *àgápe*. This is the distinctive Christian contribution to the theory and the experience of love. (Cf., e.g., Romans 5:5, 1 John 4:16; 1 Corinthians 13: 8–13; St. Thomas Aquinas, *Summa theologiae*, I–II, q. 65, a. 5c ad 1; II–II, qq. 23 [a. 1] and 26 [a. 3c].) Concerning this form of love, it would be hardly sufficient to say, with Leibniz, that it is "universal benevolence," as if it consisted just – or even primarily – in increasing a human tendency and passion – even to the maximum of its possible scope – by sheer human effort and means. Cf. G. W. Leibniz, "Preface to the 'Codex juris gentium diplomaticus'", *Die philosophischen Schriften*, ed. C. J. Gerhardt (Berlin and Halle: 1875–1890; re-issued, Hildescheim: G. Olms, 1960–1961), III, p. 387. English transl.: G. W. Leibniz, *Philosophical Papers and Letters*, trans. and ed. L. E. Loemker (2nd. ed. Dordrecht: D. Reidel, 1969), p. 421.

[123] *Ibid.*

[124] Josef Pieper, *Über die Liebe* (7th. ed., Munich: Kösel Verlag, 1992), viii.

[125] G. W. Leibniz, *loc. cit.*

[126] *Ibid.* (English transl.: pp. 421–422). The emphasis is mine.

[127] G. W. Leibniz, *Nouveaux essais sur l'entendement humain*, ed. J. Brunschwig (Paris: Garnier-Flammarion, 1966), Book II, Chap. XX, § 5, p. 138.

[128] Cf. *supra*, n. 122.

[129] G. W. Leibniz, *Nouveaux essais sur l'entendement humain*, p. 138.

[130] Cf. *supra*, p. 82 and n. 126.

[131] Cf. *supra*, n. 81.

[132] J. Marías, "Comentario", p. 225. It seems to me that, at this point, Marías can only be metaphorically availing himself of the language of agency with regard to things, for, even though things must be intrinsically capable of such "transcendence," if they are to be lovable and eventually loved, it is nonetheless true that the only agency involved in that is the lover's, though, to be sure, the beloved object would correspondingly "instance" the one in love.

[133] *Ibid.* As I hope will become apparent in what follows, it would be advisable to introduce the word "actual" in order to qualify Marías' employment of "intrinsic" in this citation.

[134] *Meditaciones del Quijote*, p. 312. (English transl.: p. 32). The emphasis is mine.

[135] Ortega speaks of the beloved object as "a virgin that is to be wooed or courted if it is to become fruitful." (*Ibid.*) The translation has been modified.

[136] *Ibid.*, p. 311.

[137] *Ibid.*, p. 312.

[138] Cf. *supra*, p. 70ff.

[139] Cf. *¿Qué es conocimiento?*, Part III, pp. 112, 120ff, 141ff, and 154ff.

[140] Cf. *supra*, p. 82 and n. 126.

[141] Cf. *supra*, n. 135.

[142] Cf. *supra*, p. 72f. One could argue that Ortega's employment of the locution "reverberations" (*Meditaciones del Quijote*, p. 311) conveys, from an external standpoint, what one would characterize, from an internal point of view, as the "intrinsic virtualities" of the beloved object (Cf. J. Marías, "Comentario", p. 226). Accordingly, one could say that Marías had in mind the essential and reciprocal relationship between the extrinsic and intrinsic dimensions of love when he asserted that "… [w]hat is important and new about Ortega's idea is that it deals with an 'all-embracing connection' cf. *Meditaciones del Quijote*, p. 316]. In other words, it does not simply deal with the subject's love of things (the philosopher being the subject in this case), nor with a hypothetical and vague 'love' that things would bear one another either" ("Comentario, p. 234). To put it otherwise: Ortega was attempting thereby to give expression to the dynamic

structure of life, which amounts to the ongoing cultivation of an intrinsic nexus of reciprocity between the self *and* the circumstance, a formula wherein the conjunction "and" points to the project and endeavor of "salvation" as essential to life. Cf. *Meditaciones del Quijote*, p. 322; J. Marías, "Comentario", pp. 266ff and *Ortega. I*, Sección III, iv, §§ 70ff; and *supra*, pp. 80–81.

143 Cf., e.g., "Renán", *OC*, I, p. 448 and *supra*, n. 19.

144 Plato, *Republic*. V. 19, 475 d in *The Dialogues of Plato*, trans. B. Jowett, I., p. 738. Cf. 480 a, p. 744.

145 Cf. *supra*, p. 80 and n. 104.

146 Cf. *supra*, p. 80 and n. 103.

147 Cf. *supra*, p. 80 and n. 108.

148 *Republic*, V. 19, 475 b, in *The Collected Dialogues of Plato*, trans. B. Jowett, I, p. 738.

149 *Ibid.*, 475 e, p. 739.

150 Cf. *supra*, p. 69 and n. 34.

151 *Origen y epílogo de la filosofía*, p. 384 (English transl.: p. 60).

152 *Ibid.*, p. 387. (English transl.: p. 64).

153 *Ibid.*, p. 386. (English transl.: p. 62).

STEPHANIE GRACE SCHULL

# KNOWING THYSELF: PARADOX, SELF-DECEPTION, AND INTERSUBJECTIVITY

## KNOW THYSELF!

At the time of Socrates, those who approached the Oracle of Delphi seeking knowledge were struck in turn with the great and seemingly impossible demand for self-knowledge, for the command "Know thyself" was etched in stone above the entrance to the Oracle. Those who asked the sage Socrates for insight would soon discover that they were instead scrutinized for the very knowledge they sought. Indeed, it was the Oracle of Delphi that declared Socrates to be the wisest of all men. Was it because Socrates knows himself? Not immediately accepting the idea, Socrates accepts the Oracle's declaration with the stipulation that he is wisest for knowing that he does not know. If Socrates knew that he did not know, then he did not know himself, for he maintained that absolute knowledge is innate, that, as Plato wrote, it is written on the soul. If knowledge is written on the soul, then Socrates is correct in claiming that we must attend to our souls and examine them. Knowing becomes a process of knowing the self. The Socratic method intends to bring to awareness that which each person knows without knowing it. It is therefore no accident that this method is a dialectical process. In particular, it can only be a lived dialogue (Socrates committed nothing to writing with good reason). Lived dialogue, between two active and engaged interlocutors, can reveal the innate knowledge that lies within the soul of each speaker. They can arrive at true propositions only in the encounter of one soul questioning another, and only in the present, as it depends on an experience of the presence of the other. Although it may seem paradoxical, it is nevertheless probable that self-knowledge can only be achieved through the awareness of another. Awareness both in the sense of recognizing a Thou (a wholly other subject) and depending upon the other's consciousness for manifesting one's own unconscious content, that is to say, the knowledge that is written on the soul unbeknownst to the ego.

Here, I have introduced a particular vocabulary into the deliberation, namely, a psychoanalytic one. Psychoanalytic metapsychological theory will be explored, as its business is the achievement of self-knowledge through dialogue with another. In particular, psychoanalytic phenomenology, as it is being explored in France, investigates these very questions of self-knowledge, consciousness, and intersubjectivity. Remarkably, the unmistakable presence



*A.-T. Tymieniecka (ed.), Analecta Husserliana LXXVI*, 99–116.

of paradox and paradoxical utterances are found whenever these questions arise. The project of self-knowledge is fraught with contradiction delineating limit and impasse, as well as paradox, paradox resulting from a leaping over the gap that bars something from its denial. Questions of self-knowledge and self-referential statements have been the concern of analytic logicians. Knowing the other is also poorly understood as it involves the paradox of going beyond oneself. As such, paradoxical descriptions abound when discussing the process of knowing an(other) and knowing the self. The presence of contradiction in these *ad limina* experiences will be explored using notions proper to logic, which is precisely the sensibility that is affronted by such utterance of contradiction. This being the case, two types of analysis will be employed in investigating the nature of self-knowledge: a logical one and a metapsychological one. Weaving together what may appear to be rather unlike disciplines need not be inappropriate, for logic is of concern to the analyst as it is a psychical process among others, and the psychoanalytic subject is of interest to the logician as only subjects utter propositions. Not only are subjects responsible for propositions, they often utter propositions that refer back directly to the subject himself. In these instances, it is unwise for logicians to bracket the context of the utterance. In response to these concerns, it will be necessary to speak of the logical notions of indexicals, negation, and contradiction alongside the psychoanalytic concepts of repression, the unconscious, and subjectivity. In the end, I will propose a resolution of the paradoxical nature of certain self-referential propositions.

## WHO ARE YOU?

The English logician Lewis Carroll uses the voice of a caterpillar in his *Alice's Adventures Under Ground*, to demand of poor, lost Alice, "Who are you?" Alice stammered back to her many-legged analyst, "I can't explain myself, I'm afraid, sir ... because I am not myself you see." What a curious locution, "I am not myself." Like many of the utterances of Carroll stories, it seems fraught with nonsense. Of course, he was also an excellent logician and maker of puzzles and riddles. With this statement, one can expect that he was playing with the law of identity as well as that of contradiction through the problem of self-reference. At the heart of the law of identity is the supposed intuition that I am identical to myself; nevertheless, inherent to the law of identity, A is A, is a notion of difference implied by the distinction of "A" as a predicate term and "A" as a subject term. In what sense can I be identical to myself, and in what sense am I not myself? Obviously, if I am to understand myself, there must be a split in my self which allows me to observe and report on the nature of the self. Herein lies the problem of self-reference. Alice's inability to know herself stems from the fact that she claimed that her I is not identical to herself. Not only is Alice not identical to herself, but part of herself is actually hidden from view, keeping her from identifying with it, naming it, or recognizing it. I am not describing Alice's predicament as unique; it is the human condition. Those looking for the truth by going to a psychoanalyst soon discover that the real project is for the analysand to know himself. Like Alice, the analysand is drawn to the couch declaring a split in the subject that the person finds desperately confusing, "I do not know who I am for I am not myself." Striking directly at the heart of identity and the law of non-contradiction is the tension inherent in the utterance: "I am not myself." Which is to say, A is not A, or $A \equiv \sim A$.

## WHO AM I?

The project to know thyself assumes that the I and the self are not identical, for if they were, there would be no need to go about knowing oneself. The subject embarks on the journey toward self-knowledge by asking, "Who am I?" Paradoxically, the necessary response denies the law of identity with the self-referential proposition, "I am not myself." My existence and my essence are not identical. The next question in the dialogical process naturally becomes, "Well then ... who are you?" The question, "Who are you?" is asked by the I at the limit where it awaits the presence of the (an)other as a you, only to find that it was an echo; the echo of the I demanding of itself who it is. It is this splitting of the subject that makes psychoanalysis and epistemology possible while, at the same time, rendering such inquiries impossibly difficult.

Dialogue creates the space for the I to appear; psychoanalysis is a very directed kind of discourse that brings the disparate parts of the psyche into view for one another. The discourse is an intersubjective one, meaning both subjects present themselves fully in the moment. The difference between being fully presented and the way in which individuals normally interact, is that by "fully" I mean the special listening that attends to the unconscious material presented, whereas "everyday" discourse is limited to superficial contact between conscious egos. To explain the dialogical structure that allows for the full revealing of the interlocutors, I will use some of the terms employed by Martin Buber in his limning of I and Thou. It can be argued that an I-Thou experience is at once external and internal. Its externality can be said to be in the intersubjective nature of the discourse that it is enaged in by an I and a Thou; its internality, however, stems from I and Thou's embracing one another's presence as their own, rendering the dialogue internal, and therefore, intrapsychic. In an intrapsychic discourse, each side of the I-Thou encounter is taken in by the other, so that an authentic We is formed of an I and a Thou. The I embraces (includes) the other in the intersubjective realm, which the I experiences as intrapsychic. Buber calls this movement into the other "imagining the real" (Buber 1965, 81).

It should be made clear that although we are broaching what appears historically to have been referred to as intuition, Buber carefully distinguishes his notion of imagining the real from intuition, specifically, from Henri Bergson's conception of intuition. One important difference between the two notions is that Bergson argues that one can intuit objects through what he calls an "intellectual sympathy" that is nothing less than, "... place[ing] oneself directly, by a kind of intellectual expansion, within the thing studied ..." (Bergson 1955, 45). Setting himself apart from Bergson, Buber is describing a phenomenon that can occur only between two subjects that are engaged in authentic discourse. For Buber, "imagining the real" is a "bold swinging" into the life of the other. One becomes an I through the experience of the other qua self. Buber continues:

> Relation is fulfilled in a full making present when I think of the other not merely as this very one, but experience, in the particular approximation of the given moment, the experience belonging to him as this very one. Here and now for the first time does the other become a self for me ... which is, however, to be understood not in a psychological but in a strictly ontological sense, and should therefore rather be called "becoming a self with me." (Buber 1965, 71)

The intersubjective realm created in the meeting of the I and the Thou is ontological, one could say, a place-thing. The epistemological counterpart to the ontological place-thing is an intrapsychic knowing. Given that the between is a place-thing, one can say it dwells objectively. It is this objectivity that bridges the gap between the I as subject and the Thou as object; likewise, it provides the basis of an explanation for the potential of subjects to arrive at objective knowledge.

Truth, and therefore knowledge, come about in the in-between, in the unique place-thing that is the intersubjective realm. Pierre Fédida in *Corps du vide et espace de séance*, writes, "The objectivity of psychic reality is the act of recognizing an intersubjectivity" (Fédida 1977, 125). Freud commented that true psychic reality is in the unconscious. Later I will develop more fully the role of the unconscious in the act of knowing oneself when I address the question of psychical structures, indexicals, and logic. For the time being, it is significant that objective truths that correspond to psychic reality will be discovered in the intersubjective realm. Buber writes:

> Human life and humanity come into being in genuine meetings. There man learns not merely that he is limited by man, cast upon his own finitude, partialness, need of completion, but his own relation to truth is heightened by the other's different relation to the same truth – different in accordance with his individuation.... Men need, and it is granted to them, to confirm one another in their individual being by means of genuine meetings. (Buber 1965, 69).

The "truth" in the interhuman exists when men "communicate themselves to one another as what they are" (Buber 1965, 77). They must fully communicate what they are by being present. Being wholly present immediately reveals them: "I-am-here-now." This absolutely self-referential utterance is demonstrably pointing to the present where the meeting of the in-between happens.

## I AM ONLY IN RELATION TO YOU

On this encounter, Buber writes, "One becomes aware of it in the between which is reconstituted in each new 'meeting'" (Buber 1965, 105). The between is ontological for Buber. The I actively moves towards the other, *ad limina*, toward the limit where the I is passive and waits openly for the presence of the Thou, to be graced by the Thou. *Ad limina* the seeker calls out, "Who are you?" The reply is, "I am not myself" for I am a contingent being and not a necessary one as my essence (who I am) and my existence (that I am) are not one. That is to say, "I am only in relation to You." While essence and existence are not the same thing for human beings, this is the case for one being, God. Only God can answer, "I am who I am." Pope John Paul II writes in a recent reflection shared at Mount Sinai, "God shows Himself in mysterious ways – as the fire that does not consume – according to a logic which defies all that we know and expect. He is the God Who is at once close at hand and far-away; He is in the world but not of it. He is the God Who comes to meet us, but Who will not be possessed. He is "I AM WHO I AM' – the name which is no name! I AM WHO I AM: the divine abyss in which essence and existence are one! The God who is Being itself!" In the same spirit, Buber, a Jewish theologian, claims that we move towards grace and await the presence of the Thou as we cannot possess it, but can only meet it. Buber writes, "Of course, God is 'the wholly other'; but He is also the wholly same: the wholly present. Of course, He is the *mysterium tremendum* that appears and overwhelms; but He is also the mystery of the obvious that is closer to me than my own I" (Buber 1970, 127). Of course for Buber, the I-Thou relationship is primordially a religious one, that is our discourse with God. One can make an argument here that would support Aristotle and Spinoza, among others, that self-knowledge comes through understanding God, for knowing oneself depends on the other reflecting back to you an image of yourself, and only God can do that wholly and completely. Buber claimed that we could approach the I-Thou relationship we have with God with others, namely, through the student-teacher relationship and the patient-psychologist pairing.

## I AM NOT MYSELF

The psychoanalytic dyad is formed when the analysand enters analysis because she believes that she does not know the truth, moreover, she does not know herself. Of course, analysis works because the analysand does know the truth about her predicament. Indeed, the analysand speaks the truth without knowing that she is doing so. The statement known without being thought by the analysand in a psychoanalytic session exhibits traits of self-deception:

(1) *I know the truth but cannot think about it.*
Or we may pose it as:
(1a) *I am deceiving myself.*
It is not the same thing to think, to know, and to say, and the analysand often finds herself in the following predicament:
(2) *I know the truth but cannot tell it.*
Which can be formulated as:
(2a) *I am lying.*
Psychoanalysis works because in fact the unconscious cannot keep a secret:
(3) *I do not know what I am saying.*
That is to say:
(3a) *I am telling the truth without knowing it.*

Psychoanalysis depends on (3) and but is necessitated by (1) and (2). Juan-David Nasio in his essay, "The Concept of the Unconscious," asserts that proposition (3) not only marks the beginning of analysis, but is likewise the "founding fact of the notion of the unconscious for Freud" (Nasio 1996, 24).

It is by adopting a special listening that the analyst can *recognize* the truth in the lie, *recognize* truth in what is not said, and lastly recognize the truth embedded in the source of a paradoxical locution. The analyst can then re-tell the patient that which she just "said without saying" so that the patient may *recollect* the truth. Why then does the patient lie to herself? Why engage in self-deception? It seems that the patient "lies the truth" because she knows how to construct the lie without revealing to herself the truth. What is this level of knowing that has access to the truth that on another level the person *resists* knowing? Freud claimed that the unconscious was the *true psychic reality*, and yet it is only part of the psyche. What are the other parts? Illusions? Lies? Deceptions? The fact that a person can lie and deceive herself reveals that she is by nature not a self without division, indeed, even that there must be sufficient separation for an internal dialogue to take place.

## I KNOW AND DO NOT KNOW THE TRUTH

The analyst hears the truth negated in the discourse and reveals to the analysand the material the analysand was repressing and yet "knew" all along. It is more palatable to hear the repressed offensive thoughts from the voice of a second person, a Thou and one having authority, for it was not, after all, the I who says it but the other I as Thou. In the course of analysis, the patient realizes that she has been lying to herself. This being the case, when the patient says:

*(2a) "I am lying,"* is she actually telling the truth? Consider that all the lies were intentionally constructed to reveal the truth; it is here that the problem of self-knowledge that is the worry of psychoanalysts broaches the worries of logicians. We must continue to look into the "I" and in what way it gives rise to an intersubjective "We" between the analyst and the analysand. Likewise, we must consider the internal dialogue of both analytic partners. The analysand has an internal discourse between the ego-superego-id that takes place on the frontier between the conscious and unconscious realms. So too in the analyst; the embracing of the analysand at an unconscious level wells up in the analyst's own internal discourse. It is along with the layering and intermingling of identities that we will look at "lying" and how it is that the analyst perceives the truth as revealed in the lie. Lying involves the spoken word and cognition, definitions of truth and reality, and notions of deceit and illusion; all of which must be investigated to see exactly where the resistance occurs that gives rise to the mentioned paradoxical locutions. Given the

complexity of these statements which are uttered in a dialogue however indirectly, who then makes the judgment about truth or falsity of the statements: the speaker, the listener, or a third party?

In analysis, the person cannot allow for their thoughts, they are repressed; they disallow that they are saying this, here, now. The analysand believes, "I am not saying this, here, now." All the while the analysand knows better and tells all to those who can listen. The subject wants the other to listen to the self and to recognize the true self behind the carefully orchestrated utterances. To listen to what is not being said requires a unique species of listening, namely, one that hears the inaudible tones of the unspeakable. It is remarkable that several analysts have couched this kind of listening in terms that broach the language of paradox and contradiction, or otherwise turn the understanding of listening on its head by re-figuring what it is to be an ear. Juan-David Nasio spoke of the analyst waiting openly to be surprised by the analysand. It is remarkable that Nasio's understanding of analytic listening makes the inconsistent demand to wait expecting to be surprised, but it is precisely this brand of inconsistency that is to be expected in attending to the truth located in paradoxical utterances.

André Green speaks of listening with the body, thereby turning the entire body into an ear that attends to communicative efforts that do not hit the eardrum, and yet, are propositions that can be heard. Green writes:

> [T]he analyst does not listen solely with his ears, but with his entire body. He is sensitive to the words, to the tone of voice, to interruptions in the narrative, to pauses and to the entire emotional make-up of the patient's expression. Without the dimension of affect, analysis is a vain and sterile enterprise. Without a *sharing* of the patient's emotions, the analyst is not more than a robot-interpreter. ... Today we know that the analyst must be able to bear the chaos of some patients, in order to enable them to emerge and build a certain ordered inner space, without which no kind of social existence would be possible. (Green 1986, 313)

Echoing Buber, Green claims that there must be a shared aspect of the experience of listening to the extent that the patient's emotions are felt through the analyst as his own.

Theodor Reik referred to this kind of special listening as the "Third Ear." As we have only two ears, the analyst must listen with an ear that is not an ear. Reik comments that the discourse that passes between the analytic pair is, "... an inaudible but highly expressive dialogue" (Reik 1949, 144). Arguing that analysts should not rely on conscious perception, as it is too limited, Reik proposes attending to one's unconscious perception. The difficulty lies in making the unconscious material conscious within the analyst himself; this is listening with the third ear. In the words of Reik:

> Receiving, recording, and decoding these "asides," which are whispered between sentences and without sentences, is, in reality, not teachable. It is, however, to a certain degree demonstrable. It can be demonstrated that the analyst, like his patient, knows things without knowing that he knows them. The voice that speaks in him, speaks low, but he who listens with a third ear hears also what is expressed almost noiselessly, what is said *pianissimo*. There are instances in which things a person has said in psychoanalysis are consciously not even heard by the analyst, but none the less understood or interpreted. There are others about which one can say: in one ear, out the other, and in the third. The psychoanalyst who must look at all things immediately, scrutinize them, and subject them to logical examination has often lost the psychological moment for seizing the fleeting, elusive material. Here – and only here – you must leap before you look; otherwise you will be looking at a void where a second before a valuable impression flew past. (Reik 1949, 145–146)

Reik elegantly argues that the analyst must capture the instant in which the unconscious presents itself, that is, when the unconscious speaks without saying, by whispering the truth. He uses the word, "here" to mark the moment in time-space that the unconscious knowledge reveals itself. Looking for it will only cause you to miss it, and any attempt at logical examination will come up empty. As such, this special kind of listening cannot be captured by logical examination, nor by conscious attention, but by catching what the ear misses.

## I AM SPEAKING THE TRUTH WITHOUT KNOWING IT

It is not by accident that the ear cannot hear the truth communicated by the analysand, for the analysand is repressing the truth; as such, the truth is forbidden to enter consciousness and by no means allowed to leave the lips of the analysand. The analysand must believe, "I am not thinking this here now." Moreover, the analysand must believe, "I am not saying this here now." Of course, in order to do so the patient must have full knowledge of that which cannot be known, so that it can be denied. The analysand tries to negate the knowledge by banishing it with the stamp of impossibility. As it turns out, the denial just postpones the inevitable, as the repressed material returns to consciousness slipping past the censor as an impossible locution. Its disguise as an impossible locution, or paradoxical proposition, is ingenious, as it feigns accepting the judgment of the superego while communicating the denied truth. For example, when Reik encountered an impasse in analysis resulting from a patient claiming to be thinking about nothing, he went to Freud for advice. Freud told him to ask the analysand to think of a thought he could never think of, an impossible thought. By calling for an impossible thought, which should yield nothing, Freud understood how to speak to the unconscious and circumvent the system of defense. The analysand made the inconsistent claim that he was thinking of nothing, yet one can never think of nothing, it is impossible. In fact, the analysand was thinking of the impossible, namely, the repressed thought. By asking directly for an impossible thought, the analysand could express the "nothing" he was thinking about (which is the unthinkable material that is repressed), while convincing the superego to let down its guard, thereby permitting himself to talk about nothing. Statements such as "I am thinking of nothing" and "Think of that which cannot be thought" appear to be poorly constructed sentences in a logical context, yet they are loaded, and often heard, utterances in a psychoanalytic context. As does, "I have nothing to say," "I refuse to think about it," "I do not know what I am saying." All these propositions imply the contradictory statement, "I know the truth and do not know the truth." In logic, this last proposition violates the law of non-contradiction and is by definition false; in psychoanalysis, however, its truth is revealed in its impossibility.

The unconscious id uses paradox to foil the logical superego and voice itself and so triumph over the superego by thwarting it with contradiction that the logical superego rejects. Freud writes in *An Outline of Psycho-Analysis*, "The processes which are possible in the id (the *primary process*) differ widely from those which are familiar to us through conscious perception in our intellectual and emotional life; nor are they subject to the critical restrictions of logic, which repudiates some of these processes as invalid and seeks to undo them" (Freud 1989, 85). The id circumvents the critique of conscious perception by presenting itself in paradox, thereby befuddling the censor, as it cannot make heads or tails of a paradoxical proposition. The material makes it past the censor but must bear the judgment that it is a non-sensical statement. Such paradoxical locutions seem to be the way of the unconscious while (intentionally?) giving our logical side fits. Imre Hermann writes, "Logic is at the service of defense, its point of view is that of a superego, that is to say it is rigorous, objective, and idealized. Logic announces a morality of thought" (Hermann 1978, 114). Logic at the "service of defense" analyzes paradoxical utterances by holding them to an objective, an idealized standard, a move that solidifies the paradox's insolubility. For this reason, Reik, among others, calls for an unconscious listening that is intersubjective as opposed to objective and is grounded in the authentic present, contrary to an idealized suspension of time.

Intersubjectivity, which will lead to an intrapsychic experience of the "between" of the analytic partners, is necessary if one is to adopt the special listening that can hear the truth in what is not being said. The ego and the superego cannot hear this dialogue, as the statements were constructed to thwart their systems of logic and defense. The ego, which is responsible for conscious knowing and the application of the reality principle, is deluded, and Freud argues, psychotic, as it is at the mercy of the superego and id. The true self is not the conscious ego; the conscious ego is psychotic (Green 1986, 21). The ego attempts to create a functional interpretation of what it is in light of the pressure put upon it by the id, the superego, and the external world (Freud 1989, 15). The ego knows it is deceiving itself, but it cannot help but misrecognize the truth and recognize the lie. Truth is in the utterance, but not where the analysand figures it to be. Therefore, the analysand is speaking the truth without knowing that she is. It seems that the id cannot keep a secret, although the superego would prefer it to be otherwise. Amongst the intense parsing and deception, only a third party can understand what is really being communicated, namely, the analyst. For this to be the case, it is not the same thing then to say and to know. W. R. Bion might have added that it is not the same thing to know and to think, for I can think without knowing it and not think about what I know. Recall the three statements made earlier that are at the heart of analytic practice and bring the analysand to the séance:

(1) I know the truth but cannot think about it.
(2) I know the truth but cannot tell it.
(3) I do not know what I am saying.

The split in the subject, demonstrated by the fact that such statements are actually made, renders self-referential propositions all the more interesting. One wonders which part of the self utters the statement, which is supposed to speak to which part of the self, about which part of the self? It is to be expected, then, that judging the veracity of a self-referential proposition will be a complicated matter.

## I AM LYING

We have discussed numerous propositions that violate the law of identity and non-contradiction. Now it is time to turn to the cases of self-referential propositions already mentioned, such as, "I am not myself" and "I am lying." The former is an example of a violation of the law of identity, and the latter is a version of the liar's paradox. In Alexander Koyré's discussion of the liar's paradox in his *Épiménide Le Menteur*, Koyré proposed that the statement, "I am lying" is one species of a family of statements that are not intended for use in a first-person context, as is the case with the species "I am dead," "I am silent," "I am asleep," and "I am absent" (Koyré 1946, 349). These statements share quite fascinating characteristics with respect to their structure and in the nature of the predicate terms. Postponing a discussion of the utterance "I am lying" until further into the paper, let it be noted here the other locutions are first-person, present-tense statements that implicitly negate an active state of being. They deny the subject's life, ability to speak, awareness (awakeness) and presence. It is remarkable that the problematic self-referential statements made above, delineate and define the characteristics of the unconscious. The unconscious is dead, without voice, asleep, and not present. In other words, the unconscious is that part of the psyche that possesses the death drive, forbidden to speak; it reigns during sleep, and is never present to consciousness, so it is absent. The conscious part of the psyche is awake, alive, and present, which comprises the predicates that were negated in the self-referential propositions mentioned above. Although the conscious is *of* the present, it is not *in* the present, for it creates its version of the present by bringing together recent past and near future. Juan-David Nasio explains in "Le Patient-limite et son rapport à l'instant présent," that the subject born in the moment is the unconscious subject and the moment of its appearance is unconscious time; one can also say that the present is the time of the unconscious insofar as the idealized past and future are the times of the ego (Nasio 1996). To Nasio's comment, I would like to add that the id and superego by virtue of being unconscious dwell in the present albeit in two distinct ways. For the id, everything is lived as though it is happening now. The superego, on the other hand, experiences an objective and idealized present much like the sense of present time logicians use to analyze propositions. The unconscious can only be present to the conscious as absence; that is to say, its presence is recognized through its absence. This may be related to the mentioned, paradoxical locutions and why they negate consciousness to reveal the presence of the true self. In first-person, self-referential, paradoxical locutions subjects deny themselves and their consciousness – is this so that the unconscious can speak?

Looking back at Koyré's discussion of "impossible," self-referential propositions, his footnotes (included in the quote) and judgements of nonsense and impossibility, unbeknownst to him (truth was in his comments, but not where he figured them to be), all act as signs posting the way to an understanding of the role of the unconscious in just such propositions. That is to say: there are cases where one would find these propositions uttered regardless of the logician's claim that it is illegitimate to do so, and likewise in spite of the egological defense mechanisms. In the words of Koyré:

> We seldom say what we mean and mean what we say. Moreover, we do not often know what we mean and intend to say. Besides, the words we speak, the sentences we hear, take their complete and full meaning only in and through context – we do not say or hear everything. Consequently, in order to interpret the words we hear we have to reconstitute and to restore the whole of the intended meaning; since, however, we are accustomed to meaningful speech [his note: Even when we speak "in order to say nothing."], to hearing sentences that have a meaning (or at least, that pretend to have one), nothing is more difficult for us than to apprehend a pure nonsense. We give a meaning even to what has none [his note: Just as we put order and logic into our dreams.]. (Koyré 1946, 349–350)

Koyré was precisely critical of the potential sensibility these self-referential statements could have. These statements, he judged, were self-contradictory and should be shunned. Any attempt to make sense of them would be nonsensical. Yet he stated that we often do not know what we mean and intend to say, and, in fact, he added that we sometimes speak "in order to say nothing." Curious that a person who forms the statement is unaware of the meaning he intends to convey, but of course, psychoanalysis understands this predicament all too well. It can be said that the ego has difficulty making sense of such statements, but that was the intent of the unconscious, to voice itself by any means necessary. It is worthy of note that Koyré specifically made a reference to dreams as being exemplary of nonsense, to which any attempt to give meaning or apply logic would be futile. The concern of the psychoanalyst, however, is to do precisely that, namely, reveal the order and logic of our dreams as unconscious manifestations of our thoughts resulting from primary processes.

Koyré, in the same article, contests that the statement "I am lying" is unlike the other self-referential utterances that otherwise share many similar traits. To quote Koyré, "Thus, the statement 'I am silent' has both a predicate and a subject; the vice is that it is impossible to link them together. Therefore it is a false judgement. … On the other hand, the expression 'I am lying' has no subject and, accordingly, it is not a judgement" (Koyré 1946, 353). He disallows that the indexical "I" could be the subject of lying. I have already shown above how the problematic self-referential statements outline the defining characteristics of the unconscious. Now I will argue that the expression "I am lying" is equally significant in the context of the unconscious revealing itself through paradoxical locutions.

Looking back at the propositions that are derived from the fundamental utterances that open up the possibility for analysis to take place, one can see their sense arising as they do from the psyche of a split subject: (1) I am telling the truth without knowing it. (2) I am deceiving myself. (3) I am lying. When the analysand utters "I am lying" the expression has a subject term, but it is not where we expect it to be. The ego must voice this statement, and it is the ego that lies; it is the ego that represses the truth. From the perspective of the unconscious, everything is happening at once, fully present, and all thoughts are lived as the eternal recurrence of the present. The repressed unconscious bears itself upon the ego at every moment, while the ego is ever vigilant of holding it at bay. The id and superego dwell precisely outside of time, but for that reason they experience everything as present, albeit in the different senses of present discussed above, while the ego dwells in the present but experiences the past and future only. Given the radical difference of experienced time amongst the structures of the psyche, any statement that speaks to this, here, now will not maintain its meaning when interpreted through the three structures of the psyche.

The difficulty logicians have with the expression "I am lying" is that it refers to itself at the moment it is uttered. A psychoanalyst could understand it as the ego making this claim, never fully present in the present, while the id triumphs, as it has revealed that the ego lies by repressing all of the banned thoughts existing before the mind at the present moment. The subject, of the locution "I am lying," becomes all the censored thoughts the speaking ego has in mind without thinking about them. The ego lies by not revealing the truths that are present to it at the moment it speaks. The ego is out of touch with the repressed material as it dwells in a synthesis of recent past and near future, pushing the paradoxical, but authentic, present into the repressed unconscious along with all other forbidden material. The superego dwells comfortably in an idealized present by ignoring the context that makes each moment unique. When the superego hears the phrase "I am lying," it judges it according to the rules of logic and thereby dismisses it as meaningless, for the predicate term has no subject. The ego and superego have difficulty with the implicit indexicals in the phrase "I am lying" which could be restated to make the indexicals explicit, including the essential indexical "I" and its interlocutor "you," as in "I am lying to you here and now."

## I AM HERE SPEAKING THE TRUTH, RIGHT NOW, TO YOU

Indexicals pose the greatest problem in translating between the superego, ego, and id. An "indexical' is a logical term used by logicians when faced with the difficulty of analyzing propositions that refer to the subject's view of that which is happening at the moment the proposition is uttered, and it is believed that certain words are the source of the problem. Examples of indexicals are the terms, "this," "here," "now," and "I." Given the pervasive nature of indexicality, it is not at all satisfactory that these seemingly benign, quotidian terms should be so problematic for logic. The self-referential nature of indexicals is at the root of the difficulty. Dan Zahavi perceptively notes:

> Indexical reference is often taken to be a perspectival mode presentation. It embodies a subjective point of view on the world. Whereas an object might be intrinsically heavy, soluble, or green, it cannot be intrinsically "this," "mine," or "here"; it only becomes so relative to the confronting subject. To think of something indexically is to think of it in relation to oneself. Our indexical reference reveals the object's relation to the referring subject, and it consequently implies a kind of self-reference. This is why indexicality has been claimed to be egocentric, why it has been said to be anchored in some kind of self-presentation. (Zahavi 1999, 24)

It is the absolute subjectivity of indexical locutions that gives logicians such great pains and likewise perturbs the egological defense mechanisms. The goal of presenting the id to the ego is that of psychoanalysis, as is made evident by Freud's famous expression, "Where the id is the ego shall be." This encounter between the id and ego, which would result in the id introducing the repressed material into the ego, is not the work of introspection, for the unconscious material would encounter the difficulties of the defense mentioned above. The repressed knowledge can only be presented to the ego from an(other), not an other as id, but an(other) as the self of an(other), namely, the analyst. How can the analyst know the radically subjective experience of an(other), fully and completely in the context denoted by the indexicals, without in some profound sense becoming the analysand?

It is through the proposed theory that the unconscious is shared and created in-between the analytic partners that the answer to this question takes shape. Nasio argued in his *Five Lessons on the Psychoanalytic Theory of Jacques Lacan* that the analytic partners meet in the space of the in-between where there is a shared unconscious that engulfs the analytic couple. Transference occurs in the space of the in-between. In transference, the analyst's unconscious is engaged with the analysand's to such an extent that the analyst assumes (takes into relation, adopts, takes within himself) the symptoms of the analysand, and at the same time, the analyst becomes the cause of the analysand's symptoms through identification. The symptoms of the analysand would include paradoxical, self-referential locutions. Transference in this sense represents a union between the analytic partners, which is nothing short of an identity, and a necessity of the analyst is to help the analysand attain self-knowledge. Here we have a profound "we" which speaks the truth, right here and now, to those present in the meeting, in the between.

Where exactly is truth to be found in the psychoanalytic séance? The analysand enters onto the analytic scene with the exclamation "I do not know." The analysand approaches the analyst seeking knowledge with the hope of coming to know what he does not know. Clearly, what is not known is the self who poses the statement, which becomes "I do not know who I am." It is then through symptom, that which is said without being spoken, that the patient understands, "I do not know the truth about who I am, and so I must be lying to myself." This statement claims that one can lie without knowing the truth, stranger still that one can lie to oneself; and the greatest paradox is that one cannot know oneself as oneself. As has already been mentioned, the analysand knows the truth without believing it and speaks the truth without knowing it. The analysand speaks the truth in the moment, but only the analyst can hear it by adopting a special listening, a listening that comes about from suspending the ego and superego.

The analysand enters into transference with the belief that the analyst is the judging, harsh superego. It is in transference that the analysand projects his ego onto the analyst, thereby liberating the repressed unconscious material and fully presenting the self. The analyst, for her part, tries to suspend the judging and defensive parts of the psyche leaving herself open to accepting the unconscious material of the analysand by making herself present through her absence; notably, the analyst sits behind the analysand and consequently cannot be seen. Through the melding of each analytic partner's unconscious into one, the analyst can know the subjective nature of indexical locutions, in this intrapsychic state, while circumventing the system of defense. The paradox of how one can know another is resolved in the between. Likewise, the paradox of a self, knowing itself, has likewise found a resolution in the same moment. With respect to the present that is the moment of the encounter, of the between, of the I and the Thou that is also the relationship between a patient and therapist, Buber wrote:

> This making present increases until it is a paradox in the soul when I and the other are embraced by a common living situation and (let us say) the pain which I inflict upon him surges up in myself, revealing the abyss of the contradictoriness of life between man and man. At such a moment something can come into being which cannot be built up in any other way. ... For the inmost growth of the self is not accomplished, as people like to suppose today, in man's relation to himself, but in the relation between the one and the other, between men, that is, pre-eminently in the mutuality of the making present – in the making present of another self and in the knowledge that one is made present in his own self by the other...(Buber 1965, 70–71)

The particular example of pain Buber mentions in this quote could easily be replaced with that of symptom, so that the symptom is shared, and the analyst feels it well up within her as her own. Indeed, the absolute subjectivity of a symptom is fully experienced by the analyst, for in transference, the analysand projects his ego onto the analyst, thereby liberating the unconscious and fully presenting the self; in turn, the analysand is made present in "his own self by the other." The analyst listens, with the third ear, and can hear clearly the inaudible expressions of the unconscious in the symptomatic, paradoxical locutions. In this unique place-thing of the between, safely hidden in the paradoxical present, the paradoxes drop their pretense of impossibility and for the first time the impossible can occur: The truth is told. But does the truth make sense? Well that depends on whom you ask....

*State University of New York at Stony Brook*

## REFERENCES

Bergson, Henri. *An Introduction to Metaphysics*. Trans. T. E. Hulme. New York: The Liberal Arts Press, Inc., 1955.

Buber, Martin. *The Knowledge of Man*. Ed. Maurice Friedman. Trans. Maurice Friedman and Ronald Gregor Smith. New York: Harper and Row, 1965.

—. *I and Thou*. Trans. Walter Kaufman. New York: Charles Scribner's Sons, 1970.

Fédida, Pierre. *Corps du vide et espace de séance*. Paris: Jean-Pierre Delarge, 1977.

Freud, Sigmund. *The Ego and the Id*. Trans. Joan Riviere. Ed. James Strachey. New York: W. W. Norton and Company, 1960.

—. *An Outline of Psycho-Analysis*. Ed. and trans. James Strachey. New York: W. W. Norton and Company, 1969.

Green, André. *On Private Madness*. London: Hogarth Press and the Institute of Psycho-Analysis, 1986.

Pope John Paul II. *Homily of the Holy Father John Paul II: Celebration of the Word at Mount Sinai*. Saint Catherine's Monastery, 26 February 2000.

Koyré, Alexandre. *Epiménide Le Menteur*. Paris: Hermann, 1947.

Nasio, Juan-David. "The Concept of the Subject of the Unconscious," in *Disseminating Lacan*, ed. David Pettigrew and François Raffoul. Albany: State University of New York, 1986.

—. "Le Patient-limite et son rapport à l'instant présent," in *Lacan avec la psychanalyse américaine*, ed. Judith Feher-Gurewich and Michel Tort. Paris: Éditions Denoël, 1996.

—. *Five Lessons on the Psychoanalytic Theory of Jacques Lacan*. Trans. David Pettigrew and François Raffoul. Albany: State University of New York, 1988.

Reik, Theodor. *Listening with the Third Ear*. New York: Farrar, Straus and Company, 1949.

Zahavi, Dan. *Self-Awareness and Alterity: A Phenomenological Investigation*. Evanston: Northwestern University Press, 1999.

PEDRO LUIS BLASCO AZNAR

# THE TRUTH OF THE I AND ITS INTUITIVE KNOWLEDGE

This study is a preliminary form of a wider study in a critical dialogue with classical and contemporary philosophy, one which will be developed in my next book, to be titled *An Anthropological Theory of Knowledge*.

My research is based on the knowledge of truth to be found in "I-myself" and in the "*I*-Other" bond in its unique, original and unrepeatable equality and intimacy.

This research is a continuation of the critical analyses of Kant and Dilthey. Kant, by means of a critique of pure reason, established the conditions of possibility of knowledge in general and of scientific knowledge in particular. Dilthey, by means of a critique of historic reason, reformulates the sciences of the spirit. It is my project to present the insufficiencies of these critiques for the knowledge of "I" and for interpersonal knowledge and, consequently, to undertake the then necessary new analysis relative to the original reality of the "I", that is each individual by means of what I call *the critique of intuitive reason*. This critique clearly makes it manifest that the personal reality of the "I" is beyond all concepts and all language. It is unconceptualizable and ineffable, but it is understandable. Its *aletheia*, its exclusive ontological truth, is revealed to me only in one particular intuition: the intersubjective personal intuition in which we are given the revealed reciprocal presence of "I" and "You" by means of their loving coexistence. My critique of intuitive reason, furthermore, reveals that love in its ontological dimension, is constituted as a gnoseological category.

* * * * *

The theory of knowledge, as viewed from the perspective of the history of philosophy, itself had a relative and homogeneous development from the Greeks up to Kant. The ancient Greek philosophers and scientists, the pre-Socratics and, above all Socrates and Plato concerned themselves with gnoseology. However, it was Aristotle who established the basis and the determinative focus that this research would have to follow in subsequent philosophy.

It is the aftermath, when the theory of knowledge underwent three important transformations in sense, that I will now set forth: transcendental idealism, historicism, and the sociology of knowledge. But I hold that it is necessary to undertake new critical research in order to complete the critique that Kant initiated. This involves a new transformation, which, in all



*A.-T. Tymieniecka (ed.), Analecta Husserliana LXXVI*, 117–131.

likelihood, does not have the epistemological scope of the previous ones. However, I should point out its philosophical relevance because the knowledge of the "I" that I now consider is not knowledge in the psychological sense; although psychology also has much to say in this respect, knowledge of the "I" does not belong to it exclusively. And because what I refer to as the anthropological theory of personal and intersubjective knowledge does not constitute knowledge alien to the discourse on knowledge of human beings in general, it has as its proper object, and in a particular way, the "human" being, the individual or the concrete person that each one is and precisely in that which constitutes his human singularity, his peculiar manner of being and nature, what an individual originally is. The anthropological theory of knowledge that I will set forth is integrated into that general theory of knowledge which has been denominated gnoseology, and which develops from the it while bearing in mind the proper characteristics of the I, and sees the I as the object of knowledge.

More particularly, what I wish to set forth here is the idea that in the same way that Dilthey had, in one way or another, to rectify or complete the Kantian basis of knowledge, it is still necessary to complete the work initiated by Kant and later developed by Dilthey.

Firstly, not only did the critical philosophy of Kant provide philosophy with a new method of reflection and research, but its particular realisation and application constituted an authentic transformation of philosophy, as is developed in K.-O. Apel's contrastive study *Towards a Transformation of Philosophy*, and are made within the orientation of the dialogic thinking of Habermas. Kant's philosophy remained a philosophy of the subject, as did all modern philosophy until Hegel, but a subject whose subjective rationality sought "to regulate" the reality of the natural and human world by means of science, law, etc. Apel's and Habermas's work, however, constitute an intersubjective philosophy, a philosophy of discourse, a philosophy of communicative action that tries to claim a rationality that is commensurate with the human world, with human life, as emancipation from manipulation and the evasiveness of mere technical-instrumental reason, or teleological reason as Weber puts it.

Kant's philosophy is a re-examination of metaphysics that refers to and demands a critical review of knowledge. It deals with a critical analysis, one confronting the prior dogmatism, of the possibility of knowledge in general and fundamentally and of the possibility of scientific knowledge in particular. The rationality of the subject is shown to correspond to the rationality of phenomenal reality, which reveals *a priori* the subjective conditions of knowledge of it and constitutes it as a phenomenal reality. These conditions expose the theoretical irrationality of metaphysics, but do not disallow its practical rationality. Kant proposed to himself a foundation, an explanation of knowledge in accord with the prevailing considerations in the philosophical tradition that polarised the relationship between knowing and the known, the relationship between subject and object that defines modern philosophy to a great extent. But Kant gave the discourse a Copernican turn that freed it from the "dead end" to which it had been geared by rationalism and empiricism, one that did not ignore for one instant the entity of the object – in this sense his philosophy is still a realist philosophy – while analysing thoroughly the *modum cognoscentis*. This determined the transcendental character of his philosophy.

For Kant the model of knowledge was scientific knowledge, the subject of knowledge was the subject of scientific knowledge, and the object of knowledge was the object of scientific knowledge.

The model of knowledge for critical analysis is that of a rationality imposing the strict conditions that delimit and define not only scientific knowledge, but all that could be the object of scientific knowledge as well; this is the monistic model that arose with modern science from its first steps in the Renaissance until Newton, and afterwards from the empiricist tradition and positivism of the XIXth century to the Vienna Circle and, practically down to today with Karl Popper, etc. This, to my understanding, excessively zealous scientific model, holding rational control of experience and experimentation to be the genetic principle of knowledge, that followed the guidelines of Aristotelian–Thomist philosophy – *nihil est in intellectu quod prius non fuerit in sensu* – put aside once and for all, all that could ontologically be the object of science and scientific knowledge and all those theories that had resonance with ontology, although at times only superficially, theories of magic, alchemy, organicism, panpsychism, etc. Undoubtedly and undeniably, this defined and determined scientific monism in the field of modern thinking traces back to the training of Kant and his academic activity. His philosophy, more than any other, appears to be a radical and coherent development of another of Thomas Aquinas's maxims of Aristotelian root: *quidquid cognoscitur ad modum cognoscentis cognoscitur*; the Kantian development of this would lead Cartesian rationalism to the culmination of the idealism of Hegel: this maxim contemplated but little the traditional demands of objective experience, of the experience of objects in so far as, and in the manner as they can be objects of experiences; and its radical proclamation by Hume woke Kant up from his rationalist dogmatic dream. And thus, the subject in Kant appears as a transcendental *res cogitans* and the object as phenomenal reality itself in all its complexity, although not well differentiated, as we will see.

The problem of knowledge establishes a dialectic of subject and object that demonstrates narrow and insuperable links between ontology and gnoseology, and which brings about a confrontation between – and goes beyond – the maxim that I have just quoted and a paraphrase that could be its twin: *quidquid cognoscitur ad modum "cognill" cognoscitur.*

Then, secondarily, Dilthey picked up the challenge and echoes of arousing faint voices, as he confesses at the beginning of his *Introduction to the Human Sciences*, and had to reconsider at length questions generally thought to have been solved since Kant. It turns out that if one has to analyse the subjective side of *cognoscitur* – the *modum cognoscentis* – he will nevertheless have to comply with the objective conditions of that same *cognoscitur* composed of a *quidquid* that patents, empirically and objectively, a reality clearly differentiated that makes the monistic model of the Kantian episteme useless for a determined area of being. In just this way the Newtonian model turned out to be an inoperative barrier as well as the limit to scientific knowledge itself, to researching and to scientifically explaining what was no longer an indistinguishable reality with the emergence of the microcosmic field of material nature revealed by the latest scientific research, which situation initially baffled the scientists themselves. Scientists could only call out disconcertedly, "How contradictory Nature is!" This was in the twenties on the shores of Lake Como (I. Prigogine and I. Stengers, *Order out of Chaos: Man's New Dialogue with Nature*).

The object also "imposes" its conditions "regulating" knowledge, that is to say, determining how knowledge in general is possible and in particular scientific knowledge and, therefore, imposing the criteria setting the boundaries of the respective individual sciences, categories or specific ways of knowledge, as enabled and required by their objects, etc.

Dilthey, in this way, corrected Kant: He incorporated the critical project of Kant, which meant putting limits on it. He did this by surveying the real field of its validity, the dominion or "natural" field of the object of knowledge, and at the same time of the object of his transcendental analysis. The object of knowledge manifests in its own being that it functions differently in the two "hemispheres" of the world of knowledge; but firstly, to my knowledge, it manifests even more and more strictly, that there are two hemispheres of the world of being, of "the real" (of what is real), of "what there is", etc. to which both hemispheres of the world of knowledge correspond. Dilthey broke the ontological monism of the phenomenal world of Kant and, thereby, broke also epistemological monism. There is, thus, a natural world and a spiritual world to which the sciences of nature and the sciences of spirit correspond respectively. The transcendental analysis of knowledge now requires the completion – not the denial – of "pure reason", of the Kantian critique with a critique of "historical reason".

It was not for nothing that it was in the second half of the XIXth century that the human sciences definitively arose, the sciences of human life as "life" and as "human": psychology and sociology primarily. It was also at this time that a clear consciousness of history as a dimension of human life arose, when full rational and social consciousness was acquired – different from the religious and eschatological consciousness of Saint Augustine and Bossuet – of the historic fact, of history itself. *The New Science* of Vico, above all the second edition of 1730, was its starting point: And then Hegel was the primary influence. Already in the *Phenomenology of Spirit*, but above all at the end of the *Sittlichkeit* in the *Philosophy of Right* (1821) he initiated his peculiar conception of history, widely expanded in the posthumous *Lessons about Philosophy of History*. Likewise, Herder made fundamental contributions to the constitution of historical science. On the other hand, this was also the moment of the development of philological studies and the philosophy of language, of aesthetics, etc. All in all, this was about the world of culture, a world having its own entity as the object of the spirit, of an objective spirit that has its own rationality and its own science. Dilthey saw it thus and also saw the subject involved in all, in its specificity as a human subject, as its original reference, as creator of that human world in which he manifests and objectifies himself, the object, in different phenomenological forms, of the different sciences of the spirit.

It is also true, and this is the third moment of innovation in the development of the theory of knowledge, that the critical analysis of scientific knowledge and the development itself of the sciences of nature, as well as of the sciences of spirit, is also conditioned by the social nature of humans, that is to say, by the social realisation of humans in all its dimensions: the creations of man reflect human nature, but they also inevitably reflect the culture that he has created. Man, and with him all his creations, are "sons of their time", including the particular sciences. Knowledge in general, as well as scientific knowledge in particular, are interceded by the subjective and objective conditions of knowledge, and also by those conditions derived from its social reality. This has been made manifest by the vast trend of the "sociology of (conventional) wisdom" or "sociology of knowledge" initiated by Max Weber, Scheler, etc.

From this perspective, the social reality of science establishes some conditions for its own constitution and development that should be taken into account by a critique of knowledge. Now, then, the sociology of knowledge discovers and highlights some conditions for its realization: such conditions affect only extrinsically the relationship of the subject with its object. For instance, it deals with the social organisation of knowledge, of the private interests that prompt the activities of the laboratories, the direction of scientific policy in public and private research centres, etc., and its practical application in technique. These are conditions whose importance has been dealt with by the sociology of the categories of knowledge, the sociology of the conditions of knowledge validity, and the sociology of science itself, and according to the considered thought of E. Lamo de Espinosa, they are different, I understand, from the subjective conditions and the objective conditions of the possibility of knowledge, and the sort of knowledge possible in each case to be derived from the ontological reality of the subject and from the ontological reality of the object and not from its social reality, which are those realities that I now wish to consider here.

Here is where I have set my sights, because I have to think that this historical process of the critical review of knowledge is still insufficient and a new correction is needed.

Kant, by developing rationalism and anticipating idealism, paid attention to the *subjective* conditions of knowledge, which simultaneously determine the constitution of an object as an object of knowledge. This path has attracted many subsequent philosophers, whose critical contributions have been, simply speaking, positive or negative with respect to the Kantian approach and critique. Dilthey, on the other hand, dealt with the *objective* conditions of knowledge, circumscribing the limits of validity of the critique of Kant and, at the same time positing a new world around the object of knowledge which, because of its ontological specificity, requires a new critique to put forth its own features as object of knowledge, and which in doing so, determines the possibility of a subject becoming the subject of its knowledge.

In this second path Dilthey still did not take into account that which is apart from and different from the world of spirit, the human world, the world of life. There is also the world of *man himself* and of the human and personal subject who lives and gives life and makes human his world, becoming an objective presence and the object of his spirit. It is evident that the critical review of Kant and the contributions that reject it or develop it as well have been extraordinarily rich: Above all there is the epistemological critique that Dilthey attempted with the copious and creative researches that it has provoked. But we are now concerned with advancing to the limit the critique that Kant initiated, and Dilthey continued analysing with greater accuracy the fields of reality and of knowledge. This means analysing them in order to delimit again the fields of reality in which their respective analyses have validity: Dilthey put limits on the scope of the critique of Kant, linking it to the natural world and the sciences of nature. For my part, I see it as necessary to also put limits on the validity of Dilthey's critique because not everything that is not of the natural world is the object of the same epistemology, that of the *sciences* of the spirit. The world of the spirit, the world of the human, in its turn, embraces two fields that are epistemologically uncompromisingly distinct because, and this is a *logice prius*, it embraces two fields of reality that are ontologically different: the *ambit of the objectivation of the subject*, of the objective and multiple realizations of the culture of each period, of the human spirit considered in all of its dimensions: artistic, religious, social, legal, etc., i.e., the world of the human in all of its manifestations, products and objectivations, from the most transcendental to the most insignificant in the everyday life of an everyday man; and the *ambit of the subject himself*, the personal ambit of the human subject, creator of all that is spiritual, human and cultural, the ambit of each man and each woman in particular, of the individual persons that each one of us is in all the multiple dimensions of our being: the ambit of the subject, of the "I", in which there is no sense in speaking about *episteme*, about the epistemological demarcation of knowledge, neither of content nor of fields of knowledge that distinguish sciences from each other; it is not even worth speaking of a gnoseology of this human subject in the same way in which one considers a general gnoseology of knowledge.

Looking back in history, from Aristotle on, at least, science is only universal science. Further, in some way, and as conceptual knowledge, it is also common knowledge of the things that surround us, of nature and the natural phenomenon, of the people that we meet, of ourselves, and in general knowledge of reality. But we know and we think about what is conceptually real, and we say and express what is real by way of language, by way of words, or linguistic expressions that are verbal expressions of our concepts. However you understand the linkage between language, thoughts, and reality, its always deals, from a linguistic, psychological and gnoseological point of view, in one way or another in conceptual and linguistic knowledge, because we know things as soon as we think about them, and we think about them as soon as we have their idea, that is to say, by means of their concepts.

Well now, the concept, although its extension and comprehension might vary, is universal and abstract; it is said and it is thought of many individual instances. But if I think on the knowledge of the individual man in particular and on his personal reality, on the knowledge of "I" singular, on his originality, on his uniqueness, on his exclusive self-identity, unique and non-repeatable, on knowledge of the "I" as to who and how he is in his intimacy, on his own being different from any other I, I find that concepts do not fit, because it is unthinkable: there is no way of thinking of it without losing it, and in the end it is unutterable and ineffable. But it can be known and it is known, only in another way.

A transcendental critique of knowledge of I – one eschewing the dogmatic of knowledge of another "I", of personal and subjective self-knowledge and of intersubjective as well as interpersonal knowledge (knowledge beyond phenomenological objectivations, etc.) reveals that the subjective conditions of the possibility of knowledge of the "I" as object or content, are very different from those that Kant thought about – and that the objective conditions of the possibility of knowledge of the "I" as object or content constitutes the possibility that the subject is constituted in the subject of its knowledge, which conditions are very different from those pondered by Dilthey. The object of knowledge whose conditions of possibility were analysed by Kant was truly different from the object of knowledge whose analysis was carried out by Dilthey. And the respective objects of knowledge of Kant and Dilthey are truly different from the object that I am analysing.

Thus we deal with a different theory of knowledge, a new theory of knowledge. Nevertheless, this is rational knowledge; knowledge of the rational subject that concurrently becomes, in itself or in others, the object of its own knowledge. Thus it concerns neither a critique of pure reason, nor a critique of historical reason, to use classical expressions, but, by means of analogy, a critique of *intuitive reason*, whose development is the object of this work whose working title is "An Anthropological Theory of Knowledge".

* * * * *

This *Anthropological Theory of Knowledge* whose principal idea I seek to develop here very concisely, is structured in three moments. The sequence of those enunciations, to give a name to each one of them, creates a *sui generis* syllogism:

1. gnoseo-ontological reflections;
2. onto-anthropological reflections;
3. gnoseo-anthropological reflections.

1. My starting point in this first moment, and in my argument as a whole, is the easily defendable thesis of a constant correlation between ontology and gnoseology in the history of philosophy. It will be sufficient to analyse this thesis from the historical perspective of the development of philosophy in order to verify that these two areas of philosophy are not reciprocally independent. In truth, philosophy has a tendency to develop as a system, and by virtue of the consistency of the system itself and its own systematic nature, none of the philosophical disciplines, nor any thesis set forth in them, are absolutely independent from each other, but rather, according to their respective content, in a closer or more remote manner, they show a logical relationship in certain respects. In particular, as far as the anthropological theory of knowledge is concerned, the ontology created by a philosopher is in no way independent from the gnoseology that he would be willing to defend; and vice versa. Even more, the indubitable relationship between them has the shape of a foundation – i.e., ontology – and of what is founded, i.e. gnoseology. The philosopher who understands and responds in a particular way to the problem of being, of reality, has already started to establish the bases of his conception of the problem of knowledge, truth, of its possibility, etc. although in fact he does not develop a gnoseology. At the opposite pole, the philosopher who has developed a gnoseology, even if he has not systematised, elaborated, or thought the philosophy of being, has already introduced ontological suppositions in his philosophy and has implicitly started from them. For instance, the way Plato conceives knowledge and its foundation is correlative, consequent, and founded on his conception of the reality of being. The philosophy of Aristotle is developed in the same way; his ontology differs very much in accord with his respective gnoseologies. This is also to be seen in Kant, Dilthey, Nietzsche, and Bergson for example, who together with Plato and Aristotle are the philosophers studied in this part of my work.

2. The second moment is the ontological analysis of the reality *man*, of the *being* who is the *human being*. It is not my intention to develop now a philosophical anthropology, a philosophy of man. On the contrary, I try to sum up a conception of man that is reduced to what is essential in this research: to collect and to expose certain features of this being who is man, with certain attributes or qualities and intrinsic elements of *human nature*. It seems to me more accurate and objective to speak about human nature, to think human nature rather than to think and speak about the essence of man. So to begin with the fewest possible assumptions, making the minimum concessions, which are perhaps inevitable, I start by trying to understand man as *human nature*. The human being is human nature. In any event, our effort tries to

highlight some specific lines of man, of the human being that distinguish him or that he does not share with the rest of human nature or with the rest of beings. There is an ontological difference between the human being and all other beings. There is a general ontology and possibly some "regional" ontologies, but there is undoubtedly a philosophical anthropology, a philosophy of man, a specific ontology of the human being containing the peculiarity or specific way of being human, the nature that man is.

3. Finally, the last moment comes with the conclusion to the syllogism previously enunciated: commensurate with ontology or consistent with ontology in a possible philosophical system; or in a conception of philosophy as a system, there is a gnoseology. And if the human being is a particular being amongst beings, ontologically different than the others, in a gnoseologically relevant way, that is to say, with respect to those constituents and intrinsic elements or features of his nature that have to do with the content of his knowledge as determinant of the objective conditions of possibility of that same knowledge, the natural constitution of his being makes him the object of an ontology equally particular and distinct from other possible ontologies. Therefore, the knowledge of the human being that each one is, in an individual, original, intimate, etc. way, the object of a type of knowledge different than knowledge of the rest of beings, and the comprehension and the analysis of this peculiar knowledge is the object of a peculiar gnoseology, of a different theory of knowledge, one that I call the anthropological theory of knowledge. To express it in other words: if we consider ontologically the individual, original, unrepeatable, etc., human being in its unity, then each person does not correspond with the corresponding being that is the object of knowledge proposed by the gnoseologies of Plato, Aristotle, Kant, Dilthey, Nietzsche and Bergson. This is the reason why those gnoseologies are not suitable or are at least insufficient for understanding and explaining the knowledge of I-myself or of I-other. We will see as well that existentialism has as well as not proposed a theory of the class of knowledge of the "I" such as I am developing; the existentialist philosophies do not consider accurately the problem of knowledge that I intend to analyse nor the responses to its problems that are here sufficient and suitable.

* * * * *

My anthropological theory of knowledge distinguishes necessarily two types of knowledge of the subject, accepting two ways of access to knowledge about the I, both of them necessary absolutes and necessarily compatible, but nevertheless very different with regard to their respective content and knowing what they provide or allow with regard to the cognitive "faculties" by means of which one has access to them and with regard to its subjective and objective conditions of possibility; that is to say, they are required and indispensable, approaching by way of the subject as well as by way of the object, so that there might be in fact such knowledge.

These are a) the gnoseological-conceptual way and b) the gnoseological-intuitive way.

a) As does all knowledge, the gnoseological-conceptual way starts from what in general we can call *doxa* or common knowledge, and by means of the appropriate depth of knowledge and relevant research it elaborates an *episteme*. At the level of the *episteme* the subject can be the object of quite varied scientific research: biology, biophysics, as well as psychology, cultural anthropology, history, or philosophy itself. This signifies that the human being is of such complexity, or by virtue of his nature of such constituted and ontological richness, that being an integrated element of those hemispheres of reality, of the real world, it is necessarily the object of research of the two hemispheres of the world of knowledge: the sciences of nature and the sciences of the spirit.
b) Nevertheless, in addition to this specialized epistemological knowledge there is a generalized knowledge, a common knowledge. This *doxa* is true knowledge, perhaps more about what man is, about what we are or how people are; it is a knowledge that arises from personal experience of contact with others, in our constant exchange with each other in the daily life. Well now, regardless of the epistemological and doxic level of knowledge that we have about human beings in general and about ourselves, this knowledge is always knowledge of the phenomenal manifestations of the I, that is to say, of the feelings and desires, of the actions, of the potentials that the person develops, of his cultural objectivations, of the way he appears and of his appearance, of his *phainomai*. But the "I" that is the human being neither reduces itself nor is it exhausted in any of its phenomena, in any of its desires, in any of its potentials, actions, and objectivations; the human being is not yielded in his mediated appearance, in the mediation of his *phainomai*. This phenomenal knowledge is a descriptive knowledge, sayable, linguistic, and it is definitively a conceptual knowledge. Therefore it can be generalized, in that each one of these qualities in different grades and shades, can be individually perceived in other subjects and is predictable of other individuals. But none of these qualities and phenomena offer or express all of the "I" as such, because the I in itself is another I, it is not

> the multiplicity of its phenomenal manifestations, but rather its original and originating unity. A description of the colours of a portrait does not correspond to the perception of the colours themselves in that portrait. Just as the description of a portrait is not equivalent to the contemplation of the portrait, so the description of a person is not equivalent to the direct and immediate and personal knowledge of that person. To say that it is not equivalent is to say that it does not comprehend the same, that perception is more truthful and more total and goes well beyond accurate description; it is to say that personal and immediate knowledge, *inmediate*, is more authentic and total and captures and knows better the truth of the subject, the I that each person is.

This conceptual knowledge of the persons around us is as valid and truthful as scientific knowledge as well as common knowledge. But starting from this conceptualisation of the "I", the gnoseological-intuitive way is also necessary, because the I is not yielded in these qualities or traits, in those concepts with which we think of them. Knowledge of persons is not exhausted by conceptual knowledge. The original, unrepeatable, intimate singularity of the particular I *unus et unicus* is neither the predicate of another I nor apprehensible in universal and abstract concepts: it is only seen and can only be captured by an immediate perception, in a *personal intuition*. The common reality of the subject, which can be decomposed or analysed and predicated by other subjects, is the object of daily conceptual knowledge, and neither the I-itself nor the I-another relationship in the actual encounter or in reminiscence can be thought unless by means of the traits, qualities, etc. contained in our ideas and concepts about this I. But the reality of the I in its individuality, singularity, unity, uniqueness, originality, unrepeatibility, intimacy does not fit into conceptual, descriptive, sayable, linguistic knowledge. Yet we know that I and we are aware of this knowledge. We know this I, but this I is ineffable and indescribable; upon expressing it and upon saying it, we know we say it, but we know that it escapes us; it cannot be seized. Personal intuition is the only adequate and absolutely truthful knowledge of the intimate and personal authentic being each human being is. Intuitive knowledge is the proper personal knowledge, because it is the only one in which this is given, which contains and captures and captivates the original, intimate, interior, and irrepeatable being that the I is, and this is why it cannot be translated into concepts and conceptual knowledge: intuitive knowledge, intuition is indescribable, for the I-itself is ineffable.

There are, indeed, other types of intuition and intuitive knowledge, but from sensate intuition to the intuitions of an intellectual in his investigations, we deal with incomplete intuitions, of incomplete gnoseological efficiency, for they cannot be translated into concepts and concrete propositions. That is why the habitual intuitive knowledge of things, of all other reality, of all truth different from the I is not intuitive knowledge itself inasmuch as it needs to be thought conceptually.

Therefore, creating a general context of a gnoseology of intuition, a philosophy of intuitive reason that describes the traits of that particular intuition that is personal, is necessary as it is discovering and explaining its conditions of probability.

Anticipating very briefly the subsequent exposition, I think that, as is so with all knowledge, intuition is given in a relationship between the subject and an object in an immediate relationship, in an immediate giving of self by the subject to the object and in a special way in a personal intuition, in an immediate and reciprocal giving of the I as subject to another I as object, of I to you and of you to me, making each other obvious, introducing to each other the entity of the truth of oneself upon abandoning the depth of concealment. The inevitable reciprocity of this intuition arises precisely from this intuitive personal relationship having a distinct grade of knowledge.

Certainly, the interpersonal relationship develops itself very extensively and gradually: starting from a mere coincidence at a stand in the supermarket, in the queue for a bus, in a redundant conversation with the employee of an office, to the relationship amongst companions at work but even more to the affectionate relationship of friendship and above all to the amorous relationship between a couple. Still in its least engaged and more external way the interpersonal relationship that is established by pure coincidence ceases to be absolutely superficial: in itself there is a corporal language and non-verbal communication, which at times is very significative. But it is above all in friendship and in love that reciprocity is a condition of the possibility of the intuitive knowledge. In the love of his friends and in the love of a couple, because there is progressively a more intense reciprocal sharing and a greater reciprocal personal transparency, because this reciprocity is inseparably linked to the special living with another that is progressive sharing and this loving living together is always dual, and because therefore, in this loving reciprocity, in the measure in which I and you are progressively more one, for love in any event is always ontologically linking, I and you are present one in the other. That is why, I say that in love in general, in the love of friends, and in the love of couples, intuition arises progressively in all of its immediacy and in all its spontaneity.

The condition of possibility for this knowledge of the I-itself and of the I-other bond is also a new type of gnoseological *epoché*: an intuitive *epoché* that consists in renouncing, putting aside, in parenthesis, but is ready for other things and for other types of knowledge, the subjective mediations, including individuality itself as totality, precisely because the individuality itself never ends in itself. I want to say that in order to truly know the "I" and to know truly the other, one has to go to their encounter unconditionally and to abide by the pure presence of each, to their unveiled and immediate presence – without the mediation of prejudices or of prior judgements, of preconceived opinions, of classifications that words constitute and of categorised concepts, without the structure of linguistic delimitation, that, like veil over veil, screen and conceal the truth of their being, making them unrecognisable and a mystery to thought.

But this knowledge, as I said before, can only be more profound, deeper, when this personal presence is reciprocal presence, in the ambit of progressively reciprocal presence. This profound reciprocal presence which is life deeply co-lived by you and me, without reservation, is the more authentic encounter of you and me and the more truthful and intimate personal knowledge.

My anthropological theory of knowledge synthetically collects something that was already known and something that it discovers itself: a study of the human being that highlights the ontological dimension of love, and a critique of knowledge of each personal I that emphasises the ontological dimension of love as well as a critique of its knowledge that highlights its gnoseological dimension, united to and founded on its ontological dimension; that is to say, love gains gnoseological status as a condition of the possibility of personal knowledge as intuitive knowledge.

On the other hand, this non conceptual and ineffable feature of the I and of intuitive knowledge, of the intuition of the I-itself and the I-other bond in the encounter of you and I, is somewhat mystic; this loving experience as an ontological and gnoseological experience of the I-other bond has certain rational analogies with mystical experience, religious mystical experience, but it is mystical experience that is not reducible to religious experience. We should bear this in mind in framing an anthropological theory of knowledge in order to understand and to determine more adequately the experience of intuition.

Finally, the ineffability of the I, of its knowledge, of its intuition, cannot be translated directly into concepts, to propositional language; it cannot be resolved or superseded in any way or form and it does not have to be resolved or superseded. But it is thrown somewhat into relief by means of literary language, by means of literary figures, for instance, by means of the metaphor. This can help to transmit more approximately than descriptive language the experience of the I, knowledge of the I, personal intuition, but it does not embrace it all, it does not understand it all. The metaphor cannot substitute for personal experience of the I. A metaphor will be able to suggest the content of an intuition in a closer way and one more appropriate than the logic of rational discourse. Although it can contain other intuitions, intuition of the I cannot be translated totally into images and metaphors. My anthropological theory of knowledge analyses this relationship, this approximation, but reveals that the knowledge of I-myself and the I-other bond is not properly metaphoric knowledge, that intuition and metaphors are commensurable, not equivalent. Because metaphors, as well as poetry, are sustained by the conceptual language of words, although constituting a peculiar game of language, the I, in its true being more itself and different from any other I, is beyond being expressed in a metaphorical way, beyond that which can be thought and said metaphorically and poetically. In fact, the metaphor of the "I" can only be understood, is only discovered and made obvious to those who have already had the loving experience of knowledge of an I, to the extent that they have participated in, lived, and known an encounter with an I and felt its reciprocal presence.

* * * * *

This study assembles with considerable accuracy my responses to the gnoseological problem of knowledge of the I which is the human being. I have wanted to study this here with sufficient precision, unless in some way that precision contradicted reality, to attempt to expose in a rational, conceptual and linguistic discourse the experience of subjective knowledge and interpersonal knowledge, in other words, the experience of intuitive knowledge.

*Universidad de Zaragoza, Spain*
e-mail: pblase@posta.unizar.es

# SECTION II

In the city of Puebla.

DIANE G. SCILLIA

# BLURRING THE BOUNDARIES BETWEEN ART AND LIFE: JAN VAN EYCK'S *GHENT ALTARPIECE* (1425–32); ALLAN KAPROW'S *APPLE SHRINE* (1960) AND *EAT* (1964)

Jan van Eyck's *Ghent Altarpiece* of 1426–1432 was a fundamentally innovative work in its depiction of naturalism: the artist presented to the viewer a rational world very much like that in which the viewer stood. Moreover, van Eyck intentionally broke down the boundaries, or rather, broke through the picture plane of his panels, posing his Adam and Eve so they seem to pierce the invisible divide between the fictive space of their shallow niches and the real space of the viewers, and he also seems to include the real viewers standing before the *Altarpiece* as an extension of the groups of figures kneeling and standing around the painted Altar of the Lamb in the lower center panel. Since the ensemble was erected in a chapel too small for it to begin with, the visitors/viewers/participants of this altarpiece were forced to stand close to it and to experience its immediacy.[1] In spite of its new location, behind bazooka-proof plastic, in the Church of St. Bavo at Ghent, it is still easy for the sensitive modern viewer of this work to imagine himself or herself penetrating the picture space, literally moving from the real world into the painted distant landscape, visually sampling the fruits arranged there for our delight.

That the innovative nature of the *Ghent Altarpiece* was also noted by its fifteenth-century viewers can be demonstrated by the fact that within twenty-five or so years of its dedication in the Church of the Sts. John (now St. Bavo) in Ghent, the townspeople used its compositional format as the basis of a *tableau vivant* erected in one of the city's squares during the *Triumphal Entry of Duke Philip the Good* (1458).[2] This tableau, played out by living figures dressed in costumes based upon those in van Eyck's panels, complete with a realistic Holy Lamb that pumped blood into a real chalice and a three dimensional Holy Spirit that descended on a guy wire from the upper stage of the tableau to the lower one, marked Ghent's return to the Duke's protection after a period of rebellion. Here we can see how a major work of art provides inspiration in contexts outside of that in which the original functioned. The work of art can take on new meaning after the death of its artist, and the traditional formats and ideas depicted in it can lend themselves to new, unforeseen, contexts when they shift from the world of religious or liturgical usage to that of the secular political sphere. The breakdown of the boundaries



*A.-T. Tymieniecka (ed.), Analecta Husserliana LXXVI,* 135–150.

between the world of the painting and the world of the viewer is now complete, although the *tableau vivant* was played out on a three-level stage, in front of onlookers.

Naturally, in this translation from the liturgical to the political, some figures – the Adam and Eve, for example – were edited out of the *tableau vivant* and the meaning of the Altarpiece as a whole underwent a change. By the end of the fifteenth century, important visitors to Ghent were shown the ensemble during their tour of the highlights of the city – even Albrecht Dürer gives us an account of his visit/viewing that reads more like what we find in the letters written home by our contemporary artists than what we would expect of a near contemporary of Jan van Eyck.[3] Also by 1500, figural groupings and structural elements derived from the *Ghent Altarpiece* were being employed in panel paintings that served no liturgical functions – that is, did not function as altarpieces – but which were hung from the columns along the nave of a church as testimonies of faith.[4] At least two such works survive, one in the Prado in Madrid and the other at the Allen Memorial Art Museum at Oberlin College in Ohio. It seems, then, that van Eyck's ideas about making the painted surface as naturalistic as possible – that is, to make it look exactly like the real world – had some effect on moving the art work away from its limited role enhancing the Sacrifice of the Mass towards a larger, more complex role certifying a personal faith or serving as a community's expression of loyalty. Later in the sixteenth century, an exact copy of Jan van Eyck's great work was commissioned by Philip II of Spain, so he could have his own "version" of the ensemble for his own devotional purposes. This copy is now in Antwerp at the Museum of Fine Arts.

While I cannot document any direct connection between van Eyck's *Ghent Altarpiece* or the *tableau vivant* of 1458 based upon it (or, indeed, any of those later panels that derive from the *Ghent Altarpiece*) and Allan Kaprow's *Apple Shrine* (1960) and *Eat* (1964), the modern works similarly use quasi-liturgical forms and symbols to provide the viewer/participant with the means of experiencing a secularized ritual that has roots in earlier religious forms.[5] Moreover, in Kaprow's works we also see a breaking down of the space of the art work and the real space of the viewer. In fact, the viewer/participant now enters the art work and becomes part of it.[6] While Kaprow's rituals and forms are purposely designed to be universal and non-denominational as well as non-sectarian, the environments themselves – in their large scale and in their use of open-ended symbols and scripted actions – recall the processions and movement of figures together with the ritualized devotional forms seen in van Eyck's *Altarpiece*. This must have been intentional.[7]

In discussing these similarities, I will review Kaprow's writings (c. 1960–72) as well as draw on my own memories of discussions with this artist dating to the mid and late 1960s. From 1964 to 1967, I took undergraduate studio art courses (a sequence of painting studios as well as a studio in Assemblages, Environments and Happenings) and art history courses (in nineteenth and twentieth century art) with Allan Kaprow at the State University of New York at Stony Brook. Rituals of a religious or quasi-religious nature had long fascinated him and are examples of "play" as outlined in Johan Huizinga's *Homo Ludens* (1944).[8] It seems natural to me that he would link ideas informing his own art works with those great art works he had himself studied while at New York University and at Columbia University and which he taught at Rutgers and at Stony Brook. After 1967, I went my own way and entered graduate school in another state and pursued a career in art history with a special interest in Medieval and Northern Renaissance painting. I have had only one further contact with Kaprow since then, in 1994. Enough time has now passed (almost forty years) and we should start to put Kaprow's environments and Happenings of the early 1960s into a larger art historical context.

Robert E. Haywood's article on the "alliance" of the Judson Memorial Church in Greenwich Village and "cutting edge" artists working in New York City, starting in the late 1950s and throughout the '60s, brought to light the fact that Allan Kaprow played a prominent role in the Judson Art Program.[9] He even served briefly as gallery director in November 1960 which, according to Haywood, "led directly to the mounting of his environment 'An Apple Shrine' in December 1960" at the church. This was one of his "passive" environments, where the audience/participator/visitor walked through, stood and observed, although Kaprow himself outlined that *Apple Shrine* was a work you "GO IN to, not LOOK AT."[10] In *Apple Shrine*, the visitors threaded their "way through narrow passages of board and wire choked with tar paper, newspaper and rags. Lights changed from very dark to very bright. At the end of the maze was a large restful space where apples – some real, some not – were suspended from a tray and signs read 'Apples, Apples, Apples, etc.' There were bright bands of color hanging from the ceiling. After taking this in the viewer made his way out of the gallery."[11]

The multi-level meanings of apples in American culture – where the fruit makes its "first" appearance (to English speakers) in the *Book of Genesis* and symbolizes the fruit of the tree of knowledge of good and evil through the more prosaic and local readings associated with the patriotic ideals of "Mom, Home and Apple Pie" or of "Johnny Appleseed" or the folkloric "an apple a day keeps the doctor away" – all inform Kaprow's *Apple Shrine*.[12] At the heart of *Apple Shrine* was an altar of apples. The word *shrine* also has several meanings, all of which, when linked to the word *apple* as in this title and seen in light of the above description of the environment, enhance its quasi-religious aura.

In the catalogue text for his environment *Words* (1962), Kaprow states,

> *Words* is an environment, the name given to an art that one enters, submits to, and is – in turn – influenced by.... In its impermanence and changeableness, it is ... fashioned from the real and everyday world, a world it celebrates, probes, comments on, perhaps, and surely dreams about. ... On one level, *Words* is light-hearted, jazzy, flip. Within this mood, there are contrasts. The larger room is public, bright and more formal in both the character and also in the placement of lettered strips, cloth-rolls, and red and white blinking lights. The small room is more subdued, private, organic, and less "arranged." On another, less obvious, level, the composition of the environment rooms within a normal room, their centrality and squareness (9' × 9' and 6' × 6'), the repeated words and phrases, the passage in gradual degrees from the outer world into an inner one, may suggest to the sensitive participant a sanctuary or tabernacle of sorts, an enshrinement of The Word. In this presence, our acts become ritual and our everyday is transformed.[13]

Although *Words* (which was presented in the Smolin Gallery in Manhattan) has been described as one of Kaprow's works in which the visitor/participant manipulated the environment – slips of paper with words on them could be taken down and new ones put up, in effect, changing the environment from day to day – it retained some of the spatial organization of *Apple Shrine*, complete with an inner sanctuary. If we substitute *logos* (the Greek equivalent of word) in Kaprow's penultimate sentence in his text, we can better understand the many levels of meanings incorporated in the quasi-pilgrimage one made through the more maze-like *Apple Shrine*. Wordplay delights Kaprow. Even in the original *Apple Shrine*, the word *apple* was juxtaposed to painted images of apples and to actual and fake apples.[14]

Moreover, as Jeff Kelley notes in his Introduction to the collected writings of Kaprow (1993), Kaprow insisted on using the word *play* to describe what he did in making these environments and Happenings.[15] It was Huizinga's conception of play, one of the crucial elements of human culture which included religious rituals as well as theatrical and other events, to which Kaprow was referring. *Homo Ludens* was on the reading list for the comprehensive examination that all Fine Arts majors had to pass before receiving their baccalaureate degrees from Stony Brook in 1967. On the oral part of this exam Kaprow and the other faculty members questioned us closely about Huizinga's book. Hence, Kelley might be understating the importance of this book in the following:

If there is one word dirtier than *copying* in the lexicon of serious art, Kaprow thinks it must be *play*. With its connotations of frivolity and childishness, play seems the antithesis of what artists are supposed to do. But Kaprow has always sought a certain innocence in his work, inviting humor and spontaneity, delighting in the unexpected. For him, play is inventive, and adults must be endlessly inventive to remember how to do it. Play is also instructive, since it imitates the larger social and natural orders: children play to imitate their parent's behavior and rules, societies to reenact ancient dramas and natural schemes. As a ritual reenactment of what Johan Huizinga calls "a cosmic happening," play at its most conscious level is a form of participation. As such, Kaprow sees it as a remedy for what he calls gaming (the competitive, work-ethical regulation of play) as well as for the ossifying routines and habits of industrial-age American education, which have less to do with learning and fun than with the "dreadfully dull work" of "winning a place in the world." With the work of art as a "moral paradigm for an exhausted work ethic," and with play as a form of educational currency that artists can afford to spend, Kaprow completes the education of an un-artist by assigning a new social role, that of educator, a role in which artists "need simply play as they once did under the banner of art, but among those who do not care about that. Gradually," he concludes, "the pedigree 'art' will recede into irrelevance."[16]

Another title on that required reading list at Stony Brook in 1967 was John Dewey's *Art as Experience* (1934).[17] Kaprow's emphasis on the non-commodification of art – that is, the denial of the market value of his works – brings us back to art that must be experienced: i.e., as environments and Happenings. Neither *Apple Shrine* nor *Words* was a *tableau vivant*, but they were works which included the movement of visitors through a space, and in which, by so moving and participating, the visitors shared some experience.

During the mornings and afternoons of the last two weekends in January 1964, Kaprow's environment *Eat* was presented.[18] Visitors were limited to twenty for each one-hour period the environment was open to prevent overcrowding and to keep free circulation in the space. The published "Cave Plan for *Eat*" (sketched by Kaprow) includes a note in Kaprow's own hand. It reads, "Note: work conceived as a quasi-eucharistic ritual. Contrasts between symmetries and asymmetries of physical things and activities [,] intended as a reciprocal rhythm between the stable and unstable. A. K."[19]

According to Michael Kirby, one entered *Eat* after walking through the old Ebling Brewery in the Bronx. This building fronted low cliffs that contained a large cave. Walking down several corridors and through doorways, the visitor finally came to the environment set in this cave. Its walls had been incompletely painted with white paint and the effect was that "age and seeping water had created a sense of decay." The entrance space was "narrow and dark" and led from the door to a low platform set beyond a stone arch. At right angles to this platform and a step down from it was another platform with rectangular wooden towers at each end. These stood about seven feet from the floor of the cave itself. On top of each tower a young woman sat motionless on a chair facing away from the entrance. The young woman on the left had a gallon of red wine and the young woman on the right had a gallon of white wine. If the visitor asked for wine, she poured some into a paper cup and gave it to him. Neither woman spoke or moved, except to pour.

Directly in front of this platform, apples hung on rough strings from the cave's ceiling. If the visitor wished, he could remove one and eat it, or if he was not very hungry, merely take a bite from an apple and leave it dangling. The cave divided into two large branches or bays of equal size to the right and left of the apples. To the right, the bay contained a forest of charred wooden beams and a young woman sitting frying sliced bananas in brown sugar over a small electric hot plate. If a visitor asked for some, she gave them to him, but did not speak to him. Nearby whole bunches of bananas still in plastic wrap hung from the ceiling. If he wished the visitor could take a banana and eat it.

To the left, the bay contained a square structure about eight feet high, constructed out of wooden beams. In the spaces between the beams and on a table inside this square enclave were loaves of sliced bread, jars of strawberry jam and a few table knives. The only way into the structure – where most of the food was located – was to climb a tall ladder propped against the side. At the very rear of this bay was another ladder leaning against the cave wall. It led to a small opening set high in the wall, in which a man sat with a large pot. "Get'em! Get'em! Get'em!..." [as in "Come and get'em!"], he called out over and over again, pausing every so often and then calling again. If a visitor climbed the ladder, the man cut a piece of boiled potato, salted it, and gave it to him. Visitors were free to wander about through the cave. Some ate and drank; others did not. At the end of the hour the remaining visitors were ushered out, the "performers" replaced by fresh volunteers and new visitors allowed to enter.[20]

Here, too, *Eat* was not a *tableau vivant*, and if its composition was more complex and its spatial arrangements were more tightly arranged than the earlier *Apple Shrine*, it still retained some of the same elements seen in that environment. The open-ended handling of symbols we saw earlier in *Apple Shrine* applies here also. Even the layout of the cave, with the indications of water flowing through it in the diagram (and the glistening of the water in the published photographs) recalls the channel connecting the Fountain of Life in Jan van Eyck's *Ghent Altarpiece* with the real chalice on the real altar table before it. The channel of water leads the visitor to the fruits arrayed in the interior of the environment: first, the hanging apples and, then, either to the right (and the banana slices in brown sugar) or to the left (and the bread slices and strawberry jam and the salted potato slices). Each of the contrasts (light and dark), symmetries (left and right), colors (red and white; white and black, brown), foods (fruits, bread and jam, potatoes), fruits (apples, bananas, strawberries), liquids (water and wine) and structures (charred beams, square towers, platforms) – and gender roles (male and female servers and male (and female?) communicants) – etc. in *Eat* can be interpreted on several levels simultaneously.[21]

In the above description of *Eat*, the visitors/participants move through the space, asking for food and drink or not. There is no set path for them, nor is there a sequence they must follow: there is no plot to this environment/ Happening although there is a narrative of sorts. The visitors individually choose to participate (to ask for food and drink or not; to move one way or another) or not. Kaprow has emphasized that in his environments and Happenings, the visitor/participant functioned like the actual paint flung by Jackson Pollock on to his canvases although these participants move of their own free will.[22] Kaprow saw his works (such as *Apple Shrine* and *Eat*) as extensions of the picture space of Pollock's abstract expressionist paintings – with an actual floor now serving as the "canvas." In effect, what we see in *Eat* is a large secular altar/shrine conceived in three-dimensional space with the figures of actual human beings moving through that space.

According to Kaprow, such Happenings were developed through a process that began when he saw Pollock's all-over canvases exhibited at Betty Parsons' gallery in Manhattan around 1950:

> the effect was that of an overwhelming environment, the paintings' skin rising toward the middle of the room drenching and assaulting the visitor in waves of attacking and retreating pulsations. Pollock confirmed this sensation when he wrote that while he was working, he was "in" his painting and wasn't consciously aware of what he was doing. ... Then Harold Rosenberg's article "The American Action Painters," in *Art News*, December 1952, completed the growing idea: why not separate the action from the painting? First make a real environment, then encourage appropriate action. The expanding scale of Pollock's works, their iterative configurations prompting the marvelous thought that they could go on forever in any direction including out, soon made the gallery as useless as the canvas, and choices of wider and wider fields of environmental reference followed. In the process the Happening was developed.[23]

With modest substitutions in this quotation we can have a description of *Eat*. Kaprow's environments and Happenings should be handled critically in a manner similar to paintings – whether abstract expressionists' works or those of the older historical schools – rather than to theatrical events. The artist has stated again and again that there were no actors and no audiences and no stage and no plot in his works, that they took place in the real world, thus they were not theatre.[24] Moreover, Happenings took place in real time. Since Kaprow himself did not perform in these early works, they were not "performance" art either.[25] Indeed, at least one theatrical historian agreed with Kaprow. Richard Schechner's published chart (in the *Tulane Drama Review* (T32) Summer 1966) analysed the characteristics of various types of "performances."[26] The note at the bottom of this chart reads: "Happenings and related activities are not included as theatre in this chart. Happenings would not necessarily have an audience, they would not necessarily be scripted, there would be no necessary symbolic reality. Formally, they would be very close to play."

This brings us back to *play* and to Huizinga's ideas. As Kelley, following Kaprow, has stated, *play* has negative associations to many contemporary artists, especially when this word is linked with what they do. We accept play's importance in the development of the individual and his socialization, but play is something children do, usually without adult supervision. In the late twentieth century adults have tried to control children's play, directing it to more "educational" forms and usually destroying for the child whatever fun there was in playing. This is a quirk of our advanced Western culture.[27] Other, less advanced, societies recognize play as a child-directed activity through which the child learns about his world and his place in it. Why is modern Western culture so anti-play?

The standard pattern for education established in the Netherlands by the early sixteenth century and in use until early in this century was the so-called "7-7-7" system (seven years of play followed by seven years of schooling followed by seven years of training). Those first seven years of play deserve our attention: Huizinga was a product of this system. He knew the importance of play firsthand: imaginative play allows the child to explore societal roles; more structured play allows the child to try out what he has learned from his parents and peers.

Invented rituals based on "adult" forms (e.g., funerals for pets) teach the child values in a way that a formal class never could. Who among us did not "play teacher" when we were children? It was not just because we missed being in school; we were trying on this role to see if it "fit." And we usually imitated a favorite adult at the same time. Huizinga is careful about linking play and art in *Homo Ludens,* but the imaginative re-creation of the world by the artist is akin to that seen in a child's playing. Notice, the word forms here are *play* and *playing,* not "a play." Child's play also has no actors and no audience and no stage and no plot. It takes place in the real world, in real time. Moreover, the sacred play of religious rituals was highlighted by Huizinga as one of the highest activities of adulthood. Yet, the twentieth century also saw the demise of public religious rituals and of religious art works.[28] Kaprow himself has stated that the church should follow the artist out into the larger world to find more current means of expressing the issues facing contemporary society and faith than the traditional forms and symbols used in earlier religious art.[29] His solution to this quandary was the form of play he called a Happening.

There is another problem in discussing these points with an audience of art historians as opposed to an audience of phenomenologists. As we have just seen, it is very easy when talking about Kaprow's works to slide from art to art history to education to theatre to philosophy, etc. In the contemporary academe, this poses a difficulty for the scholar researching Kaprow's ideas and works. Some writers take too narrow a focus on Kaprow and ignore the sources he has himself claimed for his works or they leave out the larger implications of his works simply because they have "pigeon holed" his environments and Happenings as theatre or as examples of Fluxus or Conceptual Art and refuse to see beyond their own self-imposed limits.[30] They miss the obvious by mis-classifying it, and especially in seeing *Apple Shrine* and *Eat* as theatre-like performances rather than as visual equivalents of paintings.

As modern visual equivalents of the great religious altarpieces of the past – and of Jan van Eyck's *Ghent Altarpiece* in particular, although Hugo van der Goes's *Portinari Altarpiece,* Matthias Gruenwald's *Isenheim Altarpiece,* or Rogier van der Weyden's *Last Judgment Altarpiece* at Beaune, which could also have served as "prototypes" – Kaprow's environments allowed their visitors/participants to share experiences in real time and in real space, touching upon some of the same ideas as these early great works: the commonality of communion, community, and communication.[31] Literally, all of these art works (old and new) are about belonging. We can now begin to understand Kaprow's provocative idea that the best, and most innovative, use the artist can make of the past (and the present) is misuse.[32]

*Kent State University*

NOTES

[1] For Jan van Eyck's *Ghent Altarpiece*, see Elisabeth Dhanens, *Hubert and Jan van Eyck* (New York: Alpine Fine Art, n.d.), pp. 374–381.

[2] Jeffrey Chipps Smith, "'Venit nobis pacificus Dominus': Philip the Good's Triumphal Entry into Ghent in 1458" in *'All the World's a Stage.' Art and Pageantry in the Renaissance and Baroque,* eds. Susan S. Munshower and Barbara Wisch (University Park, Pa.: 1990), pp. 258–290, includes earlier bibliography. For recent works on *tableaux vivants* in civic festivals of the Northern Renaissance, see Diane G. Scillia, "The Audiences for Israhel van Meckenem's Proverb Imagery, circa 1500" in *New Studies of Northern Renaissance Art in Honor of Walter S. Gibson,* ed. Laurinda S. Dixon (Turnhout: Brepolis, 1998), esp. pp. 87–91.

[3] Albrecht Dürer, *Durer's Record of Journeys to Venice and the Low Countries*, ed. Roger Fry (New York: Dover Publications, Inc., 1995), p. 79.

[4] I presented some of this material in "Jan van Eyck's *Ghent Altarpiece,* the 'Toog' of 1458 and the *Fountain of Life* (Oberlin): Visual Images in Translation," read at the 24th International Congress of Medieval Studies, Western Michigan University, Kalamazoo, Michigan, May 1989.

[5] In 1966, Kaprow knew he was caught in a paradox even as he called for a new kind of art, one totally unlike and unrelated to earlier art of works. In his article "Experimental Art" (1966) (reprinted in Allan Kaprow, *Essays on the Blurring of Art and Life,* ed. Jeff Kelley (Berkeley: University of California Press, 1993), he states "Painters and sculptors cannot put out their past as they would a light" (75) and "A residue of esthetics and masterpieces lies on the inside of our eyelids as patterns of semiconscious recall" (76). As a trained painter and an art historian, he could not have ignored the associations sparked by his own lectures on the "machines" painted in France in the 19th century or on the large polytychs of earlier centuries. Significantly, he had studied with Meyer Schapiro at Columbia, an art historian known for his contributions to Medieval and Modern art scholarship. Moreover, in his classes at SUNY at Stony Brook, both studio classes and art history classes, he frequently spoke of the importance of rituals, religious, secular and personal, and of our need to "invent" such rituals in order to have some control over our lives. He was not alone, in the early 1960s, in seeking to make some kind of large scale meditative art works of quasi-religious or spiritual content or intention. At least two painters linked to Abstract Expressionism, Barnett Newman (with his fourteen panels of *The Stations of the Cross,* 1966) and Mark Rothko (with his *Chapel,* 1964–1967), commissioned by the de Menil family as objects of contemplation in a non-denominational chapel at Rice University, employed forms that looked back to the late Middle Ages and the Renaissance. Louise Nevelson's *Homage to 6,000,000,* no. 1 (1964), one of her environmental shadow boxes, is in effect a large "shrine box" altarpiece for Jews killed in the Holocaust (for which, see Carla Gottlieb, *Beyond Modern Art* [New York: E. P. Dutton and Co., 1976], pp. 74–77 and fig. 5, who likens this work to a columbarium). Neither Newman, nor Rothko nor Nevelson used overt religious symbols or imagery in their works. It may be significant that Kaprow, Newman, Rothko and Nevelson had Jewish upbringings, each of them trying to "invent" a new visual, artistic tradition that merges art and religious experiences. In contrast, Andy Warhol's icons of 1960s celebrities come out of his Byzantine Catholic upbringing, but serve no spiritual purpose. Kaprow also taught (in his studio courses at Stony Brook) that because certain conventionalized forms had become so much a part of their mental constructs modern artists were unable to free themselves from these forms which are implicitly contained in the works of art they produced. One, therefore, had to be aware of the past and how it affects one's own art works, whether consciously or not. See, Allan Kaprow "The Shape of the Art Environment," *Artforum,* 6, no. 10 (Summer 1968), pp. 32–33 (reprinted in his *Essays on the Blurring of Art and Life,* op. cit., pp. 93–94) where he touches upon some of the same ideas.

[6] Robert E. Haywood, "Heretical Alliance: Claes Oldenburg and the Judson Memorial Church in the 1960s," *Art History,* 18, no. 3 (June 1995), pp. 185–212, esp. p. 208. Kaprow also defined environment, an art work one entered into, in his catalogue text for *Words* (1962), which is cited within my text below. Compare, Michael Kirby, "The New Theatre," *Tulane Drama Review* (T 30) 10, no. 2 (Winter 1965), p. 24.

[7] Throughout this study I refer to Kaprow's original *Apple Shrine* (1960) and *Eat* (1964), not the re-enactments/re-installations of 1991 at the Fondazione Mudima, Milan. Kelley's Introduction in *Essays on the Blurring of Art and Life,* xxv, and Dorothy G. Shinn, *The Duchamp Effect: The Influence of Marcel Duchamp on the Work of John Cage and Allan Kaprow* (M. A. Thesis, Kent State University, 1994), pp. 100–101 and p. 132, discuss the role of Schapiro in Kaprow's education as an art historian. *Apple Shrine* is an environment and *Eat* has elements of a happening, but both are discussed by Kaprow in similar terms. Kaprow's Happenings of the 1960s were always thought out and scripted, although he tried to leave things open in order for the chance event or unplanned occurrence that could color the experience. Unplanned events work best when everything else is tightly scheduled. We can see this inclusion of chance operations in both *Apple Shrine* and *Eat.* When it comes to themes, particularly of a ritualistic nature, Kaprow is less likely to have overlooked a parallel with traditional art works, like the *Ghent Altarpiece,* given his interest (see above, n. 5).

[8] Johan Huizinga, *Homo Ludens: The Play Element of Culture* (Boston: Beacon Press, 1955). Kaprow's name appears among the "Painters" receiving Guggenheim Awards in 1967, see *Art News* 66, no. 3 (May 1967), p. 8. He told his students at Stony Brook that on the application for this grant he wrote, "I want to play." Kaprow's ideas about play are outlined in Allan Kaprow, "Education of an Un-Artist, Part II" (1972) reprinted in his *Essays on the Blurring of Art and Life,* op. cit., pp. 113–116.

[9] Haywood, op. cit., p. 208. Kaprow is cited at least eight times in this article. That he served as art director of the Judson Memorial Church art program through the Fall of 1960 and early Winter of 1961 underscores my argument that Kaprow's *Apple Shrine* had some religious connotations, whether implicit or explicit ones.

[10] Ibid. Because there is no consistency in the published accounts of this event (even in Kaprow's publications), I shall call it simply *Apple Shrine.*

[11] Shinn, op. cit., p. 106.

[12] Multiple readings of symbols in art works is not new; see, Julien Chapuis, "Early Netherlandish Painting: Shifting Perspectives" in *From Van Eyck to Bruegel. Early Netherlandish Painting in the Metropolitan Museum of Art,* ed. Maryan W. Ainsworth and Keith Christiansen (New York: Metropolitan Museum of Art, 1998) p. 12, who cites James H. Marrow, "Symbols and Meaning in Northern European Art of the Late Middle Ages and Early Renaissance," *Simiolus* 16 (1986), pp. 151–155. Compare, Susan Sontag, "Against Interpretation" in *Against Interpretation* (New York: Farrar, Straus, and Giroux, 1966); and Meyer Schapiro, "The Apples of Cezanne: An Essay in the Meaning of Still-Life," in *Modern Art, 19th and 20th Centuries* (New York: G. Braziller, 1978), pp. 1–38. This article originally appeared in 1968 and must reflect some of the ideas that Schapiro discussed in his classes at Columbia when Kaprow was a student. See, especially, Schapiro's footnote 24a, where he argues that modern painters might find in older art works – along with exemplars to up-date – "a mask for meanings in his own works – a mask that adds to the ambiguity of the whole."

[13] Jan van der Marck, "Pictures to Be Read/Poetry to Be Seen," *The Journal of Typographic Research,* 2, no. 3 (July 1968), p. 267 (Kaprow's text is reprinted from the Smolin Gallery

catalogue). Van der Marck's article originally appeared as the catalogue introduction for the exhibition *Pictures to Be Read/Poetry to Be Seen,* organized by the Museum of Contemporary Art, Chicago (Fall 1967). Compare the description of this work that appears in Aldo Pellegrini, *New Tendencies in Art.* Translated by Robin Carson. (London: Elek, 1966), p. 243; and Gottlieb, op. cit., pp. 256–257. *Words* was one the seven environments re-enacted/re-installed at the Fondazione Mudima in Milan in 1991, for which, see, Shinn, op. cit., pp. 124–132.

[14] Joseph Kosuth's *One and Three Chairs* (1965), in which he juxtaposed a real chair with a written description and a photograph of the same chair, is somewhat similar in conception, which he addressed in his article "Art After Philosophy," *Studio International,* 178, no. 915 (October 1969), p. 135: "The art I call conceptual is based on the understanding of the linguistic nature of all art propositions." I thank my colleague, Carol Salus, for this reference. In Kaprow's *Apple Shrine* with its juxtapositions of the word apple with painted images of apples and real and fake apples (which predated Kosuth's chairs by five years), this "art proposition" is extended to include larger questions of replication, for which, see Hillel Schwartz, *The Culture of the Copy. Striking Likeness, Unreasonable Facsimiles* (New York: Zone Books, 1996).

[15] Kelley, "Introduction" in Allan Kaprow, *Essays on the Blurring of Art and Life,* p. xxii.

[16] Ibid.

[17] For Dewey's influence on Kaprow, see Shinn, op. cit., pp. 122–123, who quotes an interview she had with Kaprow in 1994; and Kelley, op. cit., pp. xxiii–xxvi. Pamela Lehnert had earlier uncovered Kaprow's use of Dewey, see *An American Happening: Allan Kaprow and a Theory of Process Art* (Ph. D. Dissertation, The University of North Carolina at Chapel Hill, 1989).

[18] Michael Kirby, "Allan Kaprow's Eat," *Tulane Drama Review* (T30) 10, no. 2 (Winter 1965), pp. 44–49. Much of the following is taken directly from Kirby, with some of his language amended to conform to current usage.

[19] The uterine shape of the Ebling Brewery cave in Kaprow's sketch bears comparison with the diagram of "GOK The Monster, a Model for a Popular Theatre," depicted in Paul Sills, "A Monster Model Fun House," *Tulane Drama Review* (T30) 10, no. 2 (Winter 1965), p. 225, and *Hon (She),* the 82-foot long recumbent female figure that Nikki de Saint Phalle constructed in collaboration with Jean Tinguely (her husband) and Per Olof Ultvelt at the Moderna Museet in Stockholm in 1966 (Gottlieb, op. cit., pp. 61–65), figs. 4a and 4b, where the work is called "She–The Cathedral"; and Whitney Chadwick, *Women, Art, and Society* (London and New York: Thames and Hudson, 1990), ill. 195). According to Chadwick (312), "Spectators entered *Hon* through the vagina and found themselves in a female body that functioned as a playground, amusement park, shelter, and pleasure palace with a milk-bar installed in one breast and an early Greta Garbo film playing elsewhere." Gottlieb gives a fuller description of *Hon* and explains exactly what Tinguely and Ultvelt contributed to the work. *GOK* (Sills, op. cit., p. 226) and *Hon* also play with eucharistic imagery, but *Hon* goes further in tweaking the idea of *Mater Ecclesia.*

[20] This description together with the published photographs in Kirby's article form the only record of the original *Eat.* In 1991, Kaprow re-created this work, and six other environments, at the Fondazione Mudima in Milan, for which, see Shinn, op. cit., pp. 124–132.

[21] Kaprow always insisted on the unrestricted meaning of every thing or object in his environments and Happenings, see, Kaprow, "Education of an Un-Artist, Part II" (1972) reprinted in his *Essays on the Blurring of Art and Life,* op. cit., pp. 113–116. *Eat* also resembles a chapel of atonement, in particular the chapel constructed and painted for the Scrovegni Family in Padua known familiarly as Giotto's Arena Chapel. Pilgrims came to that chapel to perform a devotional pilgrimage by following the sequence of painted scenes and to pray for the soul of the grandfather of the chapel's founder. Harold Rosenberg, *The Anxious Object. Art Today and Its*

*Audience* (New York: Horizon Press, 1966), p. 93, made this link between Giotto and Kaprow but did not specify which of their works are similar.

[22] Kaprow, "The Shape of the Art Environment" (1968), in his *Essays on the Blurring of Art and Life*, op. cit., pp. 93–94. This article originally appeared in *Artforum*, 6 no. 10 (Summer 1968), pp. 32–33 and included a photograph of *Apple Shrine* (1960). In his discussion of the movement of "spectator(s)" in the art environment – here a gallery – he states: "any casual meanderings on their part will thus be the formal equivalent within the exhibition floor area of, say, Pollock's drips within the canvas area." In his contribution to "Jackson Pollock: An Artists' Symposium, Part I," *Art News* 66, no. 2 (April 1967), pp. 33 and 59–61, Kaprow elaborates on the movement of people within a space vs. the movement of paint in Pollock's drips: "If an artist wanted to make the connection, he could say that some of the underground discotheques come straight out of Jackson Pollock – and he'd be right. Look at the action! It's all-over: it's intense; when you're in it you don't know what you are doing; it seems to go on without beginning or end; the noise and the lights assault you; the pulsations, changing, ever-so-slightly, come in waves; you are surrounded; it's overwhelming...." Also see Allan Kaprow, "Notes on the Creation of a Total Art" (1958), reprinted in his *Essays on the Blurring of Art and Life*, op. cit., pp. 10–12.

[23] Shinn, op. cit., p. 116; and Kaprow, "Jackson Pollock" (1967), p. 60. Recently, Bruce Hainley cited Kaprow's "The Legacy of Jackson Pollock" (1958) in his review "Out of Actions: Between Performance and the Object, 1949–1979" in *Artforum* 37, no. 1 (September 1998), pp. 145–146. According to Hainley, "Kaprow's brilliant contextualization (1958) of Pollock's practise" was the means by which "Pollock's space" was used by other performance artists. Hainley goes on to state that "By placing Pollock as the innovator, even impetus, curator Paul Schimmel [of the Museum of Contemporary Art in Los Angeles] structures all the works on display as (safely) art and brackets performance as the legacy of Ab EX [Abstract Expressionist] painting – the museum in this sense inoculating art from what in the end might bring it down (which is the point, after all)." The bracketed expansions here are mine. Kaprow's article of 1958 appears in his *Essays on the Blurring of Art and Life*, op. cit., pp. 1–9.

[24] Allan Kaprow, "Untitled Guidelines for Happenings" (c. 1965?) in his *Essays on the Blurring of Art and Life*, op. cit., pp. 59–65. Also, see Allan Kaprow, "Pinpointing Happenings" (1967) in ibid., pp. 84–89. Surprisingly, authors whose works followed Kaprow's into print cite Michael Kirby, *Happenings. An Illustrated Anthology* (New York: E. P. Dutton and Co., 1965) or the issue of the *Tulane Drama Review* (T. 30), 10 no. 1 (Winter 1965), edited by Kirby, instead of Kaprow's own words. Compare, Fred W. McDarrah, *The Artists' World in Pictures* (New York: E. P. Dutton and Co., 1961) p. 177 and figs. 175–180 showing *18 Happenings in 6 Parts* (1959) under construction; Lucy P. Lippard, et al., *Pop Art* (New York and Washington: Frederich A. Praeger, 1966), in which Kaprow is cited nine times in several different contexts; Pellegrini, op. cit., pp. 238–241; and Gottlieb, op. cit., pp. 257 and 258, where she concludes that because sound plays such a big role in Kaprow's Happenings, they are "a mixture of theatre and dance, not of theatre and art." Marilyn Stokstad, *Art History* (New York: Abrams/Prentice Hall, Inc. (2nd edition) 1999), pp. 1108–1109, a widely used introductory textbook, states "Kaprow's works and those of his followers were badly received by the avant-garde New York art critics, because they were considered theater and because spectators enjoyed the happenings so much. One critic even asked the rhetorical question, "Is artistic experience about being entertained?' Such questions about what art is and what role it plays in society have marked the richly innovative period following World War II." Compare Allan Kaprow, "Letter to the Editor," *Artforum*, 6, no. 5 (January 1968), p. 4, for his response to Jane Livingston's review of *Fluids* (1967) that touches upon some of these same issues.

[25] Since 1971 Kaprow has participated in his own art works, which he entitled Activities to keep them distinct from the Happenings of the late 1950s and 1960s, for which, see James T. Hindman, "Self-Performance: Allan Kaprow's Activities," *The Drama Review (TDR)* (T.81) 23, no. 1 (March 1979), pp. 95–102. Hindman (97) carefully distinguishes between the pre-1971 Happenings (in which Kaprow did not always actively participate) and his post-1971 Activities. Also, see Allan Kaprow, "Participation Performance," *Artforum*, 15, no. 7 (March 1977) pp. 24–29 (reprinted in his *Essays on the Blurring of Art and Life*, op. cit., pp. 181–194), which Hindman did not credit. Compare, Roselee Goldberg, *Performance Art from Futurism to the Present* (New York: Harry N. Abrams, Inc., 1988, reprint 1996) pp. 128-129, where she discusses *18 Happenings in 6 Parts* (1959) as one of Kaprow's performances. In contrast, Michael Rush, *New Media in Late 20th Century Art* (London and New York, 1996), *passim*, but especially p. 78, places Kaprow's environments and happenings as "a few examples of the multiplicity of art works" in the mid-1960s. Kaprow's own thought on "artist as shaman" (as opposed to "artist as showman") can be found in his contribution to "Jackson Pollock" (1967), pp. 59–60 and 61. Also see Allan Kaprow, "Performing Life" (1979) in his *Essays on the Blurring of Art and Life*, op. cit., pp. 195–198.

[26] Richard Schechner, "Approaches to Theory/Criticism," *Tulane Drama Review* (T32), 10, no. 4 (Summer 1966), p. 35; and *idem*, "Happenings," *Tulane Drama Review* (T 30), 10, no. 2 (Winter, 1965), pp. 229–232.

[27] Kaprow, "The Education of an Un-Artist, II," in his *Essays on the Blurring of Art and Life*, op. cit., pp. 113–116.

[28] Haywood, op. cit., pp. 191, 192–201, 205–206 and 207–208. The Judson Memorial Church sought out artists using non-conventional forms and imagery to express the anxieties facing modern mankind. Many contemporary artists have been concerned with spiritual matters in their works, but art historians have been reluctant to follow the threads these artist leave them in the various published interviews about their art or in their "manifestos" now anthologized in edited form. As I noted above in n. 5, Kaprow was not the only artist "reinventing" an art with religious significance in the early and middle 1960s. Perhaps the problem is with the over-restricted definitions of spiritual and religious in the face of non-traditional and non-denominational and non-sectarian imagery or even a total lack of imagery, as well with the tendency to emphasize novelty in these works.

[29] Allan Kaprow, "The Artist as a Man of the World" (1964) in his *Essays on the Blurring of Art and Life*, op. cit., p. 58.

[30] Allan Kaprow, "Letter to the Editor," *Tulane Drama Review* (T 32), 10, no. 4 (Summer 1966), pp. 281–283. For Fluxus, see Estera Milman, ed., *Fluxus: A Conceptual Country* in *Visible Language*, 28, nos. 1–2 (Winter/Spring 1992). Kaprow's association with this group is well known. He is included on the "Fluxchart" in Ken Friedman with James Lewes, "Fluxus: Global Community, Human Dimensions" (169) and is mentioned (sometimes negatively) in several of the other articles included in this volume. Moreover, Kaprow's collage *Self-Service* (1966) was included in the exhibiition celebrated in this volume of *Visible Language*. Kaprow would be the first to deny that he was a major Fluxus artist although he was a guest speaker at the Symposium held in connection with the Fluxus exhibition at the Wexner Center at The Ohio State University, Columbus, Ohio, in February 1994.

[31] Shinn, op. cit., p. 132.

[32] Kaprow's statement appears as the title to his contribution to "Jackson Pollock: An Artists' Symposium," op. cit., (1967), p. 33. According to Kaprow (60–61), Pollock "misused his sources by all conventional standards, taking from them rather marginal attitudes or tones of feeling,

rather than developing their central principles." He holds that "insight and growth in art lies less in certain lineal developments that are synthesized at some later point, than in a numberless range of arbitrary attractions and repulsions to and from things in and out of art," and he concludes, "...in other words, art may now be a discipline of deliberate 'misuses' (or free-interpretations and combinations) of source material." This seems to correspond to what Don Ihde called "bricolage" in his essay "Columbus: New Technologies/Old Cultures" in *Postphenomenology. Essays in the Postmodern Context* (Evanston, IL: Northwestern University Press, 1993) pp. 28–30. My special thanks go to Robert D. Sweeney, Dorothy Shinn, David Cubie, Lyneise Williams and Albert Reischuck, who served as sounding boards for many of the ideas presented here.

CARMEN COZMA

# MUSICAL ART AS ENLIGHTENMENT AND UNDERSTANDING THROUGH ETHOS: THE EXPERIENCE OF THE "HUMAN"

One of the most elevated plans of the spiritual life, so hard to catch and to reproduce through words, musical art is a "world" that needs to be known and understood. It is a "world" that deserves to be searched by the philosophical organon. Thus, we can find something of the "absolute Truth" as the "identity of the artistic-musical Beauty and the moral Good". Actually, there is here an interweaving of the aesthetic and the ethical ideal, the registering of man as human, a synthesis of values and principles for an existence in liberty and dignity, in the logos, in order, in supreme harmony.

The philosophical-hermeneutic search of music – in the "unity of composition-performance-audition" – shows us the power of this art to open forever new horizons to cognition and understanding, to the shaping of the human. Here is an extraordinary power for communication and creativity even in the terms of the ethical doctrines centered on the categories: "utility", "pleasure", "happiness" – respectively, Utilitarianism, Hedonism and Eudaemonianism – essentially, a hypostasis of the deep meanings of the human, such as aspiration, but also such as settling down in life. And, thus, a kind of achievement, the awarding of sense to our existence in its ethical dimension occurs.

It's not easy to discuss music.

The approach, by way of words, to the *musical art* is marked by consciousness of the difficulties of submitting it to investigation and trying to explain it in verbal language, owing to the ineffable and intimate and infinite in the musical experience and the constant metamorphosis by which music sustains them. Striking root in the pure intuition, having the Idea as object – in Plato's sense, as the essential and the eternal in all the world's phenomena – music is the objectifying "of the whole will which constitutes the world". (Schopenhauer 283) Owing to the features of phenomenality and temporality, music is substance and permanence, surviving beyond the phenomenal world, in the most profound structure of the whole of existence. The musical art *works into being*, in the space of the "intelligibility" – that "noetós tópos"/*νοητος τοπος* of the philosophy of the ancient Greeks.

The expression of art through sounds, *music* means penetration and striking root into essence, the redeeming of the essence, integration in "logos", in the



*A.-T. Tymieniecka (ed.), Analecta Husserliana LXXVI*,151–158.

plenitude and dynamism of the becoming. It displays a unique capacity to impersonate the most diverse aspects of reality, the complex and contradictory character of life; actually, the enlightenment of the truth of life. Especially, seeing that the musical art is a peculiar "world" – an expression of ethos/*ηθος* – a realm that strives for the achievement, the awarding of sense, and, last but not least, for assurance of the lastingness of artistic works.

Music is a fundamental presence in everything that lives. Some way or another, everywhere, we can find rhythm, harmony, and movement, correlated with the basic element of sound.

This art of sound has developed in connection with the cultural and moral evolution of man. It accompanied him, expressing him with the highest degree of accuracy, in what was always most significant to him: *his humanity*, specifically, in the felt tension between what *he "is"* and what *he "should be" and* what *he "would like to be"*.

Real and ideal are here interwoven. More than that, in this *particular way of being: music*, we find the human's passing from one living state to another, the balancing of the concrete and abstract configuration. Music is a complex experience, a perpetual work and adventure of the spirit, an eloquent sign for man as situated in "mystery" and, also, "revelation of the mystery" – as the Romanian philosopher Lucian Blaga (in his *Trilogia cunoasterii* 74 – see the note on his "trilogies" in "References" 77) conceives the "ontological mode of human being"; and, thus, music opens on the most profound dimension of being: the *moral* realm, with its axiological and normative unity.

As state-of-being-in-the-world, music is a framework for bringing in value, for projecting and recognizing the *human*; it is a mode of human communication. In music is the discovery of the human being, with his conflicted nature; here is the imagining and assimilating of the ideal he is capable of attaining. Touching man in his essence and dynamism, music elicits a direct echo in him, persuasively reinforcing the idea that the art has a high moral importance for his existence. Music gives to man the urge to make, in freedom, something noble and admirable, lasting and grand, something that speaks in the most expressive way about the man's situation in this world, about the sense of human creation.

Music is a special "lógos"/*λογος*, calling and sitting on the horizon of meaning. In an immediate relation with philosophy, the musical substance is built, generally, on those philosophical-moral ideas, simple and at the same time great: *the dignity of man, merit/areté, the value of Human Life; the meaning of life*. In the being proper to music, there is man with his capacity for (his own) creation, (his own) perfection, for permanent endeavor toward self-fulfillment; man with his capacity to give (himself) existential sense within the horizon of true values – that dialectical identity of the Ancients: *the Beautiful, the Good, and the True*, in the classical Greek paradigm of *καλλον και αγαθον, και αληθης* (*kállon kaí agathón, kaí aléthes*).

The art of sound is re-presentation, an essential one for the Human. For the one who lives the experience of music, the labor is one of sensory-emotional comprehension (first of all, while listening), of meditation on what was heard and fixed in memory, and after that, reasoning in order to judge. This path passes from initial vagueness and spontaneity to clarity and the integration of the substance. *The ethos of music* is one of becoming conscious, of amazement and interrogation, of raising problems. Contemplating the sensible musical reality, we secure access to authentic being. The route is *from* and only *through* sensible existence to the essential "*to be*".

The emotion of the musical phenomenon works in connection with the noblest living and with philosophical reflection. Thus, it becomes possible to understand the truth that *music means the creation of a world and integration into it, an independent world, logical and beautiful in an absolute way, a noble world, perfectly ordered. Music is a possibility, a promise and even an introduction to the world of something that did not previously exist and would otherwise remain non-existent, something that man aims at as an ideal being.* Passing through the senses (and only by this way), from the first impressions (in musical hearing), it overruns the natural condition: This means the affirmation of the spiritual value of man risen up to the level of meditation and understanding. Thus, there is developed his power to claim his proper *dignity/αξια* (*axía*), his *virtue/αρετη* (*areté*) – which sustains *progress/προκοπη* (*prokopé*). Therefore, in terms of a phenomenology of musical art, it is necessary to consider these: its defining features, its genesis and purpose, the message contained in it, the ways of achieving through it the "being"/man as composer, performer, listener, as one "who has Being in rapport with Being" in Heidegger's words.

The content of music is made up of fundamental experiences of being, of ethical experiences first of all, with the duality that characterizes them, with the conflict, but also with the reconciling: *the equilibrium, the supreme harmony*. Was not music designated by the Greeks: *αρμονια/armonía*?

Beyond the phenomenon of music, we can find what is general and more profound in its meaning. Here is the revelation of hidden sense, the enlightenment of an original principle, of a pure state, of intelligibility: the relation with music supposes the living of a thorough "moral experience" (see Rauh), as the *Human* one is. Subjects who approach music by themselves approach superior values. In stirring states of consciousness that are very different, music has remarkable qualities allowing the recovery of the essence of man in the right reason of being, in his most beautiful qualities. Through those, he rises to the level of "harmonious life", that *ομολογοϑμενος βιος* (*homologouménos bíos*) of Stoic philosophy (Diogenes Laertius VII). He causes to triumph the "best part of him"/*αριστος* (*aristos*), as a "regal man": wise and temperate/*σοφον και σωφρον* (*sóphon kaí sóphron*), righteous/*δικαιον* (*dikaion*), sovereign, free/*ηγεμονιχον* (*hegemonikón*), gaining some of the "great honor"/the *κοσμος* (*cósmos*) itself according to Socrates, the spokesman for Plato in *The Republic*.

Music is able to summon up acknowledgement of, and endear hearers to a referential world, a real *human* one, a world that man needs for living and through which he takes the specific measure of his *nobility* and *dignity*. Through the part that music plays in individual and social life, this art gives expression, determines and develops moral feeling, relating us to the ideal – that of the supreme Good / *το αριστον* (*ariston*). It fosters a motivation for being, a superior one: that of continuous *humanization*.

A wave that overwhelms (and, here, there is something of the mystery of music), this art of sound proves to be an unrivaled force for expressing, for exalting, for giving greatness and brightness to the feeling for life, inspiring aspiration to a superior state. Specifically, music sustains faith in a "meaning of life" as a morally supreme value, which implies an ensemble of preferences and exigencies that open the option of the *human way of being*. If it be well guided – a condition spotlighted still by the Ancients, in the theory of the "ethos of armonía" – music is an extraordinary, a privileged positive influence that leads to *Truth* and the *Good*, to *wisdom, love, freedom, duty, happiness*. Containing life's contradictions, music is a way of outrunning them, of *attaining equilibrium*. Here is what Tolstoi perceives in art, in general: "a moral organon of human life" (Tolstoi 266).

Our belief is one built on the *valence music possesses to draw out the moral sense*; herein, we pursue finding its real being.

A fine emanation of the spirit, musical art galvanizes subtly the *ethos of life*. It inspires noble, lasting, profound feelings. It orders them and lends perspective to them. These feelings in their turn have a strong influence on thinking and the will. Attitudes and demeanor are determined by them. Music is a source of moral meditation, of perpetual inspiration in the sense of the *authentic general human values*. Moral edification becomes the aim. Such a view reveals the *identity of artistic-musical Beauty and moral Good*. The lesson conveyed by music, having origin in its constitutive elements of rhythm, interval, measure etc., is that of listening to and following the rules of supreme Harmony. This seeding of man's soul, when it has crystallized the conscious understanding of "world harmony", binds the individual to that harmony. Being sensitive to the harmony of musical structures, man gets a clearer conception, he thinks more of the order and communion he is in need of. Disciplining his spirit, he can be enlightened as he searches and understands some of life's truth, of the foundational principle of his condition; he can become more conscious of the greatness and nobility he is capable of.

As a source of spiritual satisfaction and happiness, music proves to have a utilitarian character for the progress of the human being; it has an ethical usefulness, which amplifies the artistic Beauty that sustains its validity. Even the liberating function of music is one of the moral restoration of man. On the horizon of an ideal he is able to project *humanity* / *το φιλανθρωπον* (*philánthropon*). An ethical and esthetic ideal shaped by musical art, *humanity* represents that which gives power to man, the science-wisdom of assuming the inherent hardships of his existence without remaining bound by these. On the contrary, they are overrunning – with sensitivity and, at the same time, strength. To do music (to compose or only to perform it), as well as to receive music are preeminent situations for the initiation of man into the mood of *humanity*.

For the subject who opens himself to music, this art offers the paradigm of the Beauty-Truth-Good identity, *a paradigm of the ideal: aspiration towards Beauty, which represents the supreme Good and the absolute Truth* – as Plato emphasized in his *Dialogues*. The whole history of musical art testifies to the affirmation and endurance of those works that prove – beyond any haphazard condition – to accord with the desires and commands of high psycho-socio-cultural signification, having for their reference in the ensemble of these three Values – the key idea of humanism.

"Harmony", "measure", "fair proportion", "perfection in shape": *Musical art* – being produced by a noble force – contributes substantially to the fulfillment, to the flourishing of the human. Actually, the original Hellenic term: *αρμονια* meant beauty, fairness, pleasure and utility, peace, a state of "good", happiness. Music means gladness that is by occasion useful given its orientation to what is noble, to what can brighten the soul and the mind, what is exemplary and deserves effort, which impels us toward morality, stopping more easily the inclination to and contacts with "evil".

In music, we can perceive aspects of a moral *Utilitarianism*, as well as of a *Hedonism*; and more, aspects of a *Eudaemonianism*. Musical art has developed owing to its *utilitarian* character, one obvious during the oldest periods; gradually, its *hedonistic* and *eudaemonist* characters came to the fore. Mainly touching the senses but in the end an intellectual-philosophical exercise, music brings forth *pleasure/ηδονη* (*hedoné*) – in a spiritual sense, one tending towards *happiness/εὐδαιμονια* (*eudaimonía*) – ecstasy. At the same time, it is possible through music to discipline the mind, to enlighten perspectives on "reason", "just measure", "order" – those aims so much appreciated by the ancient philosophers: the *λογος–κοσμος* (*logos-cósmos*), the precept of Delphi: *μηδεν αγαν* (*medén agan*). It is possible through music to effect a change in one's soul, to change one's practical plan of behavior. This art generates and grows *spiritual pleasure*. It opens a way of *living usefully and happily*. Asking for a complex, generous and deep *understanding*, it also imparts a "science" centered on *respect for law* and consequently lays the ground for *a dignified human life*.

From ancient times till now, we recognize in music a certain utility as seen in the benefits of chant. The therapeutical, curative effect of music means, essentially, the assurance of a psychosomatic equilibrium and of a "moral health", the achievement of a balanced state and harmony (inside and outside of *man, in auto-* and *hetero-relations*). Through that harmony is structured the *human* as a value-ideal. Even only as entertainment, music represents an opportunity to forget, at least for the time being, the worries and bellyaches of existence. The musical work, the fundamental acts of music (composition, performance, audition) are real ways "susceptible to rescuing us into joy", that "joy of being" "affirmed by everyone, every moment" – according to Simone de Beauvoir (de Beauvoir 196).

Through music – through its ethos – man gives to himself the chance to experience a superior *pleasure*, one real, pure, natural and necessary *pleasure with a view to the good* – so that Plato (in *Philebus* and *Timaeus*) and Epicurus (*Letter to Menoeceus*) have granted it priority; a *pleasure* "which perfects activity" as Aristotle says (*Nicomachean Ethics* VII), owing to the purification accomplished by the music, the beauty of this art being in accord with spiritual nobility.

Man lives a special delight by music – a peculiar manifestation of the aspiration towards the Good; music as well touches the Good and also *happiness*. And this is because of the "joy of integrity and authenticity", of "balance and harmony" that are achieved as a man becomes able to learn the true meanings of music, to reach some of the "wisdom of being", given that music has an internal resonance and promotes accord between man and world as well. It is a substantial amplification of the *human*.

The aspects of a moral Hedonism and a Eudaemonianism in musical art received early recognition in the theorizing of the ancient Greeks about art in general, in Aristotle's conceptions of *mimesis*/*μιμησις*, *poiesis*/*ποιησις* and *catharsis*/*καθαρσις* (*Poetics*). We refer, certainly, to his speaking of "the songs that purify the soul", that "give us a harmless joy" (Aristotle, *Politics* 206).

Through artistic-musical creation, we find something about the meaning of *happiness, eudaimonia* – a value that has always concerned the human spirit. Musical works cement the correlation of a system of values, which is very important for the essence of human being – on the individual and social planes of existence. These are values that enable man to flourish, to maximize his state of well-being, to achieve happiness and, thus, to receive a revelation of truth (*Plato, Phaedo*). Above all, cultivation of the soul aims at a kind of refinement (*mousiké, mousikón*/*μοϑσικη*, *μοϑσικον* signifies even "culture" in opposition to *amousón*/*αμοϑσον*). The art of sounds – in its truthfulness – reveals a "world" perfectly consonant only with a perfect soul, enlightening into Truth, unable to participate in "evil" (in any hypostasis thereof). This harmony cannot at all take part in disharmony – as we read in Plato's *Phaedo*.

*Al. I. Cuza University*
*Iasi, Romania*

REFERENCES

Aristotle. *Nicomachean Ethics*; Romanian translation: *Etica Nicomahică*. Bucharest: Ed. Casa Şcoalelor, 1944, VII.

Aristotle, *Poetics*; Romanian translation: *Poetica*. Bucharest: Ed. Academiei Române, 1965.

Aristotle, *Politics*; Romanian translation: *Politica*. Bucharest: Ed. Cultura Natională, 1924.

De Beauvoir, Simone. *Pour une morale de l'ambiguité*. Paris: Gallimard, 1964.

Blaga Lucian, *The Trilogy of Knowledge / Trilogia cunoaşterii, Opere*, Vol. 8, Ed. Minerva, Bucharest: 1983, p. 74. Philosopher and artist (poet, dramatist), Lucian Blaga (1895–1961) is the author of a metaphysics of "mystery", elaborated in a monumental system in his works *The Trilogy of Knowledge, The Trilogy of Culture, The Trilogy of Values* and *The Cosmological Trilogy*. The central focus of Lucian Blaga's philosophy is on "man's singularity of being in the horizon of mystery" and "the revelation of mystery". According to Blaga, "existence within mystery" represents the "ontological mode" of existence, as "an accomplished human being".

Diogenes Laertius, *Lives and Doctrines of Eminent Philosophers*; Romanian translation: *Despre vieţile şi doctrinele filosofilor*. Bucharest: Ed. Academiei Române, 1963, VII.

Epicurus, *Letter to Menoeceus* quoted in Diogenes Laertius, *Lives and Doctrines of Eminent Philosophers*, X.

Plato, *Dialogues*; Romanian translation: *Opere*, Vols. I–VII. Bucharest: Ed. Stiintifică şi enciclopedică, 1975–1993.

Rauh, Frédéric, *L'Expérience morale*. Paris: Presses Universitaires de France, 1951.

Schopenhauer, Arthur. *Die Welt als Wille und Vorstellung*. Romanian translation: *Lumea ca voinţă şi reprezentare*, 3 vols. Iasi: Ed. Moldova, 1995, Vol. 1, p. 283.

Tolstoi, Leo. *Qu'est-ce que l'Art?* Paris: Perrin et C-ie Editeurs, 1898, p. 266.

ROBERT D. SWEENEY

# TRACE, TESTIMONY, PORTRAIT

This is a paper that deals, in part, with an aesthetic issue, the nature of the portrait. But more fundamentally, it is concerned with an issue in the philosophy of history (and thus of epistemology), namely, the question of the veridicality – the truth-value – of accounts of the past. In turn it also touches on an issue in ontology, namely, the status of the past, the "pastness" of the past. The guiding threads of the paper are taken from a recent study by Paul Ricoeur, "*La Marque du Passé*."[1] It will undertake to show that the portrait is not primarily a representation in the classic sense of a copy or facsimile, but that it fits more appropriately into the category of testimony and in that role it serves as a founding model of historical evidence and as a guide for interpretative procedures. In that context, it is hoped, some light might be shed on the complex question of the evaluation of the portrait as an aesthetic object and how this affects the ontology of the past – past-time.

The question of the portrait has a long history, of course, having been an item for discussion in both Plato and Aristotle. For Plato, understood as a type of image or *eikon*, it was central to the debate over the illegitimacy of the poet, although, as Ricoeur points out, the *eikon* was also seen as divinely inspired, and thus legitimate.[2] As Aristotle saw it, the portrait, as a type of image, had two dimensions: the graphic entity itself, and the reference to a referent, to a model.[3] In the first dimension, it was simply a visual object; but in the second, it was an example of the *eikon*, which, as Ricoeur emphasizes, was and is modeled after the striking of the seal of wax and, as such, is basically a copy. The seal, in turn, fits into the category of the trace, the generic term that covers all cases of a vestige left by a passing event. The trace, of course, has been considered the key item in historical research, which, in general, is understood to gather traces into documents and documents into archives. History has been described as a "science by traces." (Marc Bloch)

At the same time, the trace is a key component in the debates that characterize poststructuralism, coming to prominence in the deliberations of Emmanuel Levinas on the meaning of the face, the visage of the other that is the source of responsibility, of subjectivity, and of ethics as first philosophy – that says "Thou shalt not kill!" The face, Levinas claims, is among other things also a trace, that is, a present object that refers to an absent event or entity. This absent dimension is unique in the case of the face because it constitutes a "past that was never present."[4] Derrida was moved to generalize this



*A.-T. Tymieniecka (ed.), Analecta Husserliana LXXVI*, 159–169.

feature of the trace to make it the fundamental feature of all words, of all signifiers, indeed of all texts, in effect making all meaning an endless pursuit of "substitution," "postponement," and so forth. In so doing, it can be argued, the key claim of post-structuralism was brought center-stage, viz., the position that the traditional concept of truth was based essentially on representation understood as imitation ("correspondence" theory) and that representation is, in reality, an endless round of postulation and suggestion – supplementation, substitution – never completed, never "certified." The trace, Derrida says, "is not a presence but rather the simulacrum of a presence that dislocates, displaces and refers beyond itself."[5] The shadow of "relativism" hovers over this position, even if many poststructuralists, including the later Derrida, disavow that term and its implications. Ricoeur, like many others, sees the epistemic status of the trace as an enigma, since, of course, there can be no *tertium comparationis*, at least, in the case of traces of the past. But at the same time, as we know, Ricoeur has undertaken to rehabilitate mimesis, so that it becomes creative imitation.[6]

It is in this context that Ricoeur has reexamined the concept of the trace as he himself had formerly dealt with it in *Time and Narrative* III, and found that it should be displaced by testimony, which introduces "the world of the witness who reports what he has seen and demands to be believed." Instead of the enigma of "resemblance" that surrounds the trace, he describes the "fiduciary relation constitutive of the credibility of the testimony" – the credibility of a "witness whose presumed good faith can be put to the test of a confrontation of testimonies." We must stop asking, he says, whether a story resembles an event; instead we must ask "if the ensemble of testimonies, confronted with each other, is believable." This believability works in such a way, he says, that the "witness has made us *assist* at the narrated event." (MP, 14)

To underscore his position, Ricoeur reiterates his emphatic assertion that portraits are not copies. Ricoeur's general point then, is that the heart of historical meaning is not primarily in the veridicality of the trace, but in the trustworthiness of testimony. Not that veridicality is jettisoned, only that it presupposes the predominance of testimony and its trustworthiness, its "*fiabilité*." Believing is a matter of believing that, on the background of believing in; essentially, we believe in someone, not something. Central to this belief, Ricoeur says, is a "linguistic dimension, absent from the metaphor of the imprint, namely, the world of the witness who reports what he has seen and demands to be believed. The imprint that the event leaves is the seeing relayed by the saying and the believing." (MP, 13)

A key step Ricoeur makes in his effort to rethink his own position here, modifying that developed in *Time and Narrative*, is to look once again at the metaphor of the striking or stamping of the seal, which dominates the tradition of trace as *eikon*. This metaphor, he says, does have a staying power, that is, the trace has a resistance to being superseded by testimony, precisely because of its physicality; it presents a seemingly clear-cut cause and effect relationship. (MP 14) His strategy, then, is to propose an analogous relationship in the case of testimony, that is, analogous to cause and effect, and he does this by utilizing a theme derived from the work of Levinas: the "passivity" of "being affected."[7] Levinas develops this theme to support his claims as to the passivity of responsibility; Ricoeur uses it to describe a central aspect of testimony.[8] The witness who gives testimony has been so shocked, upset, startled, or even traumatized by an event that he has to speak of it. It is the suddenness, the sharpness of this "being affected," that lends it to an analogy with the "imprint" of the trace. But while the analogy may diminish the difference between trace and testimony, and thereby even reinforce the staying power of the *eikon*, it does not finally remove the difference, mainly because the physical character of the imprint is in a separate order from the emotional nature of being affected – this "pathic" quality which Ricoeur compares to the *Wirkungsgeschichtliches Bewusstsein* of Gadamer's hermeneutics. And this quality carries over to the recipient of testimony, the listener: "Through the story," Ricoeur says, "the hearer, having become a witness of the second degree, is found in his turn placed under the effect of the event of which the testimony transmits the energy, indeed the violence but also the rejoicing." (MP 17)

It should be noted as well that for Ricoeur's epistemology, testimony belongs to the same general category as does attestation, as suggested by its etymology, but more important, because it stands between "doxic belief" and the absolute certainty of episteme: "Whereas doxic belief is implied in the grammar of 'I believe-that,' attestation belongs to the grammar of 'I believe-in.' … It is in the speech of the one giving testimony that one believes."[9] And just as attestation has as its opposite, suspicion (cf. the "hermeneutics of suspicion") so also does testimony have an opposite, viz., false testimony.[10] As Ricoeur points out, much of the progress of historical research depends on weighing the relative trustworthiness of differing or even conflicting testimonials. Thus the issue of resemblance embedded in representation as copy, *eikon*, is defused, displaced. In turn, this enables him to reinforce his earlier use of *Vertretung* or place-holding (*lieutenance*) over against *Vorstellung* (representation) to describe historical intentionality. Whereas *Vorstellung* is representation in the classical sense of referring through an image to an object (i.e., by way of a copy), *Vertretung* suggests an indirect way of referring to the past and is more appropriate to historical intentionality; it is "history taking the place of or standing for the past."[11]

The importance of this position for the philosophy of history is that it responds to an issue raised by historians, namely, what is the relative value of the virtues of veridicality and fidelity? Contrary to the usual view in which veridicality is the exclusive or paramount virtue of historical research, Ricoeur concludes that fidelity is actually more important, but only if understood as being in a dialectical relation with veridicality. Indeed, he concludes his study with this phrase: "One cannot work with history without doing history" (i.e., being historical).[12] At the same time, in so doing, Ricoeur is extricating himself from "tropology" – the position of Hayden White that brings history and fiction into such a close embrace that history is reduced to allegory, resulting in the ignoring of the importance of documentation, and of "singular causal imputation." In other words, the aporia of representation, Ricoeur explains, would be insoluble if we were to stay within the theme of resemblance or copy. But again, being displaced by testimony doesn't altogether remove resemblance from the scene: Thus "it is necessary to reverse the relation of priority between testimony and portrait, and to posit that the portrait sets forth an *eikon* of its model only because it first wills itself faithful to it in the manner of a testimony." (MP 17n)

To explore testimony a step further than Ricoeur has taken it, we can say that, in addition to referring to an event, it itself fits into the category of an "event," and that "event" in this context refers to an episode in a narrative – a high or low point, a turning point, a peripeteia – that demands recording. In other words, to discern any event, one has to reconstruct a narrative in which it makes sense, in which it fits into a plot as a "synthesis of the heterogeneous." In the case of history, this plot is one that is fed by other events registered in other traces/testimonials – and this, despite the claims of the *Annales* school and others that "events" are merely surface aspects of the "longue durée" (Braudel). The portrait, then, as a form of mute testimony, stands both as a portrayal of an episode and as itself an episode in an artist's life and in the life even of a model/sitter.

## THE PORTRAIT AS TESTIMONY IN AESTHETICS

Aesthetics addresses the question of the meaning and evaluation of art. With respect to meaning, on an initial level we know about the manifold aspects of portraits when we consider those done in painting or sculpture: A likeness is assumed, sometimes without verification; we are equally ready to assume (and accept) major distortions, at least up to a point – a point not readily discerned. In either case, the issue is not a matter of accurate copying. If we call it a "likeness" we do not mean a strict copy. Of course, in photography this problematic issue is amplifed many times over. Ricoeur insists that even the photograph, (of Cartier-Bresson or Doisneau) is "no less configuration or reconfiguration than the portrait. It also aims at fidelity beyond the reduplication by copy." Both relate, he says, to "the phase of the imaging of memory and, through that process, return to the problematic of fidelity." (MP 16n)

It seems to me, in other words, that the claim that the portrait is not a copy shifts the issue of its basic nature to testimony. The implication for aesthetics then, is that what distinguishes the authentic portrait from the others that may carry the descriptive title of portrait, as well as from studies of the human figure and other such images, is testimony – testimonial quality. Assuming a high level of painterly competence within a style (e.g., realistic or non-realistic; impressionistic or expressionistic-granted, not an easy quality to establish), this would mean that testimonial fidelity is somehow embodied, is the key ingredient, in a painted or sculpted portrait, indeed even in an artistic photograph as mentioned above. How can this testimonial quality be explained, explicated? I would suggest that the narrativity of the portrait is a plausible approach. This would mean that there is a narrative suggested, if not explicitly detailed, in a portrait even though the portrait might be thought to be merely a cross-cut in time.

But there are claims that only special forms of "narrative art" embody narrativity, e.g., depictions of the Passion of Jesus (not portraits, of course). How can these objections be answered or resolved? I would say that, first of all, there are the obvious, as well as subtle, clues of narrativity in a portrait: signs of royalty, implements of a work-world, instruments of warfare, uniforms, judges robes, and the like. However, these clues would not by themselves indicate a narrative unless they were somehow integrated into a plot, a "synthesis of the heterogeneous" in Ricoeur's phrasing. This perhaps would not have to be a clear-cut plot so much as a witnessing that derives from an attitudinal aspect in the portrayal that reflects the way an individual (or a group) has negotiated a series of major life-episodes. Here I would draw on Aristotle's analysis of character (personnage) in relation to plot, as introduced and interpreted by Ricoeur:

> A character is the one who performs the action in the narrative. The category of character is therefore a narrative category as well, and its role in the narrative involves the same narrative understanding as the plot itself. ... Characters, we will say, are themselves plots.[13]

One might summarize thus: If it is true that the plot drives the character, it is also true (from another perspective) that the character influences the plot.

I would infer from this that the sitter in a portrait can be considered a character in a plot and thus identical with it. But the portrait, I have said, cannot be a full-fledged narrative – only the suggestion of one in terms of its testimonial nature. In other words, to be portrayal, the depiction should incorporate suggestions of a struggle, a test, an anticipated challenge even in the midst of (perhaps) a serene posture.

It is true that the usual designations utilized to describe portraits and their depth, as distinguished from other images, have to do with terms such as "personality," "character," "individuality."[14] But all such terms, it seems to me, are based on a background of selfhood, which in turn must be understood as a narrative self – even if the actual narrative itself is mostly unknown.[15] That is, there must be a dynamic, temporal quality in the portrait to give an authentic depth to the person involved. Not all portraits project this quality and thus many (perhaps most) do not merit our careful attention or our imaginative reception. To view a portrait, in other words, is like reading a text, in that we respond to the narrative quality – the creative "synthesis of the heterogeneous" – before we give it our close analysis. All narratives combine fiction and history to some degree, Ricoeur has shown, and here much that is missing in terms of historical documentation can be said to be filled in by imaginative reception. But critical to each such portrait would be the testimonial quality – the witnessing.

One way of underscoring the point here – in stark and over-simplified form – would be to compare Picasso's studies of humans, both realistic and highly distorted, with a painting of a great portraitist such as Hans Holbein the younger – *The Ambassadors*. Both sets of images are, of course, from very different eras, done under totally different auspices. The key difference, one might insist, is to be found in the intentions of each artist, and certainly intentions cannot be dismissed. But intentions are not mere internal dispositions; especially in the case of artists, they leave very definite external indicators. Picasso's studies, we might say, are vivid depictions of idealized representations of human types, even if they are not readily identifiable as belonging to a certain human type (stolid, phlegmatic, sanguine, energetic, electric and so forth). Although based on or derived from human models, they are related by the viewer to these models with great difficulty, and not at all to any testimony or narrative, except very indirectly.[16] Holbein, on the other hand, in, "The Ambassadors," for example, was working from a very definite narrative as is indicated by the trappings of office and intellectual interests that are included in the painting. But the personal narrative is to be found primarily in the delineation of the faces, the focus of the eyes, the stoic, reserved yet assertive, confident demeanors. We know relatively little about the historical circumstances (but more than previously);[17] still the testimonial quality does constitute historical documentation even if, sometimes, in a very limited sense.

Of course, a much broader study could be made of the issues raised here. For example, there are the many distinctions and subdistinctions to be made within the category of portrait – genres and subgenres, as it were.[18] We can mention self-portraiture; mirror portraits; mummy portraits; author-portraits; collective, political, religious, portraits; drawings; photographs; etchings; and so on. This is not even to broach the prior question of how the portrait fits into the larger category of the image – of the *imago* – and the further question behind this one, as posed, e.g., by Hans Belting[19] – must the figurative image be distinguished from art? One can walk through the National Portrait Gallery in London and be struck with the range of (mostly recent) portraits, both in terms of models and of styles. These styles can be highly interpretative – often distorting caricatures – but they reveal something of the person and of his situation. One might add – they reveal even the reception of his/her celebrity performance (celebrity is apparently the primary grounds for admission to the Portrait Gallery).

The question of authenticity can be posed here even in the case of the self-portrait. We can think of van Eyck and Dürer's very convincing efforts but be actually stunned by the intensity of the self-portraits of Rembrandt (60 or more) in a recent exhibition at the National Gallery. These (the paintings, at least – as distinguished from the etchings and woodcuts) are mostly frontal – direct, focused, "sincere," we might say with no evident irony. Of course, the explanation for the preoccupation (or obsession) with self-portraits Rembrandt exhibited can be laid to practical strategies: trial efforts to get the desired expression just right; the lack of patient, tolerant models; the possibility of retouching. It is worth noting that, although he was very much in demand for portraits of others, Rembrandt spent a period of 10 years without doing any. A plausible inference from this can be that, besides his impatience or, possibly, even a kind of egotism, there was a fascination with the evidence he could discern in himself (in his self-portraits) and could register in his painting the quality of a life and its divagations and struggles – in other words, testimony. And here, I believe, is the basis of the exceptional aesthetic merit of these works beyond their extraordinary skill-level, namely a depth, a searching quality that reveals a portion of a life narrative. Once again, even though this depth has often been related to personality or character (e.g., in Brilliant) and I would not dispute such claims, I would only argue that such features are part of the narrative self, and that this depends, in turn, on the narrative. In a recent article in the *New Yorker*, Simon Schama has underlined precisely the narrative quality of testimony in Rembrandt's portraits: "… his portrait subjects had the energy and urgency of characters caught in a historical drama – their own."[20]

There are, of course, general aesthetic qualities that are discernible in the portrait but are not specific to it, e.g., those involved in what Ricoeur has called the *humeur*, the mood, of the work and its communicability, as explicated in another recent work.[21] Although this mood is primarily the contribution of the artist, and only secondarily that of the sitter, it is also by analogy an effect perceived by the viewer: "The work, in what is singular in it, frees in the one who tastes it an emotion analogous to that which produced it, an emotion of which that individual was capable, but without knowing it, and which enlarges his affective field once he experiences it. … The mood is like a relation outside of the self, a manner of inhabiting a world here and now; it is this mood that can be painted, put into music or into narrative in a work. …"[22] Of course, such an effective quality is difficult to pinpoint or describe, but I think it is supported by the narrativity.

## PHILOSOPHY OF HISTORY AND ONTOLOGY

A brief return to the philosophy of history and its ontology might reveal a richer background and even a greater usefulness, of the portrait – mostly ignored (Belting claims) by the historian preoccupied with words.

Obviously, the nature of testimony ranges well beyond the role or function of the portrait, taken in its basic sense as an image; in particular, in history this role is in terms of the collective self (identity, etc.). As J. Barash has pointed out, the collective self must be located in between purely individual, atomic identities (Locke), and organic, substantival (totalitarian) conceptions – in what he calls "implicit memory,"[23] But the source of these is still testimonials from individuals – not outright declarations, for the most part, but lives lived in the life-world of a particular national or communal entity, in other words, in a pluralistic society.

An example here might be appropriate: The history of the Holocaust can be documented in terms of death camp records and remnants and body counts, but the basic data are derived from testimonials. Testimonials are, as it were, the basic portraits. It is true that there are gaps and discrepancies among these testimonials, and even outright fabrications done in evident good faith (cf. Wilkomerski), but these conflicts do not disqualify the overall historical picture and, in fact, can enrich its grip on us. Indeed, Ricoeur has argued that fiction relating to the Holocaust can intensify and deepen its truth as relating to universals of the human condition.[24] The portrait, I would say, can serve as a model of such truth. On the one hand, it is a transversal cut (but not, typically, a snapshot) in the temporal flow, in the narrative that displays the past and exhibits the causal connections, the "singular causal imputations" that characterize "scientific" history. On the other hand, it displays the imaginative creativity of the artist – the equivalent of the fictional aspect of every formal narrative, even history – in all of which fiction and history "intersect." And in both cases – but especially in the fictional aspect – the portrait invites the response of the audience and the imaginative recreation of the narrativity. That is, we are invited by it to reconstruct the historical narrative as best we can (and, of course, often we can do very little). Nevertheless, such narrativity delivers the singularity of the work, even if not always the specifics of the identity of the sitter.

## ONTOLOGICAL IMPLICATIONS

The ontological implications of this interpretation of the portrait are considerable, at least potentially, even if limited and only discernible in a sketch; their outlines are seen best, perhaps, if we follow Ricoeur in his critique of Heidegger's ontology of temporality that is the last section of "La marque du passé." Suffice it to say here that, while in a general way he endorses Heidegger's ontology of time, it is the question of the future as rounded off by death that Ricoeur finds too limiting in Heidegger. In other words, an open future that includes the collectivity as sharply distinguished from the individual and the "close" ("near ones") is more phenomenologically valid, Ricoeur argues.

The relevance here would be that the authentic portrait, at its best, suggests not just past episodes, but a "horizon of expectation" as part of the narrative – expectations on three levels: the ownmost, the close and the distant (i.e., the institutionalized collectivity). Testimony, that is, relates to the future as well as the past, in terms of hope, trust, potential. Indeed, in this way, Ricoeur avers, we can do justice to all three of the temporal "ecstasies" – past, future, and present – without reducing one to the other or removing the experiential balance between them. The portrait, at its best and understood in terms of testimony, displays this balance: It gives us a present experience of a past moment in a life that was/is ongoing in terms of a hopeful future of self and community.

To these already highly speculative conclusions, I hesitate to add yet one more, but I believe that the "scaffolding," the "stratification," involved here is in fact that of a hierarchy of value that incorporates the personal at its highest level. When the affective is involved, even if only in the form of a "mood" that we capture (Ricoeur), that is, wherever there is a "being affected," the personal is being activated in a way that transcends the impersonal – to put it in the broadest terms. One might then – correctly – discern an effort here to reactivate the Husserlian/Schelerian insistence on a correlation between the affective and value in general, and including aesthetic value, that is, an "affective value-response." But I might even say, in addition, that with the portrait the value involved is the personal in the form of the "other," that is, with a Levinasian emphasis on its breakthrough into our totalities of closed-in worlds and narrow perspectives.[25]

*John Carroll University*

NOTES

[1] *Revue de Métaphysique et de Morale* (Henceforth: *MP*), Janvier-Mars 1998, No. 1, pp. 5–31.
[2] Plato, *The Republic*, Book X; *The Sophist*.
[3] *Parva Naturalia*, 450b.
[4] E. Levinas, "Ethics as First Philosophy," Sean Hand (ed.), *The Levinas Reader* (Oxford: Blackwell), p. 84.
[5] Jacques Derrida, "Différence," *Margins of Philosophy* (A. Bass trans.), (Chicago, University of Chicago Press, 1982), pp. 18–19.
[6] Paul Ricoeur, *Time and Narrative* I (trans. K. McLaughlin, D. Pellauer), (Chicago: University of Chicago Press, 1984), Chapter 2.
[7] Levinas, *op. cit.*, p. 82 and *passim*. "Being affected" has a history in early phenomenology: it is suggested in Max Scheler's theories on love (*ordo amoris*) in e.g., "fellow-feeling" (vid. *The Nature of Sympathy,* trans. P. Heath, New Haven, Yale University Press, 1954), p. 49; it is explicit in D. Von Hildebrand's effort to express the receptive aspect of affectivity. Cf. Von Hildebrand, *Christian Ethics* (New York: David McKay, 1953), pp. 208–210.
[8] Paul Ricoeur, *Oneself as Another* (trans. K. Blamey), (Chicago, U. of Chicago Press, 1992), p. 23.
[9] Ibid., p. 302.
[10] Ibid.
[11] Paul Ricoeur, *Time and Narrative* 3 (trans. K. Blamey and D. Pellauer), (Chicago: University of Chicago Press, 1985), p. 143.
[12] "… Il n'est possible de 'faire de l'histoire' sans aussi "faire l'histoire," *MP*, p. 31.
[13] *Oneself as Another*, p. 143.
[14] Richard Brilliant, *Portraiture* (London: Reaktion Books Ltd., 1991).

[15] Ricoeur's concept of a "narrative self," involving a dialectic of idem (sameness) and ipse ("self-constancy"), is developed primarily in the Fourth and Fifth Studies of *Oneself as Another*.
[16] Ibid., p. 150. Picasso, it is true, did a few portraits, the most notable perhaps being his portrait of Gertrude Stein. Most commentators saw it as a poor likeness, but she is reported to have been "satisfied" with it. See Brilliant, op. cit., p. 150.
[17] Susan Foister, Ashok Roy, and Martin Wyld, *Holbein's Ambassadors* (London: National Gallery Publications, 1998). This catalogue raisonné gives new information on the nature and personnel of this multiple portrait.
[18] Cf. Brilliant, *op. cit.*
[19] *Likeness and Presence, A History of the Image before the Era of Art* (Chicago: University of Chicago Press, 1994).
[20] October 11, 1999, p. 67.
[21] Paul Ricoeur, "Aesthetic Experience" (trans. K. Blamey), *Philosophy and Social Criticism*, Vol. 24, no. 2/3, p. 33.
[22] Ibid., p. 24.
[23] Jeffrey Andrew Barash, "Les sources de la mémoire," *Revue de Métaphysique et de Morale*, Janvier-Mars 1998, No. 1, pp. 137–148.
[24] "Le temps raconté," *Revue de métaphysique et de morale*, 1989, p. 450.
[25] I have deliberately avoided the issue of the ironic portrait, as, e.g., with Duchamp, on the grounds that such portraits are parasitic on the nonironic or sincere ones.

## REFERENCES

Belting, Hans. *Likeness and Presence*. Trans. E. Jephcott. Chicago: University of Chicago Press, 1994.

Brilliant, Richard. *Portraiture*. London: Reaktion Books Ltd., 1991.

Derrida, Jacques. *Margins of Philosophy*. Trans. A. Bass. Chicago: University of Chicago Press, 1982, pp. 3–27.

Derrida, Jacques. *The Truth in Painting*. Trans. G. Bennington and I. McLeod. Chicago: University of Chicago Press, 1987.

Foister, Susan; Roy, Ashok; and Wyld, Martin. *Holbein's Ambassadors*. London: National Gallery Publications, 1997.

Levinas, Emmanuel. *Otherwise than Being or Beyond Essence*. Trans. W. Lingis. The Hague: Nijhoff, 1981.

Ricoeur, Paul. "La marque du passé," *Revue de métaphysique et de morale*, Janvier-mars, 1998, No. 1, pp. 5–31.

Ricoeur, Paul. "Aesthetic Experience" (trans. K. Blamey), *Philosophy and Social Criticism*, Vol. 24, No. 2/3.

Ricoeur, Paul. *Time and Narrative* (3 volumes) trans. K. Blamey, D. Pellauer (Chicago: U. of Chicago Press, 1983, 1985, 1989).

Ricoeur, Paul. *Oneself as Another*. Trans. K. Blamey. Chicago: University of Chicago Press, 1992.

Von Hildebrand, Dietrich. *Christian Ethics*. New York: D. McKay, 1953.

RENATO PRADA OROPEZA

# PHENOMENOLOGY AND LITERARY AESTHETICS

## 1. THE AESTHETIC *EPOCHÉ*

A present-day characteristic of the philosophers who tackle the problem of aesthetics is their particular attention to both the broad contemporary manifestations of aesthetics, above all, those that in any way have the intent of renewing those of the 19th century, and particular theories, either poetic (in its broad sense) or critical.

Nevertheless, if we pay a minimum of attention to the earliest commotion of a particularly renewing literary theory like Russian Formalism inspired by the contributions of linguistics, to the theses of the theorists belonging to the linguistic Circle of Prague, as well as to those expressed by the keenest representatives of the New Criticism expounded some years later, we find in them some common postulates which can be grouped into at least two classes:

i. The conviction that literary discourse in general is a particular *language*, different from common speech or from language in its practical function, but undoubtedly a *verbal fact*. The decision to consider its particular manifestations seriously just as they are brings out the constitutive values characteristic of aesthetic work, from the point of view of both the form of expression and the manner of arranging content;
ii. As a consequence of the foregoing, there is avowed a need to pay attention to individual aesthetic configurations without seeing therein continuity with what classical aesthetics understood as sensibility, on the one hand, and, on the other, to realize the following Hegelian insight: if the aesthetic discourse has something to tell us, if it has a meaning, that will be one that is its very own. In regard to this, it is important to pay attention to the contemporary aesthetic *praxis* (in painting, the manifestations that go from expressionism to the so-called abstract painting of a Kandinsky, for instance; in music, the works of Schönberg, Ligeti, Takemitsu, Lavista, among others; in narrative, those of Beckett, Borges, Cortázar, Elizondo Hishiguro, Calvino, Saramago, et cetera), to name those that demand a special aesthetic *praxis* from aesthetic receptors.

If we take into account these contributions – both of aesthetic *praxis* and of the theories and criticisms of art that have been concerned with making it clear – the operation of an aesthetic *epochè* shows itself quite clearly. It



*A.-T. Tymieniecka (ed.), Analecta Husserliana LXXVI*, 171–182.

would seem that these pursuits unintentionally achieved what would be a hard task without them: a rigorous *phenomenological epoché* enabling us to characterize what constitutes the *eidos* of narrative-literary discourse.

In some cultural sequences both the story and the very *praxis* that characterize its discourse take charge of bracketing those elements that are extrinsic to their characteristic manifestations; this philosophy does with the world of common sense, of mythology, of the factual and physical-mathematical sciences; religion does it with mythological tales; literature with the author-person, by referring to the daily or common sense world, with language as a practical system for communication. All this aesthetic theory does not specify; it requires an additional effort to consider these achievements as constituents of a manifestation "in person," with their identity and distinctiveness. The *epoché* places us before an autonomous – if not independent – aesthetic manifestation within a network of other manifestations: *autonomous with regard to the first understanding, which should be considered as if it were "alone" in the world*, in order to pay due attention to its very own values. The first step toward constituting the sense of a discourse is taken in the domain of relative but inevitable immanence.

If we bracket the world of common (daily) sense, the referential relation of an object (of a preexisting place to space through a toponym in a narration, for instance, Paris, Capital of France, in relation to the "Paris" described in Balzac's novels; or a narrative program fit for a role, for example, that of a middle class head of the family, or that of a university professor, et cetera), of the author-person (his/her possible experiences as found in biographical data that "explain" a short story or a novel, or some of his/her values; his/her interpretations of the narrative discourse "created" by him/her, et cetera), there seem to remain, as a result of this decanting, the values inherent in the aesthetic discourse, like beauty and those related to it, which traditional aesthetics considered the characteristic values of an aesthetic discourse, although a more careful attention to this *praxis* will note that even these "values" are decanted. It is also the task of the aesthetic *epoché* to bracket, at least for the sake of minimizing the suspicion of responding mostly to prejudices that have never been well delimited with regard to the aesthetics of everyday life (that of the senses): What happens if for the time being I am not obsessed with finding the beauty in manifestations of discourse and I let the discourse present itself without these garments? Have all the discourses called aesthetic throughout history had the same axiological charge with regard to their so-called beauty? Did beauty have exactly the same value for Greeks as for the Romans? Does the beauty of a Renaissance painting – of a

Da Vinci, for example – correspond exactly to that represented in Brueghel the Elder's work? If we consider that the sense of a value or of one of its elements is established in relation to others of its own own discursive unit or series, we can at least conceive that both beauty and its counterpart, ugliness, have not always had the same meaning: at least along the path from what it had in Greek culture to its gradual secularization in the Italian Renaissance, to its mythicization in Romanticism, to its total loss in the great exemplars of expressionism and cubism, to the manifestations of the decomposition of figurative elements and its problematic presence in abstract painting. Given all this, a revision of its rule and even of its *effective* presence in current manifestations seems to be legitimate.

Let us reduce our phenomenological description to the updating of the aesthetic effect in aesthetic narration; for this purpose, let us consider a narration by Rulfo, "El Hombre," and a couple of novels from cultures quite akin to ours: the closest, *Ensaio sobre a Cegueira* by José Saramago (a Portuguese novel) and *The Unconsoled* by Kazuo Ishiguro (an English novel).

Rulfo's short story offers a narration very well adapted to the thread of a *diégesis* (story) with a fragmentary *intrigue* based on the occurrence of passing incidents, most of them analeptic, which is very well established if the reader follows the development of the fable and is able to reconstruct it fragment by fragment: the intercrossing of offense or harm inflicted and reparation or revenge by two initially unidentified characters. The unraveling occurs in a new break, this time in the order of the registry of the discourse: it passes from an impersonal relationship to an engaged one, told of by a witness, a poor sheep herder (the explicit narrator) to a *teller* (a justice official), whose interruptions of the sheep herder's discourse are just to be inferred from the answers of the latter. As we can see, we are in a literary narration with a complicated diegetic construction and one that concerns a double revenge. If the implicit reader loses the "thread" of the *diégesis*, he/she does not have the literary competence to understand the story, as it becomes confused and does not tell him/her anything. Likewise, he/she should pay attention and integrate the words of the characters into a discursive totality and a specialized sense: the mountains and the river as obstacles to the escape of "the man" who is then killed by Urquidi, his pursuer, whom the implicit narrator likes calling "the one who pursued him," when he refers to these events.

The great concentration that the narration demands, as well as the richness of its network of elements that weave this fine and complex discourse lead the implicit reader to pay attention to its elements, or better its diegetics, as we have said, to be able to construe what the discourse tells him/her, that is, its *symbol*. In this encounter with the discourse itself, the complication of a narrative construction offered as an aesthetic challenge for the reader arises to restore the meaning that he/she has to actualize, a meaning that is totally offered in the symbol articulated by the story and that responds to the total integration of the discourse as such, as soon as it *says something* in that same manifestation. This is the experience we have of the *praxis* we were committed to from the moment we started the reading with the initial sentence: "The man's feet sank into the sand, leaving a shapeless footprint, as if it were the hoof of some animal [...]".

As in the case of painting with regard to the figurative tradition, whose dismantling takes place in different aspects and from different aesthetic perspectives, there occurs in narrative a dismantling of realism and its codes, a process that began more than a century ago and continues up to the present, and which has resulted in a decanting of the narrative-literary discourse itself, but without achieving the purity of a total and originally "abstract" narration (i.e., as *pure*, as was the ideal of some poets and writers). The very nature of the substances which are reformulated – language and the daily world of action and common sense values – does not allow that discourse's total reduction to elements lacking or completely losing their semantic values, although it does allow its *reformulation*, that is, its integration into a semiotic element different from the sign of common discourse (dominated by its referential function, which makes it very practical and useful in daily communication): the *aesthetic symbol*, a discursive conformation that is particularly susceptible of various possible interpretations.

"El Hombre" obliges us to bracket the linearity of time as it is experienced, as well as the causal-logical relation of actions; the said linearity becomes the line of its break with realistic narration, and aesthetic *praxis*, as far as the reader is concerned, is restricted fundamentally to responding to the required intentionality.

Something different happens to us upon reading the remarkable novel *Ensayo sobre la Ceguera* by the Portuguese writer José Saramago. Here *diégesis* is not a major problem. We may say that in the narrative sequences the plot matches the fable. The relationship of the events is linear without breaks in continuity that demand more concentrated activity from the reader of the narration. However, the narrative's programs become unusual with respect to those of common sense, since the novel narrates precisely a particularly exceptional human situation, the progressive and actual blindness of humanity. The characters have to confront situations in the face of an uncommon world, whose axiological and semantic configurations are also very strange – "Dantesque," we might say – with a metaphor that would run the risk of "literalizing," hiding what the possible world of the novel reveals when instituting it. Aesthetic *praxis* here focuses on the constitution of this unusual world and of the narrative programs that the characters have to invent and develop to face it. Thus, we are, as receptors of the narration, obliged to bracket the values of the daily world, the programs of the roles the world offers to its members (husband, concubine, friend of the family, et cetera), as they have been marginalized by the same discourse. The world in which blind persons act and move is "another" world, totally different from the one that surrounds us; also the world of their actions is different. If we succeed in updating all the virtualities that the text displays in the diverse configurations and relationships among its actors, we can the more successfully integrate the symbolic value of what the implicit author calls an "essay" (with the entire denotative and connotative value that this term has as opposed to "narration" or "literary story") on human blindness, instituted in the discourse.

*The Unconsoled*, the English writer's novel, also presents an uncommon world, albeit one less spectacular and tragic than that of the *Ensayo sobre la Ceguera*, and its stress lies on an almost total break with the expectations of the narrative programs of its characters, even in uncommon situations, since these often appear embedded within other uncommon situations. In addition, the semantic value of the actors, as well as their narrative identification as characters, is always destroyed, either because they are not given a full semantic dimension, or because their behavior neither denotes nor connotes a common semantic dimension. The bellboy does not play the role of Ryder's father-in-law at any moment, nor does Ryder make it evident, through his actions and language, that he is the son-in-law of the famous orchestra director, who seemingly arrives in a town where nobody knows him, but where he not only has a wife but also a child. The reader of the narration has to follow the actions of the characters, considering all the time that they may suddenly change: A man, seemingly strong and firm, may move in a way that reveals him to be an invalid with an orthopedic leg. Everything is possible, though we are not in a wonderland, but in a "common" and anodyne one. As in *The Castle* by Kafka, Ishiguro's novel ends with the narrative program totally dissolved, without a finished realization in the realistic sense, as if the narrator had forgotten it. (This does not happen in the novel by Saramago, where in spite of the chaos and total confusion that prevail in a world inhabited by the unseeing, the narrative program of the focused upon group's survival does not vanish in a nebulous complete oblivion.)

In conclusion, let it be said that with all of these works, reading, as the *praxis* that builds the text, leads above all to the formation of a unique semantic universe all its own, one irreducible to common, daily experience. Each of the discourses sets up its own world, autonomous in this sense, with its organization of characteristic, unalterable, semiotic values, although the establishment of a symbol proper to each discourse is accomplished thanks to estrangements of greater or lesser magnitude regarding the world of common sense and to ungrammatical utterances and semantic distortions of the discourse of common language. The adjustment of its elements has as its discursive intention the constitution of a characteristic world of meanings thanks to the same alteration, whether of the values of language or the values of common sense. This adjustment is carried out through conventional narrative procedures that converge, as well as through the particular codes of each manifestation, with all working toward the establishment of an appropriate value, one different from the sign, which we call *symbol*. For that reason, our interpretation of an aesthetic discourse should not become independent from the symbol by casting it into oblivion as a simple means of communication, for it "lives" only within the discourse and because of the aesthetic discourse in question, and it is precisely that which provides food for thought, leading to aesthetic reflection.

## 2. THE *EIDETIC* REDUCTION AND AESTHETIC EFFECT; THE SECOND UNDERSTANDING

The great concepts of the idealistic philosophical tradition bear a burden that corresponds not only to the diachrony of their development, or better, to the inertia that has maintained them through centuries and different cultures, but also to the inertia of common sense, which they were able to suffuse. Hence, concepts such as *essence, spirit* or *spiritual, idea* and *concept* itself cannot be conceived without a heavy ideological charge which confers notes of transcendence, universality, permanence and a certain kind of eternity or infinity on them as opposed to the finitude and limitation of their contrasting concepts, and of what seems to be the characteristic of the material, everyday world. Husserl himself does not make such a radical renovation of the philosophy that he conceives the *eidos* within the phenomenological reduction that he proposes, as something without pretense to universality, absolute rationality and eternity; although this *eidos* may be of the thing itself, this very thing deserves consideration or phenomenological description where it touches the universal and transcendental metaphysical realm, when we have really stopped considering it in its finitude and limitation. However, even as we conceive that the transcendental subject does not demand identification with the radicalized *Cartesian ego*, but with the exercise and assumption of a social subject that manifests itself *in* and *for* the world that constitutes it, of which we are a part thanks to our interiorization of it, in the same way we think that it is possible to propose a constitutive *eidos* (or essence) as the characteristic core of a discourse, of a practice or of a work in general, without that meaningful and distinctive core becoming established as a thing (in-itself) outside the sociocultural *praxis* in which it acquires its value and standing. The people who synthesize a sociocultural reality are able to interiorize it and then to conceptually represent what characterizes it as its more or less complex constitutive core – the *eidos* of a discourse, of a work, even of a "thing" that has a particular meaning only within that manifestation and with regard to manifestation. In aesthetic discourse this *eidos* may lose its nuclear and characteristic elements in the course of history to a point that it can no longer offer the significance that it used to have in a given discourse: a literary work, such as an epistle, possessing aesthetic and essayistic values that are nuclear to it will eventually retain only the value of being an informational discourse. On the other hand, thanks to technology – something that *belongs* to man, or more specifically, to the *world* that man constitutes and, therefore, *also* constitutes him – the great aesthetic manifestations of photography, the cinema and televised serials have arisen. These should be dealt with by taking into account their own *eidoi*.

We think that the decantating effected by the *epoché* of the narrative-literary discourse enables us to apprehend and to intuit a core that characterizes said discourse, its particular conformation (an intentional and premeditated distancing from everyday discourse), its reformulation of the elements it "takes" as substantiations of expression and content in new forms of expressing content, characteristic only of narrative manifestations – at the first level, a particular *diégesis* that does not correspond to the "history" of our actions in the daily world, where causality and linearity are manifest. We are also enabled to discover that reformulation leads this discourse to the presentation of narrative programs and spatial semantic configurations, or characters, that are upheld not by their referential value or rather, their relationship with the world of everyday sense, but precisely by themselves. For this reason, trying first to identify what they might represent in our daily world and then to find their value, their own *eidos*, is not a good reading technique. Art, any art, is not an illustration, reflexion or reproduction of the values of language or of the daily world as such, or as they are "represented" in instances that would be incomprehensible without the aid of having the previously existent everyday world as a point of reference. Both the actions and the configurations, as well as something that we did not mention previously, namely, the *characters* (the individuals who assume or suffer from the actions of other individuals on their own plane) are *discursive values*, elements that weave – and, therefore, intimately belong to, in their very semiotic constitution – the *text*, which constitutes a possible world that demands to be considered in itself, without reducing it to another world that would turn out to be its explanation, in order to definitively constitute its meaning: the meaning of the aesthetic discourse is, in the first place, given by its internal elements as is generally the case with all discourses. Furthermore, these discourses do not "live on their own," they do not appear in isolation as islets that do not communicate with one another in the "system of systems" that the living culture of a society constitutes. This *transcendence* of the aesthetic discourse is measured in its full importance within what we call the second understanding: an aesthetic intuition of its meaningful character that arises as supported by analysis and interpretation (*explanation*).

## 3. AESTHETIC INTUITION AND THE VALUE OF THE SYMBOL

The intuition proposed by Husserl is only exercised by interiorizing the "thing *in person*," that is, through the contact we establish with its "self-sameness" – through "intimacy," if we like; this is the capacity by which human beings penetrate into the very sense of a "bundle of perceptions" and confer on it unity and persistence as soon as we can refer to it, either as a referent or as a primary instance which allows mediation, in our case, of an aesthetic type. The discourse here is not a set of statements connected in a chain, nor by any means a conglomerate of words, but a unit with full sense moved by a precise *intentionality* in a given culture. The aesthetic discourse establishes its sense as a value constituted by a characteristic intentionality, precisely that of producing an *aesthetic effect* in the receptor. Moreover, the aesthetic symbol compels this particular reception of the aesthetic discourse; just as there is no language without symbols, similarly, there is no narrative discourse (story, novelette, novel) without the establishment of a proper element by that same discourse providing it with unity and meaning within the network of the discursive relationships of a culture, thus giving it its distinctive value and supporting its communicative function. We are again in a relationship of mutual dependence, a tense dependence which is maintained and supported by a particular dialectic, undetected by a third party, in an instance that overcomes the tension.

In our book *Análisis e Interpretación del Discurso Narrativo-Literario*, we considered the particular semiotic constitution of the literary symbol that reformulates the potential for meaning of a language and of common sense or the daily world. Both these particular and ordinary semiotics "come in," and are taken into account by the narrative-literary discourse as substantiations of their ways of expression and their content; but, when submitting these to new relations that precisely establish the new forms of expression of content, which is characteristic of this particular (literary) semiotic, revelations of sense arise that do not coincide with the ones offered by the first semiotics, those of a language and of the common sense world. This reformulation establishes its own value, as we said before: the *literary aesthetic symbol*, which though derived from the signs of primary semiotics is not reducible to them. If we can say in a general way that semiotics operates in discourses that are real units within a cultural network, and that the network of said discourses constitutes the sociocultural world in which we live, then discourses form integrating and meaningful parts of our sociocultural reality; aesthetic discourses, however, do so in a manner different from those of the two primary discourses (a language and the world of common sense). They function differently than does practical daily communication but without interfering with it: they have their own *space* or *domain*.

In our work *Literatura y Realidad* we paid special attention to the ontological law of the literary aesthetic symbol, particularly the narrative, in relation to the constitution of our sociocultural world. For that reason, we dedicated a great part of the book to the lucubration of what meaning those *semioses*, those *praxes* of discursive organizations that do not have a practical function (neither when they are used for such a purpose in the daily world, nor when we expect from them a greater dominance or control of our immediate surroundings) have for us, but that man, as such, ceaselessly realizes again and again and in different ways.

Let us now expound on two pairs of characteristics that seem distinctive over against the other symbolic discourses – myth, religion and ritual in general – pairs that each have an antinomic constitution, a kind of irreducible aporia, but one whose tense dialectic confrontation becomes for us a source of the very sense of and matter for thought on the twinned characteristics and, therefore, can not be omitted or disregarded.

*i. Uniqueness/plurality:* Each particular narrative manifestation is constituted as a singular, unique discourse. Even when two stories have an apparently common theme and even should they be by the same author-person, we encounter *two* different manifestations that form two distinct – even opposite – possible worlds. A single motive that circulates from one narration to another, enters into different relationships and acquires a different meaning. For that reason, there is nothing more deceptive and fallacious than a "thematic" anthology: for example, what death or infancy symbolizes in two stories may be totally different, even contradictorily different. This uniqueness gives to a theme its identity, notwithstanding the tradition and its multiple interpretations.

*Plurality:* This last statement characterizing uniqueness brings up the presence of an antinomic element: there is no constraint in aesthetic discourse that imposes a *one-and-only* interpretation. The very value of the semiotic element it establishes (and constitutes with regard to aesthetic discourse), *the aesthetic symbol*, is not decoded in a single semantic direction. Being polysemic, it not only allows but *imposes* multiple readings or interpretations, as long as they do not *use* the text and are not aberrant. Moreover, although the aesthetic "schools" give the impression of being "chapels," as it were, small churches, this is so only with respect to the relationship each with the others, given the application it takes to establish their presence in a social context. The discursive manifestations of these aesthetic confraternities do not impose any aesthetic dogma on the model reader: *in art no orthodoxy or dogmas are possible*. Perhaps this is the artistic discourse's most outstanding distinction with respect to the religious symbol, which has its "official interpreters," although we can also say regarding its uniqueness or particularity that it is different from religious discourse: that each Catholic Mass be celebrated in a particular way does not imply that a variation in one of its elements or components, for example, the priest's race or the configuration of the church, makes it eidetically different from other Masses; this would not be so unless the introduction of truly foreign elements into the discourse at a key moment of the ritual or in some of its essential matter should occur.

*ii. Aesthetic distinction / the reformulation of that which is foreign to the aesthetic*. To characterize the first aporia, we will gloss a proposal of Gadamer's: The work of art enjoys an abstract yield. It obliges us to leave aside everything constituting the foundation of a work, such as its original and vital context, the functions of certain elements to which it has recourse (religion, customs, rituals, et cetera). In this way it becomes evident as "a work of pure art." The abstraction compelled by an aesthetic work is positive for its constitution. "It discovers and allows thc self-existence of what constitutes a work of pure art." Although a narrative discourse may include in its semantic configuration many other discourses, or parts or elements of the same, it does so in subordination to its unique aesthetic intentionality, to its "aesthetic distinction."

As in the case of the previous antinomy, the last part of the first aporia already states the presence of the second. Properly speaking, this discourse is characterized by reformulating, "appropriating," under its own parameters and discursive rules, subject to its aesthetic intentionality, other manifestations: a political or religious discourse, a custom, et cetera.

To finish this brief essay, it seems worthwhile to quote some observations of a philosopher whose cultural proximity to phenomenology in its original version is unquestionable, namely, Hans-Georg Gadamer, a disciple of Heidegger, whose preceptor in turn was Husserl:

> Basically it is to phenomenological criticism of nineteenth-century psychology and epistemology that we owe our liberation from the concepts that prevented an appropriate understanding of aesthetic being. The critique has shown the erroneousness of all attempts to conceive the mode of being of the aesthetics in terms of the experience of reality, and as a modification of it. All such ideas as imitation, appearance, irreality, illusion, magic, dream, assume that art is related to something different from itself: real being. But the phenomenological return to aesthetic experience (Erfahrung) teaches us that the latter does not think in terms of this relationship but, rather, regards what it experiences as genuine truth. Correlatively, the nature of aesthetic experience is such that it cannot be disappointed by any more genuine experience of reality.[1]

*Universidad Veracruzana*

NOTES

[1] *Truth and Method* (New York: Continuum. 1997), pp. 83–84. *Note bene*: the English translation is not as precise as the Spanish edition's translation of Gadamer's thought. For these reason we offer another English rendering of this passage: "Substantially, the liberation from those concepts that constituted the main obstacle to an adequate understanding of the aesthetic essence came about owing to phenomenological criticism of the psychology and epistemology of the nineteenth century. Through this criticism it was possible to demonstrate the folly of trying to grasp the essence of aesthetics by starting from the experience of reality and conceiving the former as a modification of the latter. Concepts such as imitation, appearance, unfulfillment, illusion, enchantment, fantasy, presuppose reference to an authentic essence from which the aesthetic essence would be different. In contrast, the phenomenological return to the aesthetic experience shows that the latter in no way reasons from such a frame of reference but, on the contrary, perceives the authentic truth in what it experiences. Owing to its very essence,

therefore, the aesthetic experience cannot feel disappointment over having undergone a more authentic experience of reality." (Consult *Verdad y método* [Salamanca: Sigueme, 1977], pp. 123–124.)

GERALD NYENHUIS

# LOS CUASI-JUICIOS

## LAS "CUASI-MODIFICACIONES" DE LAS ORACIONES EN UN TEXTO LITERARIO

Lo que acabamos de decir en relación con los varios tipos de conexiones que afectan el carácter de la nueva entidad – poema, cuento, etc. – nos sirven de pista para indicar que existen algunos índices reconocibles para distinguir un texto literario de otros tipos de escrituras, de un artículo "científico", por ejemplo. Las oraciones que se hallan en una obra literaria no son puramente declarativas ni son afirmaciones genuinas. Las afirmaciones en una obra literaria tienen un carácter solamente "cuasi-afirmativa", es decir, tienen la apariencia externa de afirmaciones, aunque no son intencionadas como juicios proposicionales genuinos. Antes de procurar explicar la teoría de Ingarden, tenemos que repasar algunos temas que vimos antes.

Una proposición es un juicio por el cual afirmamos (o negamos) un conjunto de circunstancias, expresándolo como una oración declarativa. Cada afirmación propone decir la verdad, pero hay diferencias entre proposiciones en cuanto al grado de certidumbre con que son afirmadas, en conformidad con la relación entre el intencional conjunto de circunstancias y uno objetivamente existente. Hablamos de una aseveración cuando el conjunto de circunstancias que se afirma está intencionado a ser idéntico con un objetivamente existente conjunto de circunstancias. Una afirmación, o una aseveración, como empleamos estos términos, se presenta como la verdad, aun cuando el juicio propuesto resulta ser erróneo. El sentido de una afirmación genuina se deriva de su propósito de decir la verdad, y no de la prueba disponible para mostrar que los dos conjuntos de circunstancias sean idénticos – el puramente intencional y el objetivamente existente. En este trabajo, entonces, no vamos a preocuparnos con la verificación de las aseveraciones, sino con las distintas formas de afirmar la verdad. Una afirmación siempre se expresa en forma de una oración declarativa, pero tenemos que recordar que la oración declarativa no necesariamente es la verdad, aunque la forma en que aparece sea la de una afirmación. Por eso, seria razonable preguntarnos si hay pautas que pudieran ayudarnos a reconocer la diferencia entre una oración declarativa que es un vehículo de una genuina afirmación y una que no la sea. ¿Cuál, entonces, es el sentido de una oración puramente declarativa y de las oraciones declarativa que llegan a ser vehiculos de los varios tipos de afirmaciones?



*A.-T. Tymieniecka (ed.), Analecta Husserliana LXXVI*, 183–193.

Sabemos que el correlato de cada oración es puramente intencional. Sabemos también que una oración no es una formación independiente y que normalmente está en relación con otras oraciones. Aun cuando parece ser aislada del contexto, casi siempre es pertinente a la situación en que fue emitida. Sin embargo, cuando hacemos una oración declarativa pura, la hacemos sin tomar en cuenta referencia alguna, sin propósito referencial. No tiene otra functión que la de ejemplificar las funciones de los sentidos verbales dentro de las unidades de sentido, y el carácter puramente intencional de su correlato. Además, podemos notar la diferencia entre la función de un verbo cuando ocurre en una oración puramente declarativa, y su función en una oración declarativa que propone ser, al mismo tiempo, un vehículo de una afirmación en algún contexto. La diferencia es uno de sentido; que, como podemos ver, depende del sentido que intencionamos para conferirlo a una oración, a la luz del propósito que debe servir. Finalmente, también sabemos que el factor intencional de un sentido verbal apunta siempre hacia el correlato puramente intencional que proyecta. Pero, si, en una situación dada, una oración declarativa (El teléfono está sobre mi escritorio.) se emplea para afirmar un independientemente existente conjunto objetivo de circunstancias, entonces intencionamos también el factor intencional direccional de "el teléfono" a extenderse más allá del correlato puramente intencional y alcanzar también el objeto independiente, pero también intencional. Este es un "momento"[1] en el sentido de una afirmación genuina.

Ahora bien ¿qué pasa cuando, debido a esta nueva extensión del factor direccional, hacemos referir la expresión nominal a un objeto independiente referencial?

Si consideramos solamente el conjunto de circunstancias puramente intencional del correlato de la oración declarativa, sabemos que el correlato como una totalidad queda puramente intencional aunque el carácter existencial del objeto y del conjunto de circunstancias es del tipo de objetos reales y de los independientes conjuntos de circunstancias. Sin embargo, cuando el factor intencional direccional de "el teléfono" se intenciona para alcanzar el objeto real, independiente y referencial, entonces, la capacidad del total conjunto de circunstancias, dentro del cual el correlato de la oración pertenece a la puramente intencional proyección de la expresión "el teléfono", con eso está transpuesto en la esfera existencialmente independiente del objeto real referencial y así intencionada para ser identificada con el conjunto de circunstancias que pertenece a este objeto. Si queremos evaluar y entender esta transposición[2], tenemos que considerar varios puntos.

La caracterización existencial no trae de por sí la transposición. Meramente asigna a los objetos y a los conjuntos de circunstancias la característica de ser un cierto tipo de objeto real o de conjuntos de circunstancias independientes. El acto intencional de transponer "intenciona" que el puramente intencional objeto/correlato del sentido verbal (que es el sujeto de una oración) llegue a ser, junto con el conjunto de circunstancias que pertenece e él, una presentación revelador de del objeto independiente referencial y ser idéntico con él. Esto lo hace porque el factor intencional se extiende para alcanzar, al mismo tiempo, el objeto puramente intencional presentado y el objeto independiente referencial. Los constituyentes y los otros determinantes dentro de la "capacidad" del objeto puramente intencional presentado, entonces, son intencionados para ser esencialmente identificados con aquellos del objeto independiente.

Como consecuencia de ello, la capacidad de este puramente intencional conjunto de circunstancias presentado, que es del correlato de la oración, y que pertenece al objeto puramente intencional, también es intencionada para ser esencialmente idéntico con el conjunto de circunstancias independiente que pertenece al objeto independiente. Obviamente, el conjunto de circunstancias del correlato de la oración nunca deja de ser intencional durante el curso de transposición, pero su capacidad es intencionada para conformarse al conjunto de circunstancias independiente de tal grado que el puramente intencional independiente conjunto de circunstancias presentado y el conjunto de circunstancias independiente referencial deben ser considerados idénticos, y por eso ser identificados como lo mismo.

Debido a esta identificación, el conjunto de circunstancias puramente intencional del correlato de la oración llega a ser como transparente. Nuestra atención está clavada en el conjunto de circunstancias independiente referencial mientras el puramente intencional desaparece de la vista.Tendemos a dejar a un lado la presentación intermediaria por la cual un objeto o un conjunto de circunstancias se revela. Tenemos que recalcar, sin embargo, sobre el hecho de que el conjunto de circunstancias del correlato de la oración persiste en su intencionalidad pura a pesar de su conformidad al conjunto de circunstancias independiente referencial con que esta identificado.

Si necesitamos más prueba de ello, podemos mencionar que la pura intencionalidad de un correlato de la oración se ve fácilmente si consideramos algunas oraciones afirmativas, como "cada cuerpo es extendido" o "este teléfono está en el escritorio". En un conjunto objetivo de circunstancias no hay cuerpo que es "cada", ni teléfono que es "este". Solamente nuestra intencionalidad, que se dirige hacia un objeto, que produce tales distinciones. Pero estas distinciones desaparecen de la vista porque los factores direccionales de cualquier de estas palabras como también como también los de "cuerpo" y "teléfono", etc., convergen en el correlato del sentido verbal nominal. ¿A qué grado podemos hablar, entonces, de "conformidad" y de "identidad" esencial entre el conjunto de circunstancias presentado puramente intencional y uno objetivo referencial? Lo podemos hacer, como ya hemos notado, con respecto a los constituyentes y determinantes formales y materiales dentro de la capacidad de objetos y de los conjuntos de circunstancias que puedan ser transpuestos. No obstante, tenemos que excluir la existencia puramente intencional y las características que no puede transponerse en la realidad independiente.

Claro, una afirmación genuina no puede cumplir con su función de transponer sin la función que cumple el factor direccional intencional que intenciona alcanzar el blanco objetivo. Tampoco lo puede hacer sin la conformidad, basada en la identidad esencial, de la capacidad del conjunto de circunstancias puramente intencional con el conjunto de circunstancias objetivo referencial.

Ahora bien, sabemos que no hay una identidad esencial de dos entidades a menos que las dos caen dentro de la misma idea general. Consecuentemente, la transposición puede ocurrir sólo si la idea general del conjunto de circunstancias se hace particular en el correlato puramente intencional de la oración y en la realidad objetiva. Para decirlo de otra manera: si el momento cualitativo constante de esta idea general esté concretizado en el correlato y realizado en la realidad objetivo, la transposición puede ocurrir. Cuando esta identidad esencial se aprehende y así la transposición se logra, la oración afirmativa cumple con los requisitos para aseverar que es verdadera. Sin embargo, para que la oración se aprehenda como afirmativa, la función del factor intencional direccional tiene que inducir esta identificación de la puramente intencional con el conjunto objetivo de circunstancias.

En cada oración declarativa el predicado cumple con la función de desdoblar con el verbo el conjunto de circunstancias. El predicado es, en este sentido, el "núcleo verbal", sin el cual no hay conjunto de circunstancais, y es el desdoblamiento que hace el verbo lo que hace posible la "declaración". La oración declarativa sigue cumpliendo con esta misma función aun cuando la oración declarativa es intencionada como un vehículo de una afirmación genuina. También, como resultado de la afirmación intencionada, el predicado tiene que cumplir con otra función.

Debido a que reclama ser la verdad, la afirmación genuina confiere al predicado la función que podemos llamar su "posicionamiento existencial". El predicado procura afirmar que el conjunto de circunstancias transpuesta existe verdaderamente como un hecho. El reclamo de ser la verdad que hace la oración afirmativo descanso sobre su intención de transponer la capacidad de un conjunto de circunstancias puramente intencional a la esfera de lo independientemente existencial. Depende también del efecto en la transposición del factor intencional direccional, y asimismo depende del "posicionamiento existencial" establecido por el predicado.

Ahora, dirigimos nuestra atención a la oración afirmativa modificada, tal como ocurre en el texto literario. Esta modificación es el efecto del hecho de que en la obra literaria la afirmación es, en esencia, una *cuasi-afirmación*, un *cuasi-juicio*. Debemos hacer énfasis en el hecho de que una transposición y un posicionamiento existencial genuinos tienen que estar presentes en cada afirmación genuina. No obstante, como vamos a ver, hay ocasiones en las cuales un cierto tipo de transposición no involucra un posicionamiento existencial genuino.

Ingarden distingue varios distintos tipos de oraciones cuasi-afirmativas. Las mide en una escala, de acuerdo con su aproximación a oraciones declarativas puras, por un lado, y con las afirmaciones genuinas, por el otro. El correlato de una oración declarativa pura no tiene, como sabemos, la intencionalidad derivada que intenciona una afirmación que reclama ser la verdad acerca de un conjunto independiente de circunstancias. El factor intencional direccional de un sentido verbal nominal no tiene la intención de apuntar hacia un conjunto independiente de circunstancias. Por lo tanto, el conjunto de circunstancias que pertenece al objeto puramente intencional (que es el sujeto de la oración) no está intencionado para ser transpuesta en la esfera independiente de existencia, y su conformidad con un conjunto independiente de circunstancias no se intenciona. Aunque el conjunto de circunstancias proyectado se caracteriza como del tipo de conjuntos independientes de circunstancias, la oración puramente declarativa en si no reclama ser la verdad, y su predicado no tiene un posicionamiento existencial.

Hay textos literarios, del tipo que podemos llamar "puramente ficticios" (o "pura ficción, que no hacen pretensión alguna de que sus proyecciones hagan referencia a la realidad externa. No obstante, en ellos las oraciones "cuasi-afirmativas" crean una ilusión de realidad. Esto se logra por una transposición simulada, que no se efectúa por los factores intencionales direccionales sino por unos ciertos índices materiales, tal como el colocar el conjunto de circunstancias en una ubicación específica. Pero, aun esta ilusión se mantiene débil porque los conjuntos de circunstancias en estas obras no pretenden alcanzar conformidad con el mundo "real". Retienen, entonces, su carácter de "puramente intencional" y apenas logran esconder el hecho de que son proyecciones de pura intencionalidad. Por ende, no están transpuestos a la esfera de existencia independiente, más bien, están transpuestos en su propio mundo, el mundo de una realidad ilusoria en el cual están "puestos tal como son", como puramente intencionales, y ahí se quedan, por así decirlo, suspendidos.

Se hallan las transposiciones y las "posicionamientos" también en los textos en los cuales los correlatos de las oraciones cuasi-afirmativas tienen que adaptarse, en general, a las características sociales, políticas, económicas y ambientales de ciertas épocas históricas. Aquí tampoco se logra una conformidad a los individuos y a los conjuntos de circunstancias porque la transposición, tal como es, no está efectuado por los factores direccionales intencionales. Sin embargo, hay aquí, por lo menos, una perceptible adaptación a los aspectos típicos de un periodo con los cuales están dotados los conjuntos de circunstancias que en todos los otros sentidos son puramente intencionales. Datos y detalles individuals, tales como nombres de personas, lugares, movimientos y periodos, se introducen en el texto para otorgar verosimilitud a esta simulada transposición hacia la realidad. Esta "realidad", producida por la intencionada adaptación a los aspectos, tipicos y generales, no es igual con la realidad de los conjuntos de circunstancias a los cuales los conjuntos de circunstancias de los correlatos de las oraciones genuinamente afirmativas quieren conformar y con que pueden ser identificados.

No es difícil reconocer el carácter cuasi-afirmativo de las oraciones en los textos literarios, de los dos tipos que hemos mencionado, y ver que la transposición y el posicionamiento son meramente simulados. Pero, ¿cómo distinguimos el sentido de las oraciones cuasi-afirmativas en los textos literarios que se presenta como "históricos" del que es de las oraciones afirmativas genuinas? Este caso se presenta en las obras de "erudición", en que las oraciones "cuasi-afirmativas" reclaman ser fieles representaciones de personas y situaciones abstraidas de la historia.

Aquí tenemos que comentar un tema, muy relacionado, que Ingarden introduce en su comentario sobre el siguiente estrato, el de los objetos (re)presentados. Pero, antes tenemos que repasar algunos de los pasos que hemos tomado y ampliar nuestros conceptos. Sabemos que la unidad de sentido de una oración declarativa proyecta un conjunto de circunstancias puramente intencional. En el desdoblamiento de esta proyección encontramos una "presentación" (o representación)[3] que revela los elementos constitutivos y los determinantes de un objeto que es el sujeto de la oración. Cuando intencionamos una oración para ser genuinamente afirmativa, esto quiere decir que el factor intencional direccional del sentido verbal (que es el sujeto de la oración) alcanzará no solamente el correlato intencional de aquella palabra sino también el objeto referencial cuya existencia está en la esfera de la realidad. Transponemos, entonces, en esta esfera la capacidad del objeto puramente intencional y del conjuntos de circunstancias que pertenece a él.

Durante del proceso de afirmación, queremos también que el predicado cumpla con su función de posicionar la capacidad transpuesta del conjunto de circunstancias en la realidad como un dato, un hecho. Podemos decir que el propósito del desdoblamiento del conjunto de circunstancias proyectado por la oración declarativa es la (re)presentación que revela un objeto puramente intencional por medio de un conjunto de circunstancias.

Cuando intencionamos una oración declarativa como una verdadera afirmación, la transponemos y posicionamos su objeto puramente intencional y su conjunto de circunstancias en la esfera independiente de la realidad de la manera que ya hemos descrito. Queremos que la capacidad de el conjunto de circunstancias transpuesto sea idéntica con un conjunto de circunstancias independiente. Cuando aprehendemos aquella identidad, identificamos los dos conjuntos de circunstancias. Por medio de esta identificación, el conjunto de circunstancias puramente intencional y el conjunto de circunstancias independiente ambos son directamente presentados. De hecho, debido a la función del factor direccional intencional en una afirmación genuina, el conjunto de circunstancias independiente muevo, por así decirlo, al primer plano y esconde la capacidad del conjunto de circunstancias puramente intencional junto con la intencionalidad que lo proyecta. Esto quiere decir que, debido a esta identificación, el conjunto de circunstancias puramente intencional en una afirmación genuina no se aprehende como una producción de un modelo individual independiente,ni, por ende, es aprehendido como una representación del conjunto de circunstancias independiente que le pertenece.

En un texto, literario, las oraciones declarativas no son, ya sabemos, intencionadas como afirmaciones genuinas. También, como en todas las oraciones declarativas, sus correlatos son (re)presentaciones de objetos y conjuntos de circunstancias puramente intencionales. Al mismo tiempo, sin embargo, dependiendo del tipo de obra literaria percibimos varios implícitos reclamos de ser la verdad. En las obras que mencionamos primero y las lamamos "puramente ficticias", la presentación de los conjuntos de circunstancias puramente intencionales está en una relación muy tenue con ciertos indicios materiales en los cuales están transpuestos (e insertos o posicionados), aunque sólo aparentemente. El carácter cuasi-afirmativo de estas oraciones claramente preserva la pura intencionalidad de sus correlatos, pues aqui ciertamente no hay pretensión de que el mundo creado en la obra literaria sea, a la vez, una presentación del mundo real e independiente.

De la misma manera, en las obras del segundo tipo (obras que *se adaptan* a ciertas circunstancias o periodos) las presentaciones claramente no están intencionadas como simultaneas (re)presentaciones de objetos indivduales independientes y sus conjuntos de circunstancias. Sin embargo, están intencionadas como representativas de tipos generales de objetos y conjuntos de circunstancias, en ciertas épocas y lugares. Así que, aquí también, las oraciones cuasi-afirmativas proyectan un mundo puramente intencional cuya "transposición" existe solamente en apariencia y es el resultado de su semejanza con algunos aspectos generales de la realidad objetiva. Podemos concluir que en ninguno de los dos tipos encontramos una transposición efectuada por los factores direccionales intencionales,y por eso, tampoco notamos una presentación simultanea de los conjuntos de circunstancias puramente intencionales y los conjuntos de circunstancias independientes; tampoco pasan desapercibidos los conjuntos de circunstancias puramente intencionales ya que no se hacen transparentes por medio de la identificación mencionada arriba. Como consecuencia,las características de las afirmaciones genuinas faltan y las oraciones son por eso meramente cuasi-afirmaciones.

Ponemos nuestra atención ahora en las obras literarias que se llaman "históricas", o sea, las que se presentan como obras históricas (o sociológicas, etc.). En estas obras las oraciones cuasi-afirmativas se presentan como si fuesen ser afirmaciones genuinas. En el texto de obras de este tipo, *algunos* de los objetos individuales y los conjuntos de circunstancias proyectados por *algunos* de las oraciones son intencionados para corresponder con *algunos*, pero no todos, de los objetos individuales y los conjuntos de circunstancias independientes (casi siempre del pasado) de tal manera que logran cierta conformidad con ellos. Ya hemos visto que una presentación de un conjunto de circunstancias, por lograr una conformidad tan fiel (de su capacidad) a un conjunto de circunstancias independiente que virtualmente se idéntico con el. Esta es una de las condiciones para que una oración declarativa llegue a ser afirmativa.

La otra condición es que, en base a su identidad esencial, los dos conjuntos de circunstancias sean identificados a través del alcance extendido del factor intencional direccional. Debido a que la primera de las condiciones, mencionada arriba, en muchas oraciones se cumplen, todas las oraciones en las obras que pretenden ser "históricas" dejan la impresión de ser afirmaciones genuinas. Sin embargo, solamente algunos de los factores intencionales direccionales tienen el propósito de alcanzar los individuales objetos independientes. No hay, entonces, una "transposición" de todos los conjuntos de circunstancias proyectados. No hay, como consecuencia, una identificación en todos los casos. Algunos de los correlatos, entonces, si presentan conjuntos de circunstancias puramente intencionales que son virtualmente idénticos con algunos correspondientes conjuntos de circunstancias, pero el contexto normalmente excluye una simultanea presentación de todos los conjuntos de circunstancias proyectados por el texto.

Por esta razón, el mundo presentado en el texto puede ser sólo una representación del mundo independiente. Podemos decir que en la afirmaciones genuinas intencional el virtualmente idéntico conjunto de circunstancias puramente intencional y el independiente conjunto de circunstancias referencial se identifican mutuamente debido a la función crucial de todos los factores intencionales direccionales. Los conjuntos de circunstancias de los correlatos se hacen transparentes, mientras los conjuntos de circunstancias independientes mudan conspicuamente en la presentación.

En las cuasi-afirmaciones, sin embargo, los conjuntos de circunstancias puramente intencionales se presenten prominentes, y no son intencionados como idénticos con los con los conjuntos de circunstancias independientes, porque el contexto no justifica la pretensión de que todos los factores intencionales direccionales tiene la intención de extenderse más allá de los correlatos puramente intencionales de los sentidos verbales nominales. Las presentaciones de los correlatos de las oraciones cuasi-afirmativas en las obras históricas, entonces, son meras reproducciones de algunos modelos individuales independientes y representaciones de sus independientes conjuntos de circunstancias que se hallan en la realidad. Ya que falta una identificación consistente, a pesar de la realidad simulada, los correlatos de las oraciones cuasi-afirmativas crean su propio mundo, separado de la realidad independiente. Ingarden piensa que las cuasi-afirmaciones del texto literario es lo que principalmente, aunque no exclusivamente crea para nosotros la ilusión de la realidad.

Hemos tenido mucho cuidado de poner las oraciones cuasi-afirmativas en el texto literario, más bien que en la obra literaria. Ahora tenemos que hacer otra distinción que posiblemente pueda aclarar mejor el punto. Ingarden llama "texto" el complejo de oraciones que presenta el mundo puramente intencional de la obra literaria. Cuando esta distinción se emplea, podemos llamar el texto el "texto presentativo". En el caso de una narración impersonal ("objetiva") el narrador es anónimo y el "texto presentativo" narrado presenta un mundo por medio de lo que podamos llamar una "proyección singular" por el intermediario de las oraciones cuasi-afirmativas. En este texto presentativo el narrador no se da, no se proyecta en el mundo presentado, pero aunque no se proyecta, como veremos más tarde, no está totalmente ausente.

Damos por sentado que en cierto sentido el narrador se da, aunque sin identificación específica. En esta caso, el narrador proyecta un mundo (re)presentado, pero el mismo necesita presentarse por algún texto presentativo (que tiene que ser identificado), a fin de que llegue a ser parte del mundo presentado. El narrador puede ser comparado con uno de los personajes en la obra de una novela; de hecho, puede ser uno de ellos, uno que con todo seriedad hace afirmaciones, preguntas y da ordenes. Sus intervenciones se toman en serio por los otros personajes presentados en el mismo mundo presentado. Tenemos que preguntarnos, entonces, si las oraciones afirmativas, las interrogativas o las imperativas que pronuncia el narrador proyectado (como cualquier otro personaje proyectado en el mundo proyectado) no sean también cuasi-afirmativas, cuasi-interrogativas y cuasi-imperativas.

Vamos a suponer, entonces, que un texto presentativo breve e impersonal, que suele ser sólo implícito, introduce y presenta el narrador. El mismo texto presentativo también proyecta el acto de narrar del narrador: en otras palabras, el narrador y la circunstancia de narrar (a través de su duración) constituyen el primer mundo presentado que es proyectado por aquel texto presentativo. Todas las oraciones presentativas que explícita o implícitamente afirman la presencia del narrador presentado y la presentada circunstancia de narrar son cuasi-afirmativas.

Dentro de este "primer" mundo presentado, el narrador presentado proyecta y presenta la materia narrativa, el "tema". Luego, tenemos en un plano el texto presentativo original (posiblemente sólo implícito) que presente el mundo del narrador y la circunstancia de narrar. Dentro de este "primer" mundo presentado, el narrador proyecta y presenta (o representa) objetos, conjuntos de circunstancias, problemas, ordenes y situaciones que constituyen, en un segundo plano, un "segundo" mundo (re)presentado.

Lo que acabamos de presentar es una doble proyección que resulta en estructuras de *incrustación*. La materia narrada presentada por el narrador está incrustada dentro de la circunstancia de narrar, que, a su vez, está incrustada en la estructura más amplia del texto presentativo original.

La importancia de la doble proyección radica en el hecho de que la proyectada circunstancia de narrar proyecta, al mismo tiempo, otro mundo diferente, un mundo presentado por medio de oraciones afirmativas, interrogativas, imperativas, etc. Cada vez que un texto cumple con la función de proyectar un mundo (re)presentado, en cualquier plano en la estructura de incrustación, con eso cumple con la función de un texto presentativo. Las oraciones que constituyen el texto, entonces, sufren una modificación (o una *cuasi*-modificación) por la cual una oración afirmativa llega a ser cuasi-afirmativa, una oración interrogativa llega a ser cuasi-interrogativa, etc. Cuando las oraciones, de cualquier tipo, no se toman en su función presentativa, sino como si pertenecieran al mundo presentado, donde se toman con toda seriedad, esta llamada "cuasi-modificación" no se realiza dentro del mundo presentado. Para hacer claras las distinciones, podemos llamar estas oraciones afirmaciones "ficticias", interrogaciones "ficticias", etc., ya que sus aparentes afirmaciones, interrogaciones, etc. se toman en serio dentro del mundo presentado. S puede observar muchos ejemplos de ese doble provección, la estructura de incrustación y las modificaciones de oraciones cuando leemos una obra de teatro, como distinta de la obra que vemos en el teatro. Un problema que nos queda es el de la manera en que un texto presentativo proyecta oraciones presentadas.

*Universidad Iberoamericana*

NOTES

[1] Uso "momento" no en el sentido de un corto lapso de tiempo; sino, mas bien, en el sentido de la física, como en "un *momento* de fuerza", o "un *momento* de inercia".

[2] Ingarden usa la palabra alemana "*Hinausversetzung*".

[3] Ingarden usa la palabra alemana "*Darstellung*", que a veces traducimos (re)presentación.

MIHAI PĂSTRĂGUŞ

# ILLUSION AND TRUTH IN THE WORK OF ART

## THE ILLUSION IN THE PSYCHIC PARAMETERS

The concept of *illusion*, very much used but less analysed in its complex content, has been a word that most people use with doubts and with a kind of mystical suspicion, loading it with a lot of negative connotations. First of all, we have to mention that this notion is present, consciously or not, in most human activities, and second, that it covers from an operational point of view a series of psychic activities of the affective and cognitive types. That is why this notion is correlated and articulated with a series of neighbouring notions such as appearance, imagination, representation, perspective, and so forth. By their content and scope, these notions belong either to philosophy or to the arts or technology. The association of illusion with truth can be reviewed in the sphere of the technical-pragmatic sciences, yet not in the sphere of the arts and philosophy. What would cinematography be if we took away the possibility of using techniques of illusion, and what would music or literature be?

An analysis of illusion as to its psycho-physiologic basis, however brief, is demanded, within certain limits. From this point of view, illusion is understood first as an emotion, more or less transient, of the thinking processes, resulting in a distortion of their results (intellectual perceptual images, emotional states and motor reactions). Most often this emotion occurs within normal limits, producing itself due to a combination of objective characteristics of the environment in certain situations (exhaustion, agitation, etc.). Parts of illusions are mostly due to some environmental phenomena that furnish wrong information to the analyser (the mirages that are the result of the phenomenon of total reflection; the "breaking" of an object immersed in water that is the result of refraction; meteorological illusions; the optic illusion the moon presents when on the horizon, so that it is perceived as having a larger diameter than does the moon at its zenith, Aristotle's tactile illusion of perceiving two balls when only one is placed between the extremities of the crossed forefinger and middle finger). Other illusions are due to a psycho-physiological state existent at a certain moment: high emotions, an intense wait for an event, but also some illnesses that in fact produce a fake image which the subject considers true.

Considering the nature and the level of the psychically distorted result, we can distinguish the following types of illusions: perceptive, motor,



*A.-T. Tymieniecka (ed.), Analecta Husserliana LXXVI*, 195–212.

intellectual, emotional, and mixed. The most studied are the *perceptive* ones, and among these there are several types of illusions: *optical, auditory, ponderous, tactile, skin, gustatory, olfactory, mixed.* Among these, *optical illusions* have received most interest, and most representative of this category are ambiguous forms and *optico-geometrical* illusions.

Briefly analysing the psycho-physiological basis of illusion, it is not hard to realize that philosophy and the arts especially are entitled to use the notion of illusion without restriction.

In Indian philosophy, especially with the idealist philosophers of the Sankara School, Maya designated the phenomenal side of the world, of reality, the deceiving appearance, the illusion. In the philosophy of the Upanishads, where the unique reality of Atman, equivalent to the Parmenidean One, is assiduously cultivated, illusion means diversity, i.e., the nonentity of a pluralist world external to Atman. "Only Purusha is this whole world, and what was then and what will be in the future it lasts" (*Rig-Veda*, p. 167), "but the people do not know; they see as real not the trunk of the tree, but what does not exist, the crown of branches covering it" (*Rig-Veda*, p. 167), and the Indian philosopher Sankara, the founder of Vedantic absolute monism *advaita* (9th century) also considered the sensorial world to be a simple appearance, an illusion, Maya. It is not only Eastern philosophy of Indian origin that has had this predilection for Maya; it was also present in another form in a great number of the ancient Greek philosophers: Plato (*The Allegory of the Cave*), Heraclitus, Pythagoras. We may say that the influence of this concept was transmitted and preserved in a most original form by the German philosopher Arthur Schopenhauer in his famous book *Die Welt als Wille Vorstellung* (The World as a Representation).

In this well-known work, the world was reduced to a simple illusion created by the subjective will. The forces of nature that have an important place in his thinking are particular forms of a blind will that evolves from spontaneous elementary forms to conscious ones. Human life, one of these forms, involves an endless effort of personal individualisation and preservation, always followed by disillusion; the human being can only temporarily get out of this state with the help of aesthetic contemplation, especially with the help of music (we can here perceive the influence of Plato's thinking on Ideas). He considers that the world appears differently to us than it is in reality, and that is why it is an illusion wherein the will is reflected as in a mirror, meaning that life accompanies the will as a shadow accompanies the body. His philosophy was an ontology, but more a justification of the necessity of art in human life, in order to give art a top position among the constituents of social life. The idea of appearance or illusion was constantly used as a pretext by the modern philosophers of Germany. Let us remember Schelling with his philosophy of myth, Fichte with the philosophy of the "I," Hegel with his theory of appearance that in its essence is also a theory of illusion after another model. Criticizing Kant, Fichte, and Schelling's idealism, which move the whole variety of the real world, the thing itself, into appearance, Hegel shows that by this transformation no problem is solved. The content of an appearance is given directly. For Hegel an appearance is an objective moment and not a construction, a creation of consciousness. An appearance has a direct presupposition, an independent side as compared to an essence, but it is all the same linked with an essence. We may speak of appearance as an aspect, but on the other hand the appearance is nothingness, a nonentity; the appearance is thus the essence itself, but the essence in a certain determination, in a certain moment, in a certain form and degree. An appearance, says Hegel, is an essence in its negativity, i.e., in its becoming, in the succession of forms and aspects that appear one after another and negate themselves in movement. The "negation" of an appearance, a reflection, is a tough phenomenon, a violent crash, a powerful paradox, a type of movement of the general through the particular, and not a paradox on a precipice as the relationship between the Phenomenon and the Thing in Itself was with Kant. Hegel understood better than Immanuel Kant the relationship between phenomenon and essence, between real and imaginary, between absolute and relative. For Hegel, the objectivity of the Absolute Idea and at the same time its reality are aspects implied only through the mechanism of the dialectic movement. That is why with Hegel everything is possible, as opposed to Kant. In order to justify the limits and the causes of the impossible, Kant framed the antinomies.

If, as we can see, almost all the philosophical systems, explicitly or not, have operated using the category of illusion or one of its substitutes, and if they used it in order to make clear representations or models of the world or of a segment of it, then the term itself should be liberated of negative connotations and more boldly given currency, with specifications concerning the significations and the semantic context in which it can operate.

Almost the full complement of the arts show no prejudice when it comes to this term; on the contrary, we can assert that, with no debate over semantics, in modern aesthetics, beginning with Berkeley and Hume, illusion has been assimilated as a constituting element of any image capable of confering on a work the power of suggestion, expressive intensity, evocative force, etc. Usually, from the point of view of the perception of the real, the premises of psychology help the recognition in an illusion of all the sensorial data that give a specific character to the vision, and underline the deformation of the imaging character of the represented or signified appearance, the deviation from a standard considered to be the real prototype of that appearance.

The term illusion is mostly linked with the theoretical capitalization of the artist's imaginative activity, and in the sense of actualization of the evocative capacities of the imaginary, and also in that of an objectivated game in which we identify the metamorphoses of the real, the re-melting of its data in a new structure where the essential becomes contingent and its significance becomes novel.

The emotional effect and the symbol richness of an illusion often take shape in the spontaneous creation of the artistic image and in its aesthetic perception. But in special conditions an illusion can become a procedure intentionally used in order to get a certain effect or expression by the use of various techniques of deforming representation by which a new value is introduced into an artistic value, increasing its symbolic richness and enlarging its emotional resonance. To select some examples of the intentional use of illusion we certainly recall the painter's use of perspective in depth, the "trompe l'oeil" method from the spatial arts, special effects in cinematography, the technique of the illusion of incoherence in discourse, metaphors, irony, paradox, inverted quotation, chimerical personification, timbre transpositions in music, false counterpoint, and so forth.

## THE ARTS AND THE ILLUSION

From what we have said up to now, it results that the technique of illusion, when it is artistically used, has great power to influence the production of the artistic image in the different arts. The plastic arts owe their existence to these techniques of producing illusions of colour and form in space. These arts always insistently sought those methods by which their representations are close to the originals, to the reality, or to the artistic program of a school or a trend. This practice became a science and a subject of research for artists. Even if the art does not clearly admit it, artists know very well that if they do not master the laws of perspective, no matter what, their success is uncertain, and their creation is more difficult to realize. Artists like Cenini, da Vinci, and others wrote special treatises either to guide their students, or to show the public the difficulties met by the artists in order to reproduce the image of things in colour and mass. This science, called perspective, has had an evolution over time, like any other science, so that it has advocated several perspectives, some of which are not used anymore. The science of perspective developed mainly beginning with the Renaissance, when Paolo Uccello and Leon Battista Alberti had an important role in the specification of its difficulties. There are now several perspectives used over time in the plastic arts. Vertical perspective was mainly used in ancient Egyptian painting. This was also used in the Pre-Renaissance period. This type of perspective means the representation of depth planes on levels, and that is why it is also called multi-level perspective. Another type of perspective is inverted perspective, called that because the value of the planes is inverted, the depth planes becoming more expressive and more important. The realization of such effects was a mark of Byzantine painting. The Renaissance invented and used lineal or conic perspective, in which a more plastic reproduction of the depth and form of objects was obtained, a function of alterations optically perceived, supported by the contour lines of the objects due to distance and their location in space. Another perspective, "à vol d'oiseau" (bird flight), was used mainly to reproduce landscapes from a certain height. This perspective and the cavalier one, in which objects are supposed to be seen from a point situated in the infinite, were used both in the Renaissance and in the following periods, up to the present day. But the painters and sculptors nowadays are using alternately or in combinations other perspectives, to reproduce more clearly the images they are realizing: aerial, practical and even axonometric perspective. These suggest the distances between objects through differences of colours, or they reproduce objects in the form in which they are seized by the eye, without any theoretical processing of their representation. Reproduction through axonometric perspective uses even deformations of angles and edges with the purpose of achieving a more plastic image of the objects represented in space. This last perspective is that most used in architecture. Until now we have analysed the systems invented by the arts linked with space in order to reproduce the image of objects more exactly, more plasticly and more expressively. We can see that also from a compositional point of view. Space is the place where the work of art is born, and not only that, the work of art uses space depending on its own necessities; it defines it, and even creates it. This space in which there is life is a given to which the space of art is subjected; it is a changing plastic material subject to the laws of perspective before anything else. We will notice first that it is not possible to analyse the matter abstractly and to reduce it to a certain number of general solutions that will require detailed applications. It makes a difference whether the form be rendered in architecture, sculpture or painting. No matter the difference of techniques, how decisive the authority of one over the other would be, the form is first qualified by the special field to which it belongs.

However, there is an art that seems capable of transporting itself with no difficulties to a set of techniques and this is the art of ornamenation, maybe the first alphabet of human thinking. This stands as a field of vast speculations. Before being rhythm and composition, the simplest ornamental themes, the flexions of a curve, a drawn leafage involves a future symmetry of alternation, dual or unified; this already prepares the space, the place where the ornament will be a novel existence. Reduced in an insignificant, sinuous way, it already becomes a frontier and a path. Now not only does it exist in itself, but it also creates its own environment to which this form gives a meaning. In the system of a series made of discontinuous elements, clearly analysed, powerfully rhythmical, defining a stable and symmetric space to protect those elements against unforeseen metamorphoses, a labyrinth system realizes itself by a mobile synthesis, in a bright space. "In the interior of this labyrinth, where we feel the road without recognizing it, rigorously lost by a lineal mood that sneaks to get to a secret purpose, a new dimension elaborates itself, and it is neither the movement, nor the depth, but it creates the illusion." (Focillon, p. 28)

The arts, on the whole, have the power to produce by different means the most unexpected effects, approaching even the fantastic. Through these means the creation of an inexistent universe can be realized by reflecting reality to the scale of imagination, or by mixing realistic elements with the supernatural ones. Approaching the issue of fantastic symbolism, Hegel ascertains that a difference exists between significance and the form of representation. He ascertains that the fantastic is the premise of religious artistic representations, and also of folkloric ones. The fantastic also expresses the degree to which is exercised the creative imagination in the effort of reconciling the form of representation and significance, pushing the lines of a specific figure beyond their precise limits. "So, in this field of confusion, we are not allowed to look for the real beauty, because in the ceaseless and hurried jump from one edge to another we see, on the one hand, that the scope and the force of general significance is tied, in an improper manner, to the sensible element, received either in its details, as in its form of elementary appearance." (Hegel, p. 342–343)

The idea of illusion is promoted by the ancients too. Roman style was no less sensitive to the need for illusion, even if the Romans, on the whole, were considered pragmatic in their times. The Antonine era represents that period in which the condition of humans, as they were understood at that time, were the happiest and the most prosperous, writes Gibbon in *The Decline and Fall of the Roman Empire*. This favourable state of things was attributed to the Roman genius in the respect paid to law and order, to the Roman spirit of tolerance and justice, and to a capacity for realizing wise leadership. This greatness rightly should have been reflected in art, to provide a memento for all times. Trajan's Arch commemorating Benevento and Trajan's Column were such means of expression creating the illusion of prosperity and of a greatness specific to the Romans and to their emperors. The Arch commemorating Benevento honoured an important technical achievement, and its main constructor, and incited the Romans' imagination to see overseas lands, great victories over other people considered barbarian (see the Gauls, Germans and Dacians). Trajan's Column was built to commemorate the defeat of the Dacians, considered a very dangerous people. It was imagined as a tale, the teller hero being Emperor Trajan himself, who describes – in an original manner for those times – how tough the battle with the Dacians was, how dangerous it was for the Romans and how great was the bravery of the Romans in defeating the barbarian Dacians. The beginning of the campaign is placed on the banks of the Danube River, in a Roman camp. While the Romans are crossing the Danube on a pontoon bridge, a god who is the personification of the river comes out of a cave and shows his good will by supporting the bridge. From this point forward, the action is favourable to the Romans.

In their way of informing, the bas-reliefs come close to literature – a visual tale. The method used by the Romans in these works was considered "simultaneous" and "continuous." The first category is identical to that method used by the Greeks on the frieze on the Eastern front of the Parthenon, where the entire action takes place in a moment "frozen" in the sculpted form. The second method, called "simultaneous" (separating), respects the classical unit of action, of time and of place. The continuous method – or cyclic – was created by Romans precisely to reproduce the succession of episodes along the way in Trajan's wars.

In spite of its having the content of a direct tale, the style is not realistic. To obtain effects in building the illusion, the artist based it on a set of symbols, as carefully elaborated as the words writers use. A row of wavy lines indicates the sea, an outline notched at the horizon indicates mountains, a wall may indicate a town or a camp, a feminine figure with folded curtain, half-moon shaped, indicates the presence of night. We find that the frieze from Trajan's Column prefigures the imagistic symbolism of early Christianity and of the Middle Ages. William Fleming dares even to say that "the continuous way of visual tale was adopted directly in the case of the catacomb paintings of the first Christians, adopted further in the illuminated manuscripts, the religious sculptures and the stained-glass windows of the medieval era, and we can still find it, well developed, in modern-day cartoons. Even the cinema owes something to this technique, created in the second century." (Fleming, p. 138)

If art, in general, creates illusions, that means images are more or less flitting between realism, the fantastic and the fabulous, and we cannot precisely indicate which one of the arts produces more illusion; analysing art history's moments, we can understand that the Baroque excels in this way. For a long period, the baroque was held up to define bizarre productions. The history of this style, but also a historical moment, has a lot of turns and meanings that are not part of this study. We have to show that the philosophy that stood at the base of this movement, although having roots that go back to its anticipations, is that of the end of the 16th century to the second half of the 18th century. The Baroque appeared as a reaction against the stylistic asceticism advocated by Reformation movements. Its pomp, grandiloquence and decorative luxury created the illusion of a world with surrealistic features, antagonistic to the present one. This model of art gradually replaced the Renaissance one. In order to produce the intended effects, it abundantly used contorted and curved forms; it neglected symmetry and regularity generally; and it suggested, by all its means, the impetuous dynamism that overcomes order. Baroque art preferred agitated, centrifugal contours, the pictorial play of light and shadow on surfaces, compositional elaborations in diagonals and in whirls, the chiaroscuro technique, levitations, etc. In literature, a image of aristocratic ambience was cultivated and characterized by charges of metaphorical elements, complicated expression, an invasion of the feelings that go beyond the borders imposed by reason through excess and paroxysms of passion and impatience. The philosopher of this trend was Leibniz himself, who in his work, *Nature et de la Grace, fondés en raison*, expresses philosophical principles that fully favor the realization of this trend of thought, and especially of feeling in the Post-Renaissance world. The central idea of his theodicy is that of a preestablished harmony. Each of his posited monads behaves in accord with its own created purpose. These windowless monads, each following its own purpose, form a unit of the ordered universe. Even though each is isolated from the others, their separate purposes form a large-scale harmony. It is as though several clocks all struck the same hour because they kept perfect time. Leibniz compared all these monads to:

several different bands of musicians and choirs, playing their parts separately, and so placed that they do not see or even hear one another. ... Each monad then is a separate world, but all the activities of each monad occur in harmony with the activities of the others. This way it can be said that each monad mirrors the whole universe (but from a unique perspective), in the sense that if anything were taken away or supposed different, all things in the world would have been different from what they are like at present. Such a harmony as this could not be the product of an accidental assortment of monads, but must be the result of God's activity, whereby this harmony is pre-established (Stampf, p. 248).

Beside the idea of preestablished harmony, which had deep meanings in the Baroque art and ideology, there is also the concept of *grace*, having first a theological-philosophical meaning, then extending in meaning abundantly in all the artistic creations of the trend. Leibniz understands theological-metaphysical grace as a perfection of nature, as that special way of being beautiful that means perfection in nature and in all the works of imitation of nature made by man. For this kind of harmony mirrors are the most used metaphor. For Leibniz the mirror is an element that reflects and multiplies in itself all the universe; it is the most proper instrument to create the illusion of man's location in another world, in the best of possible worlds.

A sample of the Baroque style is the church of Birnau (Germany), built between 1746–1750 by the architect Peter Thumb at the order of Salem's Cistercian abbey, with Gottfried Göz as painter. This church intended to raise us "higher than ourselves," as Leibniz said. An actual mirror is in the hands of one of the allegoric figures floating above the choir, at the feet of Mary, Mater pulchrae dilectionis, represented as the Woman of the Apocalypse in a vault apparently painted using the technique of visual illusion, to which Leibniz often referred. While the painting on the apparent vault presents the sky as if floating above the space of the church, through the intermediation of the mirror a direct, symbolic and visual rapport is established with the one who looks on. Reflected in this mirror, he finds himself simultaneously in the inner space of the church and also transported to a certain realm of meanings. This similarity between the mirror of Birnau's theological grace and the mirrors of profane grace of various palatial Baroque and Rococo rooms corresponds perfectly to the continuity between nature and grace, with differences of degree in an ascending sense between rich worldly grace and theological-metaphysical grace, as aspects of aesthetic grace (metaphysical-aesthetic). It is not at random that in Rococo rooms the mirrors are not placed in a line (as at Versailles), but are placed opposite one another on all the walls, so that anyone can see himself from his own point of view and at the same time, from all others' points of view.

"Leibniz's image of the mirrors was to be extremely favorably received by the mid-18th century aristocracy, especially by the rulers of small realms. This philosophy promoted by Leibniz and then continued in other forms by the Enlightenment and Romanticism could be also called Rococo philosophy." (Assunto, p. 90) The aesthetic program of this philosophy, as is proved by the facts, is to erase the border between reality and illusion, the identity between significant and signified. The truth in this case, which is presented as a fantasy, a fantasy that becomes reality because of the feelings stirred, stands on the emotion of the artistic or even natural sublime, or as Kant called it, the mathematical sublime.

If Leibniz's aesthetics strongly promoted illusion and matched perfectly the intentions of Baroque art and the Rococo style, it is also true, even if not explicitly so, that the present arts use illusion too as one of the most authentic instruments for building images suggesting best the truth in the vision of art, and this is not to be condemned. Artistic trends in the realm of the plastic arts, literature, music, and even architecture intensely seek such effects in their creations. Thus, illusion is not a purpose, as it was for Baroque and Rococo art, but an instrument and an immediate image effect, an expression of artistic truth.

From the artistic trends of our century we could say that expressionism started to use the subjective in order to create images with a high illusion potential. In its first stage, between 1885–1900, the trend served the creation of images that, by the presentation of personal destiny, fully reflected the social question. In the second stage, the abstract one, the trend developed its own capability to directly express the contorted human feelings owing either to the war that had just ended, or to the one that had to start, i.e., anxiety before the major themes of existence. The real purpose of this art was to reveal the "hidden real" and to express it. Then expression receives a dynamic function, an active correcting role and eventually arrives at the nonfigurative expressionism that pretended to liberate itself from the burden of matter, preserving only the action of the moving forces of hidden truths. In this hypostasis art is only a transfer station between exterior nature, visible or invisible, and the interior nature that comes to be identical with the expression.

The way by which an expression realizes its active function of practical knowledge is through spontaneity and impulsiveness. In this context expression is growth that confirms the act of absorption of the organized energy of matter, and which has a tendency to permanently remake its vital dynamism's harmonies.

Far from meaning a hypertrophied ego, the concept of expression has the extensive character of a knowledge of the universe in its most hidden reality, while at the same time being a most organized one that can also be understood. From this aesthetic point of view, matter is not nature, but just a proper environment, an intermediary by which the fundamental forces of nature act to become nature. The expression belongs to the action, i.e., it contributes to the matter's organization and harmonization, giving the "absolute" the possibility of taking shape through action. From this we learn the path of realization of the image of this art: living-expression-living. These are taking shape by making monumental forms, simplifying them down to their skeleton, intensifying the acts dematerializing the form, narrowing or reducing "matter" down to abstract effects, with no plastic-sensual function. These techniques of the spiritualization of form lead eventually to the identification of the real with the expression, as Benedetto Croce wished for the written art. But this process of losing the individuality of the figurative element and its absorption in an active field of forces will go from expressionism to constructivism and suprematism, becoming one of the basic concepts of that artistic movement. In what concerns the chromatic conception, a new attitude is emphasized: from the symbolism of expressionist techniques, that take the color of things as a gate to their truth, up to the absolute and cold reality of a color as a superficial and illusory cover, tightly separated by the deepness of things. Without passing through all artistic trends and through all the historic stages of art, we can liken art to a labyrinth to which illusion is the most dextrous instrument for explaining and supporting it. Wherever some mystery exists, illusion is quite suitably present. They exist in each work. These are things as real as other things, as true and natural because they have been created by human hands. A painting is less mysterious than a shell. But illusion is somewhere else. On the one hand, it is in the work: real; on the other hand, it is in the artist: he knows the work of his hands and the dream that will guide them. Art is a work before being a religion, a trade; it is a trade before being a mystery. The artist knows it well, he who measures his effort and his tiredness. Poussin explained his masterpieces in this way: "I have neglected nothing...." The genius becomes modest through work, but the empty spectator forgets this, and the work as work becomes the work as miracle. This way, illusion is born from lazy contemplation. It is born and in that moment it becomes double. As in politics, art is the production of illusion.

The first illusion: the objectivity and the universality of the beautiful. A work that one admires without understanding it, is so beautiful that if it imposes itself on one, one therefore has the feeling that will impose itself on everybody else. The beauty of a work is lived as though it were universal, immortal, absolute – and truly present in the work that we like or that we admire. Kant stated the essential regarding this subject. To consider that a thing is beautiful does not mean only to recognize the pleasure it gives to you, because then it would be only something pleasant, but also to pretend the objectivity and universality of this pleasure.

The second illusion follows from the first one. We can hardly see the individual and definite origin of this universal work (universal, through the pleasure we suppose it offers). It is unbelievable that what is absolute could be born out of the relative. It seems that it is immortal; how to believe then that it was possible to be born? We can imagine that it preexisted in a particular mode during the time of its creation, and that this should have appeared in one form or another, being already registered somewhere and having no chance in ways more or less mysterious of not making its way through to us. Creation? Rather, revelation. It is hard to believe that the Ninth Symphony of Beethoven could be in any way other than it is, neither that it could not exist. In a certain way, it is necessary and immortal, not only before but also after. The artist does not invent, but discovers; he does not produce, but uncovers; he does not create, he discloses. Like the great mystics, he sees what others do not. Like other artists, he has the obscurity of his night and pain, like his illumination. With all his doubts, the artist always has the plan of his work clearly shaped. In a way, we can say that it preexists the creation itself. Here the birth of the final illusion takes place. It consists in believing that the work is ruled by an idea that precedes and realizes it, an idea that is none other than the work itself, but prior to its birth. In this way, the work is always its own cause. In this vision, the painter paints because he already knows the painting that he wants to create and which exists in his consciousness even before it is painted; in the same way, if the poet writes, it is because he already knows what he wants to say. Here we can say that the subject, the artist, is the cause of the work. But there is also another cause – the work itself – what we can name "its final cause," and this ushers the work to its final stage; and without it, it could never exist. Taking into consideration that the work of art goes always to itself, in this sense its ultimate form is, as Kant said, "an ultimate without end." In this statement is captured the essence of the so-called illusion of finality.

Another problem that generates illusions is that regarding the origin of the work of art. Logic would say that the origin of work depends on the artist, and this, without doubt, is true. There still is a problem however. The artist quite often sees his work's creation as a gift from without, that he receives it and does not produce it. This is an illusion. Talent alone is not enough. The talented man is not always inspired. That is why Goethe said to Eckermann that "a masterpiece depends on nobody," that it hovers over earthly forces. These are inspired gifts, which man receives by divine grace and which he should greet, says Goethe, with veneration and gratitude. The audience seems to share this point of view, that the work in its magnificence, sometimes misunderstood, seems to exceed human capacities. Even Beethoven seems small compared with the Ninth Symphony. Hence the illusion or maybe the truth of the theme of inspiration, to which we often reduce the mystery of artistic creation. If the work is superior to the artist, then from where does this superiority and greatness, its value, come, if not from the artist? What assurance have we that it comes from somewhere else? And if it comes from without, why does it come only in a certain way and to a certain artist and not to all artists? The doctrine of finality in art, as in other fields, totally overturns the nature of things. It considers as effect what is in fact cause, and reverse. With all the illusions that it produces, this theory has its importance. Without the end it would not be possible to explain inspiration. Nothing of the work exists before its creation. Nothing gives it the right to existence. Nothing from the absolute substantiates it, nothing from the transcendental justifies it. The work has to be explained not only through inspiration, but especially through the artist's solitary work, which reveals nothing, which uncovers nothing. He, the artist, is the one that puts into value his hard to guess and unlimited forces, those forces that emerge from ideals, from imagination, from his enthusiasm, informing the gifts of talent and vigour. The most complicated relief of the most beautiful landscape is nothing but an opposition of up and down, and if there exist three dimensions, these are caused by a horizontal juxtaposition of all the points and not by a vertical multiplication of directions. As we well know, "up" and "down" are instrumental explanations, having no meaning outside an illusionary, geocentric point of view. In infinite space, all the directions are equalized (leading to a horizontality of existence), and there does not exist on any side – a true side up or a true side down. More, the opposition of an up and down would suppose that there be a centre of the universe and that our planet be flat with the result that all the verticals have the same value, i.e., as, parallels. The opposition of an absolute up and down supposes, in the first place, an egocentrism – an up that means what is over me, and a down that means what is under me. In this case, our bodies would be an axis that transforms all the verticals into parallels, but this is only an illusion about our position. This kind of illusion exists in all fields of existence, including those of the spiritual life (philosophy, morals, politics, etc.). This illusion is used, unconsciously, by politicians, moralists and philosophers when they frame concepts, and when they build up or "suppress" a world using them (as seen for example, in the case of the philosopher Berkeley, with his principle "Esse est percipi").

## THE VALUE OF ILLUSION RELATED TO THE ARTISTIC TRUTH

Maybe at this point in our explanation it is easier for us to understand what is an illusion and what is its relation with truth in art. The illusion here consists in permanently presuming to be real the presence in a work of art or in an intelligible world (reference system), of the universality of the Beautiful, specific to our desires, yes, and to the transsubjective objectivity of our unconscious. What gives greatness to art is its creation, for to create means to produce what does not exist. Art is the illusion of truth, but also the truth of illusion, because an illusion is the starting point in art and in part even a way of being for it, but not a way of existence. Art creates sense and, simultaneously, value.

Hence we arrive at a conclusion that we once considered to be a premise – that art can be likened to a labyrinth in its characteristics, among which is illusion. Or, as we saw, the illusion is that it appears to us, or in art everything is produced under the sign of appearance. The illusion does not forbid the truth, but it prepares it and substantiates it. The values cannot be overturned. It is known that the Sun's rising out of the East is but an illusion, but it would truly be illusory to believe that it could rise from the West; on the geocentric illusion – with some transformation of the parameters – we can build this illusion too. That is the way in which it is possible to reverse the relations between illusion and truth, if we change the positions of the values we are operating with and the senses that we are building.

An important aspect of this theory is the *Power of Illusion*. It is well known that literature begins with the tale. A tale represents certain events by means of auditory and visual signs. The events thus represented are mental events in the narrator's mind. His motive is the urge to communicate these events to others, to make them relive his thoughts and emotions; it is the urge to share. The audience may be physically present, or be an imagined one; the narrator may address himself to a single person or to his God alone, but his basic need remains the same: He must share his experiences, make others participate in them, and thus overcome the isolation of the self.

To achieve this aim, the narrator must provide patterns of stimuli as substitutes, for the original stimuli of the experience is actual occurrence. This, obviously, is not an easy task, for he is asking his audience to react to things which are not there, such as the smell of grass on a summer morning. Since the dawn of civilization, bards and story-tellers have produced bags of tricks providing such stimuli. The sum of these tricks is called the art of literature.

The oldest and most fundamental of all tricks is to disguise people in costumes and to put them on a stage with masks or paint on their faces; the audience is thereby given the impression that the events represented are happening here and now, regardless of how distant they really are in space and time. The effect of this procedure is to induce a very lively "bisociated" condition in the minds of the audience. The spectator knows, in one compartment of his mind, that the people on the stage are actors, whose names are familiar to him; and he knows that they are acting for the express purpose of creating an illusion in him, the spectator. Yet in another compartment of his mind he experiences fear, hope, pity, accompanied by palpitations, arrested breathing, or tears – all induced by events which he knows to be pure make-believe. It is indeed a remarkable phenomenon that a grown-up person, knowing all the time that the faces on a screen are shadows projected by a machine, and knowing furthermore quite well what is going to happen at the end – for instance, that the police will arrive just in the nick of time to save the hero – should nevertheless go through agonies of suspense and display the corresponding bodily symptoms. It is even more remarkable that this capacity for living in two universes at once, one real, one imaginary, should be accepted without wonder as a commonplace phenomenon. In this context we can speak also of the *Dynamics of Illusion.*

In the comedy, the accumulation of suspense and the subsequent annihilation of it in laughter take place at distinctly separate stages. In the tragedy, on the other hand, excitation and catharsis are continuous. Laughter explodes emotion; weeping is its gentle overflow; there is no break in the continuity of mood, and no separation of emotions from reason. The hero, with whom the spectator has identified himself, cannot be debunked by slipping on a banana peel or by any sudden incongruity in his behaviour. The gods of the Greek and Hindu pantheon might change into any shape – a swan, a bull, a monkey, a shower of coins – and yet their paramours would lovingly surrender to them. On the bas-reliefs of Indian temples, Shiva is often seen making love to Parvati while standing on his head, without appearing ridiculous. When the events in epic or drama take an unexpected turn – Odysseus' companions are transformed into swine or chaste Ophelia sings obscene songs – emotion, being provocatively vitiated, refuses to perform the jump and explodes in laughter; but if a sympathetic presentation is maintained, it will follow the hero through all vicissitudes. An abrupt change of situation which required an equally quick reorientation of the mind to a different associative context, led in the first case to a rupture between emotion and reason, in the second to a transfer of emotion to a new context whereby its harmonious co-ordination with reason is preserved. Thus, incongruity – the confrontation of incompatible matrices – will be experienced as ridiculous, pathetic, or intellectually challenging, according to whether provocation, identification, or a well-balanced blend of scientific curiosity prevails in the spectator's mind. Don Quixote is a comic or a tragic figure, or a case-history of incipient paranoia, depending on the panel of this triptych in which he is placed. In all three cases the matrices of the reality of windmills and the delusion of phantom-knights confront each other in the reader's mind. In the first case they collide, and malice is spilled in laughter. In the second, the two universes remain juxtaposed and reason oscillates to and fro between them, compassion remaining attached to it and being easily transferred from one matrix to the other. In the third case, these two reactions merge in a synthesis.

As Arthur Koestler emphasized in his book *The Act of Creation*, "... compassion, and the other varieties of the participatory emotions, attach themselves to the narrative told on the stage or in print, like faithful dogs, and follow it whatever the surprises, twists, and incongruities the narrator has in store for them. By contrast, hostility, malice, and contempt tend to persist in a straight course, impervious to the subtleties of intellect; to them a spade is a spade, a windmill a windmill, and a Picasso nude with three breasts an object to leer at." (Koestler, p. 305)

Illusion, then, is the simultaneous presence and interaction in the mind of two universes, one real, one imaginary. It transports the spectator from the trivial present to a plane remote from self-interest and makes him forget his own preoccupations and anxieties; in other words, it facilitates the unfolding of the participatory emotions, and inhibits or neutralizes his self-asserting tendencies.

This sounds like an escapist theory of art; and in spite of its derogatory connotations, the expression contains a grain of truth – though no more than a grain. The analysis of any aesthetic experience requires, as I said before, a series of steps; and the escape offered by transporting the spectator from the bed-sitter in Bayswater to the Castle of Elsinore is merely the bottom step of the ladder. Nevertheless, it should not be underestimated. In the first place, if illusion offers escape, it is escape of a particular kind, one sharply distinct from other distractions such as playing tennis or bingo.

The aesthetic experience depends on the delicate balance that arises from the presence of both matrices in the mind; on perceiving the hero as Laurence

Olivier and Prince Hamlet of Denmark at one and the same time; on the lightning oscillations of attention from one to the other, like sparks between charged electrodes. Thus the creation of illusion is in itself of cathartic value – even with a product, judged by sophisticated standards to be of cheap quality; for it helps the subject to actualize his potential for self-transcending emotions, a potential thwarted by the dreary routines of existence. Liberated from his frustrations and anxieties, man can turn into a rather pleasant and poetic creature; when he changes into a dark suit and sits in a theatre, he at once becomes capable of taking a strong and entirely unselfish interest in the destinies of the personae on the stage.

To revert to Aristotle, the cathartic function of a tragedy is to, through incidents arousing horror and pity, accomplish a purgation of those same emotions. In cruder terms, a good cry, like a good laugh, has a more lasting after-effect than this particular occasion would seem to warrant. Taking the Aristotelian definition at face value, it would seem that the aesthetic experience could purge the mind only of those emotions which the stage play has created; that it would merely take out of the nervous system what it has just put in, leaving the mind in the same state as before. The aesthetic experience inhibits some emotions, channels others, but above all, it draws on unconscious sources of emotion which otherwise are only active in underground play.

Thus the concept of catharsis assumes a twofold meaning. First, it signifies that concentration on the illusory events on the stage rids the mind of the dross of its self-centred trivial preoccupations; in the second place, it arouses in a person dormant self-transcendent potentials and provides them with an outlet until they peacefully ebb away. This peace, of course, does not necessarily mean a happy ending. It may mean unearthing an individual tragedy in the universal tragedy of the human condition – as the scientist resolves a problem by showing that a particular phenomenon is an instance of a general law. Tragedy, in the Greek sense, is a school of self-transcendence.

*The Value of Illusion.* But where do beauty, aesthetic value, or art enter into the process? To answer that requires taking several steps. The first is to recognize the intrinsic value of illusion in itself. This derives from the transfer of attention from the Now and Here to the Then and There – that is, to a plane remote from self-interest. Self-assertive behaviour is focused on the Here and Now; the transfer of interest and emotion to a different time and location is in itself an act of self-transcendence in the literal sense. It is achieved through the lure of the heroes and victims on the stage, who attract the spectator's sympathy, with whom he partially identifies himself, and for whose sake the temporarily renounces his preoccupation with his own worries and desires. Thus the act of participating in an illusion has an inhibiting effect on self-asserting tendencies and facilitates the unfolding of the self-transcending tendencies. "In other words, illusion has a cathartic effect – as all ancient and modern civilizations recognized by incorporating various forms of magic into their purification rites and abreaction therapies." (Koestler, p. 303)

It is true that illusion, from Greek tragedy to horror comics, is also capable of generating fear and anger, palpitations and a cold sweat, which seems to contradict its cathartic function. But the emotions thus generated are vicarious emotions derived from the spectator's participation in another person's existence, which is a self-transcending act. Consequently, however exciting the action on the stage, the anger or fear which is generated will always carry a component of sympathy, an irradiation of unselfish generosity, which facilitates catharsis – just as a varying amount of high voltage current is always transformed into heat. At a later stage, when the climax of the drama has passed and the tension ebbs away, the entirety of the current is consumed in a gentle inner glow.

## REFERENCES

Assunto, Rosario. *Universul ca spectacol*. Bucuresti: Editura Meridiane, 1983.
Fleming, William. *Arte şi Idei*. Bucureşti: Editura Meridiane, 1983.
Focillon, H. *Vie des Formes*. Paris: P.U.F., 1964.
Hegel, G. W. F. *Prelegeri de estetică*, Vol. I. Bucuresti: Editura Academiei Române, 1966.
Koestler, Arthur. *The Act of Creation*. New York: Macmillan, 1964.
*Rig-Veda* 10, 90, 2. Cf. Paul Deussen, *Filosofia Upanisadelor*, Bucuresti: Ed. Tehnica, 1994.
Stumpf, Samuel Enoch. *Philosophy: History & Problems*, 3rd ed. New York: McGraw-Hill, 1983.

# SECTION III

The Puebla conference in progress.

W. KIM ROGERS

# TRUTHFULNESS IN SCIENCE AND ART

For the philosopher of science Karl Popper, the first of the only intellectually important goals was the formulation of problems. "Science begins with problems, and ends with problems." (1974, p. 132) His meaning is clearer when we speak not of "science" but instead of "scientific research programs (Lakatos). The same approach to knowledge was expressed more explicitly by N. Postman and C. Weingartner: "Knowledge is produced in response to questions. And new knowledge results from the asking of new questions; quite often new questions about old questions." (1969, p. 23) Such a "questioning system of thought brought us to today's world," according to the author of *The Day the Universe Changed*, James Burke. (1985, p. 17)

When non-scientists think about scientific knowledge, they are apt to see it as merely a record of facts, as a repertory of answers. But this is certainly a mistake. The facts themselves do not push us towards setting to work on problems, and answers cannot be taken as givens. Only good questions can lead to good answers. As Postman and Weingartner put it, one needs "relevant and appropriate and substantial questions" (1969, p. 23) in order to arrive at relevant and appropriate and substantial answers. Answers are strictly correlated with questions already asked, and their truthfulness is relative to the same questions. Scientific knowledge and truthfulness cannot then be the result of any accidental encounters with the facts but make their appearance within a search for specific answers in specific enquiries.

Jacob Bronowski, scientist and poet, historian of science and culture, viewed science and art as very similar in their demands for truthfulness. "Science," he wrote, "takes for ultimate judgment one criteria alone, that it shall be truthful." (1963, p. 123) At face value, one can hardly disagree with that statement, but its whole worth depends upon what is meant by "truthfulness." "We do not mean by truth," he continued, "some chance correspondence with the facts." (Idem) Doing science does not mean producing a record of facts "but the search for order within the facts. And the truth of science is not truth to fact ... but the truth of the laws which we see within the facts" (Ibid., p. 130).

> We are not merely observing and predicting facts; and that is why any philosophy which builds up science only from facts is mistaken. We know, that is, we find laws, and every human action uses these laws, and at the same time tests them and feels toward new laws.... The laws of science, like those we use in our private behavior, remain helpful and truthful whether they contain words like "always," or only "more often than not." What matters is the recognition of



*A.-T. Tymieniecka (ed.), Analecta Husserliana LXXVI*, 215–222.

the law in the facts. It is the law which we verify: the pattern, the order, the structure of events. (Ibid., pp. 129–130)

The facts – incomplete and ambiguous as they must always be when we separate them out from future experiences and past interpretative schema – are science's point of departure for its enquiries rather than the goal towards which its search aims. But that *is* their authentic significance for science, facts *are* something to be problematized, to be questioned about the regularity or laws of their connections and differences.

Science must be described as an activity and that very much like what we call having an adventure. To have an adventure is doing and experiencing something out of the ordinary. It is the shattering of the inert, insistent, and oppressive sphere of everyday reality that imprisons the leaps of the imagination as if it were a bowl of glass. Each adventure is the incorporating into the familiar round of our days of the unforeseen, unthought of, the novel. Science, like the having of adventures, combines the elements of certainty and uncertainty in one's life.

The scientist, even as the adventurer, is ignorant as to whether he shall arrive at the point for which he sets out, and the knowledge that he does not know the outcome – the answer to his enquiry – is the very condition of his action. The scientist asks questions of reality because he does not know the answers, or indeed, if all of his questions are answerable. To the non-adventurer and non-scientist, to the "practical man," this way of proceeding may appear to be madness. (Perhaps that is one of the reasons why the "practical man" has nightmares about the irresponsible and mad activities of scientists.)

Scientists begin with the effort, not to invent a novel idea or vision of the world, but to rightly see, to unveil or discover some regularity or law in the world which is to be found through their enquiries. To paraphrase Bronowski, science does not teach us to worship what is known but to question it. (Cf. 1973, p. 360.) But scientists find themselves at the start in an inherited situation with acquired patterns of behavior appropriate to this situation – the world appears to them already in terms of some unreflected-upon interpretations of what exists. Their "research programs" are projected into the world in terms of what they understand to be the kinds of affairs which make up that world. For example, modern science sought to discover that which, as regards the separate, autonomous, and self-contained individual "things" which they believed made up reality, is always the same in their relationships to each other.

These interpretations of affairs, which are, in fact, beliefs – the repertory of our active convictions as to the sort of things that exist – constitute the very ground of our sciences. Yet with the repeated experience of some significant degree of non-conformity between the way these affairs are related and how one expected them to be related, this ground of beliefs can shift, and cracks open up. Then scientists may turn their efforts towards the discovery of new sorts of relations and may uncover new kinds of affairs. That is to say, they begin to ask new questions.

"Science," wrote Bronowski, "is a language." (1963, p. 131) I want to amend that to read instead: "It is a conversation," that is, science is to be viewed as an interaction between the enquiring scientist and the responding world. Ortega y Gasset had it right: by our searching questions we call out the thing which is not there before us and the thing responds, makes itself manifest. (See 1971, p. 304.) But let us not overlook the limited nature of this exchange, for whatever other responses the world could make to unasked questions (if indeed we knew what to ask) are now enclosed in silence. Truthfulness comes from reality's self-disclosure of its order, but always in terms of when and where one meets it, from one's historical and cultural position. Note that different qualities and different quantities of information about the ordering of the world are thus afforded by different points of view.

As Ortega wrote:

> That a science is "true" for the very reason that its doctrine is changeable, flies in the face of the traditional idea of truth, and can be cleared up only by renewing *a radice* the general theory of truth itself and by making us see that, as this is a human matter, it is affected by man's condition, which is that of being *mobilis in mobile*. (1960, p. 37)

Does this mean that truthfulness is something merely relative to the observer? Certainly not! As Ortega said, "We must recognize variations in thinking not as changes in yesterday's truth which convert it into today's error, but as changes in man's orientation which lead him to see other truths that are different from those of yesterday. It is not truths that change, but man...." (1960, p. 26) Is this then to say that truths about the ordering of the world are only our subjective interpretations of them, and so to be led to an Idealist understanding of meanings? Not in the least! Truthfulness is a quality that a human being's relations to affairs takes on through the latter's self-presentation. In as much as affairs can manifest themselves in different ways in conjunction with a human being's diverse ways of acting, so truthfulness belongs to all of these.

Truthfulness comes into our relations with others through our actions and the response of the affairs towards which our actions are directed. But truthfulness in science also "rests upon an act of free human judgment," according to Bronowski. In science "every act of judgment is a division of the field of our experience into what matters and what does not." (1963, pp. 131, 132) That is, it divides the world into what we regard as relevant and what we regard, for the purpose of finding answers to our questions, as irrelevant. And the moment one does that, one is bound to be satisfied with what is only an approximation. This can certainly lead to good laws, that is, if what we judge to be irrelevant is not very relevant, they will be good laws. But it does not follow that they will then give you a complete picture at all of the ways in which the world is ordered. Thus there is no absolute truth accessible to us, no God's eye vision of reality.

Bronowski makes this point nowhere else as strongly as in the chapter titled: "Knowledge or Certainty" in his *Ascent of Man*. Paintings, he wrote, "do not so much fix the face as explore it ... each line that is added strengthens the picture but never makes it final. We accept that as the method of the artist. But what physics has done now is to show that this is the only method to knowledge. There is no absolute knowledge." (1973, p. 353) There is progress in science, Bronowski held, "because it has understood that the exchange of information between man and nature, and man and man, can only take place within a certain tolerance." (Ibid., p. 365) Further he implored us to remember that "science is a tribute to what we can know although we are fallible. In the end the words were said by Oliver Cromwell: 'I beseech you, in the bowels of Christ, think it possible you may be mistaken.'" (Ibid., p. 374)

The truthfulness of science depends upon our acknowledging and recognizing the unity of its parts. We do this, Bronowski said, "by a highly imaginative creative piece of guesswork, but we finish with something which is only a gigantic metaphor for that part of the universe which we are decoding." (1978a, p. 70)

For Bronowski, a metaphor has an imaginative likeness to something in reality, but such a likeness is rather what our attention is drawn to by a simile. I say that a metaphor is rather "a path through woods which we take to a clearing" where some affair may reveal itself that would have otherwise escaped our view. At the same time a metaphor is our call to that affair to manifest itself, as Ortega wrote. "*Name* is that which serves to *call* someone. The word *calls* to a thing which is not there before us, and the thing ... responds, makes itself manifest" and "*the metaphor is the authentic naming of things....*" (1971, pp. 303, 304)

In Bronowski's view, "what impresses us as truth is the orderly coherence of the pieces. They fit together like the characters in a great novel, or like the words in a poem. Indeed, we should always keep that last analogy by us always." (1963, p. 131) "There is a common quality in science and poetry – the quality of imagination." (1978b, p. 5) He further declared that "we have had to push out the boundaries of the relevant further and further. Every time we do so, we have to revise the picture totally ... by an act of the pure imagination." (Ibid., p. 60) "All created works, in science and art, are extensions of our experience into new realms.... When it matches our experience and at the same time points beyond it, this is the meaning of truth that art and science share." (1978b, p. 32)

Science was described by Ortega as "pure exact imagination because it is clear that nothing can be more exact than a fantasy ... something invented *ad hoc* so that it may be exact...." (1971, p. 299) Lest you think that this is pure sensationalism on Ortega's part, compare what he said above with the words of that most rigorous of rationalists, Descartes, who in his *Discourse on Method* recommended "assuming an order, even if a fictitious one, among those [objects] which do not follow a natural sequence relatively to one another." (1955, p. 92)

Agreeing with Bronowski, Ortega wrote that "scientific theory neither more nor less than poetry, of which it is the twin, belongs to the unreal world of fantasy. The real aspect of science is its application, its practice." (1971, p. 300) Elaborating upon Ortega's theme, we could say that modern science is concerned with generating modern myths, which it calls theories. These myths give exact, and consequently clear and certain orderings of imaginary objects, and this, Ortega wrote, "makes possible an *unequivocal* comparison between the order of imaginary objects and real phenomena; one that discovers whether these latter allow themselves to be arranged in a system or series isomorphic with the former." (1971, p. 32)

Let us take as examples Newton's and Einstein's interpretations of the effect of gravity upon a moving body. For Newton, gravity's effect is conceived in terms of a force of attraction that is proportional to the product of the masses of two bodies and inversely proportional to the square of the distance between them. Einstein held, however, that it is the shape of space-time that determines the motions of objects or of light crossing it because their gravitational fields distort the space-time around them.

Both of these theories of gravitational effects deal with many of the same observed events, and the figures they give for such instances of motion are sometimes nearly the same. Would it make sense to say, though they share nothing in common except for starting with the same facts and in some instances the forecasting of similar results, that both are (almost) equally good or bad accounts of what the world is like? Can one of these theories be better than the other because it can be shown to have greater correspondence with the world and more predictive power?

From Popper's perspective, these are both the wrong sort of questions. We should ask instead, "what makes theory interesting or significant?" And he answered, it is its "relation to preceding and competing theories; its power to solve existing problems, and to suggest new ones." (1974, p. 25) What then makes a theory scientific? It is "its power to rule out, or exclude, the occurrence of some possible events ... the more a theory forbids, the more it tells us." (Ibid., p. 41) Scientific progress consists "in moving towards theories which tell us more – theories of ever greater content," that is to say, in the advancing of a theory "which can be more severely tested." (Ibid., p. 79)

Yet truthfulness does not lie in a theory's being more interesting, or more scientific or better than its competitors past and present, as desirable as these qualities may be. Bronowski wrote: "Of course a good theory has practical consequences and forecasts true results [i.e., those that pass our tests] but these successful forecasts do not make the theory true." (1978b, p. 31) Truthfulness, as was said above, comes from affairs' self-disclosure of order in response to our enquiries. We call out to the world, seeking the pattern, the structure, the unity which is not there before us. When in response, this order both meets us in experience and at the same time points imaginatively beyond that, then our relation to the world takes on the quality of truthfulness.

Gabriel Marcel understands truth or truthfulness as "a value." To elucidate this description he suggested that we ask what we mean when we say that we are "guided by the love of truth, or that somebody has sacrificed himself for truth." (1984, Vol. I, p. 58) Only in so far as truth is a value can truth "become 'something at stake' ... and [which] must in spite of everything be maintained." (Ibid., pp. 58, 59) Is this truth to which one must remain faithful a different truth from that with which a scientist is concerned?

Perhaps we should ask Roger Bacon or Copernicus, or Galileo or even Bruno whether being guided by the love of truth matters. What should we say of a scientist who finds that the answer to his question is not to his liking and who therefore falsifies his report? Or suppose that a scientist is called upon by the State or the Party to deny or renounce some conclusion to which his researches has led him. Let us further suppose that he refuses to recant and risks being sent to a concentration camp because he will not betray the truth. "For him," Marcel wrote, "it *is not himself* that is at stake, but truth." (Ibid., p. 72)

We also talk "about people having the *courage* to face the truth." (Ibid., p. 68) When we say, as Marcel wrote, that we must "'face' the truth, the expression we use is extraordinarily full of meaning." (Idem.) In as much as a human being, whether scientist or not, is concerned about the truth, he accepts the obligation to maintain that "openness" which continues to allow affairs to present themselves and their order to him. This holding oneself open is like committing oneself to be a "light" so that the affairs and their order can become "visible." (Cf. 1984, Vol. I, pp. 62–65) Truthfulness in science and art as well requires that an affair or affairs are faced, met and so may present veridically their ordering to one in one's transactions with these affairs.

Opposition to truthfulness is the "hardening of one's heart," "du refus à invocation," "choosing to live in the dark," being closed to a new meeting with an affair. And let us not forget that the contradiction of truthfulness is deceit, bearing false witness, lying to oneself and to others.

Truthfulness cannot be accomplished in increments, it is not divisible into parts. That is, a "conversation" with an affair either occurs or it does not. The results of this "conversation" depend on what expectations and questions one has brought to this meeting with the affair or affairs. Such a meeting can fulfill, partially fulfill, or disappoint any expectations. As these expectations vary from one culture, society and historical period to another, so what one gains from these meetings will also vary, while there still can be in all these cases a genuine "conversation."

Truthfulness is not a quality of something said or not-said, it is not a lexical quality as such, but rather a matter of having participated in a "conversation" so that one can testify to what one has seen and heard. It is akin to bearing true witness. Truthfulness is practiced with others, and so is a cooperative enterprise. It comes to expression in and through usually repeatable encounters shared (or shareable) with others. Science and art in their different ways are "dating services," that is, they arrange such meetings for those who seek to meet face-to-face with hitherto unknown or imperfectly known affairs and thus have opportunities to converse about their ordering.

*East Tennessee State University*

REFERENCES

Bronowski, J. 1963. *The Common Sense of Science*. Cambridge: Harvard University Press.
_____ 1973. *The Ascent of Man*. Boston: Little, Brown & Co.
_____ 1978a. *The Origins of Knowledge and Imagination*. New Haven: Yale University Press.
_____ 1978b. *The Visionary Eye*. Cambridge: MIT Press.

Burke, J. 1985. *The Day the Universe Changed*. Boston: Little, Brown & Co.
Descartes, R. 1955. *The Philosophical Works*, Vol. I. Boston: Dover Press.
Marcel, G. 1984. *The Mystery of Being*, Vol. I. Lanham: University Press of America.
Ortega y Gasset, J. 1960. *What is Philosophy*? New York: Norton.
_____ 1971. *The Idea of Principle in Leibnitz and the Evolution of the Deductive Theory*. New York: Norton.
Popper, K. 1974. *Unended Quest: An Intellectual Biography*. La Salle: Open Court.
Postman, N. and Weingartner, C. 1969. *Teaching as a Subversive Activity*. New York: Delacoarte Press.

JULIO E. RUBIO

# PHENOMENOLOGY AND LEVELS OF ORGANIZATION IN SCIENCE

## INTRODUCTION

According to the most accepted theory of the origin of the universe, cosmological evolution has generated an impressive multiplicity of entities out of an original state of indistinguishability. Some of those entities have kept a structure that is stable enough to be discerned: atoms, planets, organisms, etc. One of the conditions for entities being discerned is their limitation to a *single scale range* (Salthe 1991). Whenever we identify something as an individual object, we associate it with a specific range in the scale of complexity (e.g. humans belong to a range of complexity which is different from that of molecules). It is a specific single scale range that we will identify with a level of organization. However, the former characterization is intuitive and dependent on the concept of complexity, which we will not address directly in this work, not only because it is a particularly difficult concept, but also because we are exploring a phenomenological approach. From such a perspective, natural science begins with a phenomenological condition: the recognition of a particular phenomenon as belonging to a general class that could be called a *natural class* in the sense that it can be identified unambiguously by members of a relevant epistemic community. Our interest in this work is to establish the basic conditions levels of organization must meet to become valid objects of scientific inquiry.

## I. PHENOMENA AS INTERSUBJECTIVE OBJECTS OF SCIENTIFIC KNOWLEDGE

Natural science begins with phenomena. The construction of scientific knowledge assumes the existence of an objective reality in a phenomenal sense, that is, one structured by specific events in space and time. Recognition of phenomena marks the methodological beginning of natural science. If we want to study reality we have to address what is accessible to empirical experience[1] and not what is accessible only to metaphysical speculation.

The achievements of modern physics pose some interesting problems for the theory of knowledge that are sometimes referred to as counterexamples to realism. Our common sense about the nature of reality is put into question by the quantum statistical laws or the relativistic behavior of bodies moving at



A.-T. Tymieniecka (ed.), Analecta Husserliana LXXVI, 223–231.

very high speeds. However, even in modern physics, the conclusions are sustained by empirical evidence.

To make scientific work, the relevant phenomenon *must be clearly and distinctively recognized* by the members of the epistemic community. We could associate this idea with the notion of *intuition*, as it has been postulated to be one of the foundations of knowledge by several philosophers (e.g., Kant, Descartes). The concept of intuition has been conceived in the philosophical tradition as a way of direct perception of an object and its relationships. Nevertheless, it is not the cognitive and subjective process of perception that we are trying to address now but the process of construction of knowledge that begins when a subjective recognition is validated via an intersubjective agreement.

For phenomena to become objects of scientific inquiry they ought to be demarcated through a process of intersubjective recognition. Phenomena become objects of study only when the epistemic community can confirm the individual perception of a certain kind of event. Objectivity is achieved, in this case, through communication and identification of experience. This does not mean, however, that subjective experience is completely communicable. On the contrary, subjective experience of phenomena is fundamentally incommunicable in its wholeness,[2] as happens in aesthetic, religious or other kinds of experience. Agreement can be achieved because the subjects eventually share a state of mind. Anyway, in contrast to what is the case with aesthetic or religious experience, scientific phenomena should be accessible to any individual properly trained (Villoro 1998, Ch. X). Therefore, scientific phenomena cannot be any arbitrary realm of reality but only those realms whose perception can be validated intersubjectively. Science is a social activity in a very fundamental sense.

## II. LEVELS OF ORGANIZATION AS PHENOMENA

How it is the case that we humans are able to perceive the world under certain phenomenal categories is a problem of the theory of knowledge that is not considered here, but it is that recognition that offers the point of departure for doing science, at least since Aristotle's time. Our approach to the validation of levels of organization in science is considered phenomenological not because it can be framed in the terms of the philosophical usage of Husserl or any other philosopher, but in the naive sense that it assumes the existence of certain specific phenomena whose objectivity can be defended because a long scientific tradition has identified the same foundational phenomena.

Let us point out some general kinds of phenomena which could be the basis for a proposal of a hierarchy of levels – whose epistemological validity we shall defend with certain detail in another paper. The following phenomenal spaces are generally recognized and used as a basis for scientific work:

i. There are *physical* phenomena, which correspond to the natural kinds studied by physics (e.g., gravitational, electromagnetic or chemical classes of events).
ii. There are *biological* phenomena, characterized by perception of the environment and functional-autopoietic organization.
iii. There are *psychological* phenomena, characterized by reflexive consciousness (i.e., consciousness of the self).
iv. And there are *social* phenomena, characterized by the occurrence of symbolic communication.

The usage of the term "level" implies the existence of a hierarchical framework. We consider that there is such a hierarchy that some thinkers have defended (Wimsatt 1976). Nevertheless, the epistemological nature of the hierarchy will not be discussed here because that is not the purpose of this paper. Instead, what we want to do is to establish how the phenomenal spaces – identified with levels of organization – can become scientific objects of study. Our hypothesis is that levels of organization become proper objects of scientific analysis when they are translated into certain kinds of observational and linguistic structures. Let us begin the defense of this idea in the next section.

## III. EPISTEMIC STRUCTURES

An extreme way of dealing with the levels of organization *as structures* associates a hypothetical phase space with each level. In the words of William Wimsatt: "If the entities at a given level are clustered relatively closely together (in terms of size, or some other generalized distance measure in a phase space of their properties) it seems plausible to characterize a level as a local maxima of predictability and regularity." (Wimsatt 1976, p. 218) Wimsatt's mathematical metaphor helps us to establish some of the conditions involved in identifying a level of organization. The phase space is a mathematical tool that is widely used in physics to represent phenomena. Its use begins with the identification of the relevant properties of the object of study and their translation into state variables. The state variables configure a mathematical space in which the phenomena can be represented and their behavior generalized in terms of mathematical functions.

In the case of levels of organization we can identify the structural properties of a certain level, but they will constitute a set of properties that are particular to that level and, therefore, incommensurable with respect to any other level. It is possible to think of a phase space – in a sense which is not restricted to mathematics, that is, as a set of properties – as being proper to a given kind of phenomena; however, it is not possible to think of a general phase space wherein different levels are represented and identified through their localization in that space. We can illustrate the just mentioned situation by reference to the dynamics of a general complex system that is capable of evolution, that is, a system able to generate novel properties (e.g. any living system) as a result of dynamic processes. Rosen describes the situation as follows:

A complex system will have a multitude of partial images of the Newtonian type, which can in some sense "approximate" to the behavior of the system. But this approximation of complexity by simplicity is only local and temporary. This means that, as the complex system develops in time, any such simple approximation ceases to describe the system … the discrepancy between what the complex system is actually doing … and the behavior of the simple approximation … grows in time. When the discrepancy becomes intolerable, we must replace our initial approximation by another. The discrepancy between the behavior of such a complex system and any simple approximation is, depending on the context, called error or emergence. (Rosen 1985, p. 193)

The same reasoning can be applied to a general frame where our reference is *Nature*, manifested in layers of different complexity. The picture we can apply to a level is inadequate for representing an emergent level because different properties distinguish both levels. No single representation space is enough for two or more different levels.

There is a possible exception to the argument just made. *Complexity* could be the generalized measure distance by which to represent levels of organization within the same phase space, but at the current time there is still no definition of complexity that is general and precise enough to be used for that purpose. Furthermore, if an operational definition of complexity were available it would give us a way to establish a hierarchy and the relationship among levels, but it would not distinguish the specificity of levels given as sets of particular properties, because the purpose of such a general measure would be precisely to avoid distinctions. Such distinctions are not only necessary for the scientific enterprise, they are the very point of departure. A scientific discipline begins through the phenomenological recognition of the specificity of a natural kind. We recognize, for instance, that there is a phenomenon we called *life* before we proceed to its classification and study. It is once we recognize a phenomenon that we try to establish its relationship with other phenomena. Therefore, any definition of the levels of organization should indicate phenomenal specificity because that is the key to identifying the structural properties that characterize a given level.

## IV. OBSERVABILITY

It is not enough to accept the existence of a given phenomenon to make it a valid object of study for science; it is also necessary to obtain the structure of the phenomenon and to represent it in a symbolic space capable of giving the conditions for observation and critical discussion. That is why Wimsatt's characterization of levels of organization as maxima of predictability and regularity gives a very important clue to the question. There are two sides of this characterization, which have different but equally important and necessary conditions for the construction of scientific knowledge: the first is the translation of phenomena recognition into observation mechanisms; the second is the translation of phenomena behavior into a representation space. Let us set forth some remarks on observability first, and review the problem of representation in the next section.

For the scientific analysis of a given phenomenon it is necessary to establish modes of observation that provide the tools for anyone properly trained to recognize the same kinds of phenomena if confronted with the same empirical situation. It is not enough to recognize life – going back to our former example – as a field of reality for doing biological science. We can approach reflection about life from religious, ethical or other perspectives, but for doing biological science, the phenomenon of life has to be analyzed in terms of observable events such as physiological, metabolic or biomolecular processes.

Scientific observation is a core activity in science, which is not limited to passive perception of natural events. On the contrary, natural phenomena have to be placed under close control (Hacking 1983) in order to assure their classification in a certain space of representation. This fact has persuaded more than one contemporary thinker to defend the idea that Nature is not the direct object of study for science but instead an object of manipulation and reconstruction. For an example of such thinking, let us quote Hans Rheinberger on the subject of experimental activity: "Nature as such is not a reference point for the experiment; it is even a *danger* ... if Nature is fractionated, unfractionated Nature has to be excluded from the space of representation" (Rheinberger 1992, p. 392). But, if Nature is excluded, what kind of things are then the phenomena which are under observation in scientific laboratories? The answer is that scientific observation is not the passive perception of natural events but the active construction and stabilization of phenomena that are not themselves present in unfractionated natural reality. Science does not depend only on the phenomenological recognition of fields of reality but also on the systematic construction of phenomena. Then, for levels of organization to be the epistemic basis of scientific work, they must be mapped into highly structured fields of observation.

## V. REPRESENTATION

Predictability and regularity are structural conditions for the observation of phenomena that make scientific statements intelligible as *representations of the world.* Let us now ignore the problematic relation between representation and reality; it is a prerogative of a scientific-phenomenological perspective to not address the Kantian *noumena* as its foundation. The point is that in order to have material for scientific work it is necessary to construct a symbolic structure. As with the matter of observation, without a structure, a phenomenon could – presumably – be the object of religious or poetic reflection but not an object of study for science. The structure of scientific phenomena is obtained from regularity patterns that can be translated into scientific representations. Predictability and regularity are both expressions of structure, but they can and should be reduced to regularity only, because predictability is also a regularity condition and because there are scientific disciplines for which prediction is not an essential condition (e.g., the theory of evolution, archaeology, etc.).

Scientific knowledge has a representationalist vocation that is scarcely doubted by scientists themselves. It is only philosophers who doubt it. Besides, contemporary scientific representation does not intend to be trascendental or methaphysical. In fact, "methaphysical" is almost a bad word if we are trying to speak scientifically. *Representation* is a pragmatic condition for any scientific work that can be confirmed through the observation apparatus of any particular discipline. But observation itself is possible only when the phenomena are encoded in representation spaces. Despite the efforts of logical positivism, during the first part of the twentieth century, it was not possible to construct an observational language absolutely deprived of theoretical content. Observation and theory are intimately entrenched in scientific semantics (Hanson 1958). Observation is the mediating operation that translates a phenomenon into an object of scientific knowledge. The translation implies the use of scientific codes with certain properties. In particular, the code should give the conditions for *closure*, that is, the elements of representation should be clearly identified and their meaning restricted as much as possible – even if that meaning is changing.

The concept of *phase space*, as it is used by Wimsatt for his characterization of levels of organization, is certainly suggestive and undoubtedly adequate for scientific representation if we are talking about physics, but its mathematical specificity makes it too restrictive for other scientific fields. Let us propose instead the term *representation space* to designate the symbolic translation of a scientific object of study. A particular representation space is formed by a set of semantic variables that ideally are explicit and unambiguously defined. Returning to the example of life, we could speculate that the general structural properties of the phenomenon are *metabolism, autoreplication and perception mechanisms*. A semantic space constituted by these axes could be constructed as the framework wherein all specific manifestations of life would then be encoded.

The relationship of scientific representation and phase space is, however, a very strong one. This is because the paradigmatic model of modern science is Newtonian physics. Since the culmination of the scientific revolution in the seventeenth century, physics has been the model of scientific work. Even the cultural or humanistic disciplines are involved in a controversy about their epistemological status wherein the naturalistic approach is confronted with a human-centered, historical or hermeneutic approach. Whatever the result of this controversy might be, the matter illustrates the persuasive strength of the naturalistic model of science, which is itself based on the paradigmatic case of physics.

The Newtonian mode of translating phenomena uses phase space as its basic mode of representation. Despite the idealization of phenomena involved in this kind of epistemic construction, the explanatory and predictive power of this mathematical tool is the historical and epistemological root of the technological revolution we are experiencing. We are sure this is a good part of the reason for its having such a great influence on the development of other disciplines. However, phase space is one among many possible spaces of scientific representation sharing basic conditions of closure. They are:

i. *Logical closure*, that is, maximized coherence and consistency of the system of propositions.

ii. *Linguistic closure*, that is, limitation to a finite – even if dynamic – set of epistemic variables whose nuclear meaning is shared by the members of the epistemic community.

iii. *Observation closure*, that is, the availability of well-defined mechanisms for empirical intersubjective validation employing a body of relevant empirical results.

By "closure" we do not mean the fixation of the logical, linguistic or communication systems. On the contrary, closure is considered here as a property of dynamic systems of knowledge that are transformed by continuous readaptations to internal and external conditions. The condition of closure is reached when a system has a structure wherein the elements are related in such a way that they support each other and enable the whole structure to maintain its continuity. This condition does not reject the possibility of change. As in biological systems, the structures of systems of knowledge evolve from an original condition through variations in the process of autoreplication (Luhmann 1996). This, however, is a topic for another discussion.

## CONCLUSION

Epistemological validation of a given level of organization as an object of scientific inquiry requires that three conditions be met:

1. that there be a basal natural kind of phenomena that is intersubjectively recognized by the relevant epistemic community;
2. that there be structured and well-established mechanisms of observation that translate the *basal phenomena* into stable and controlled *scientific phenomena*;
3. that there be spaces of representation with logical and semantic closure wherein the phenomena are converted into a symbolic object of analysis.

To conclude, we propose the following operational definition of a level of organization: *A level of organization in science is a phenomenal space that operates univocally as a structural maxima of closure in observability and representation.*

*Tecnologico de Monterrey, CEM*

## NOTES

[1] By "empirical" we do not mean necessarily a direct experience of phenomena; as in the case of the origin of the universe or quantum phenomena the empirical evidence could be indirect.
[2] The reason for the subjective experience's being incommunicable is that it requires the translation of experience into language. While some structures of experience can have linguistic analogies, there is always a part of an experience that is not a structure but a phenomenal perception and it is therefore untranslatable into language.

## REFERENCES

Hacking, Ian. *Representing and Intervening*. Cambridge: Cambridge University Press, 1983.

Hanson, Norwood R. *Patterns of Discovery. An Inquiry into the Conceptual Foundations of Science*. Cambridge: Cambridge University Press, 1998.

Luhmann, Niklas. *La Ciencia de la Sociedad*. Trans. J. Torres. Mexico City: Universidad Iberoamericana, 1996.

Rheinberger, Hans J. "Experiment, Difference, and Writing II. The Laboratory Production of Transfer RNA," *Studies in History and Philosophy of Science* 23: 3 (1992), pp. 389–422.

Rosen, Robert. "Organisms as Causal Systems which Are not Mechanisms: an Essay into the Nature of Complexity," in R. Rosen (ed.): *Theoretical Biology and Complexity*. Oriando: Academic Press, 1985.

Salthe, Stanley N. "Formal Considerations on the Origin of Life," *Uroboros* 1: 1 (Mexico City 1991), pp. 45–65.

Villoro, Luis. *Creer, saber, conocer*. Mexico City: Siglo Veintiuno Editores, 1998.

Wimsatt, William C. "Reductionism, Levels of Organization, and the Mind-Body Problem," in G. Globus et al. (eds.): *Consciousness and the Brain: a Scientific and Philosophical Inquiry*. New York: Plenum, 1976.

MILAN JAROS

# MACHINIC INSCRIPTIONS OF FRAGMENT OBJECTNESS

## AN OVERVIEW

Human have acquired the ability to make heterogenous bodies and networks. This renders obsolete much of the conceptual ammunition legitimating Galilean modernity and the meta-narratives of Progress. In particular, in the course of the last decades of the 20th century the link between material reality and representations (artistic, literary, mathematical) of reality grounded in Kant's Critiques has been fatally weakened. The crisis of representation became the crisis of the critical function of representation, Those attempting to retain the critical function at any cost have turned their barbs against the very assumptions that gave rise to this function. Whether as a result of Greenbergian modernism or Derridean textual postmodernism, the real has disappeared.

The turn away from Kantian dualism legitimated by Husserl's phenomenology inspired two generations of brilliant thinkers (e.g., Heidegger, Levinas, Derrida) who developed Husserl's thoughts well beyond his terms of reference. They effectively closed off the debate concerning the relevance of the Galilean paradigm and prepared the ground for a new agenda. How do we account for the authenticity of human experience sited in the body without restricting its access to the real, to its grounding context in the material condition of humanity today? For geo-philosophers Deleuze and Guattari this is a call to radically reappraise human experience in its new heterogenous habitat. Their "disciple" philosopher-engineer of intelligent machines Smith offers an ambitious model of "thingness" embracing the full range of human experience today. The real reappears via the dynamics of the act of registration in the rising and decaying assemblages of quasi-objects.

If a system of signs (e.g., "art") is to regain access to the enchanted real of the rising and fading of assemblages, the mind must be freed from the divisions of the material world deposited in the unconscious by the Galilean centuries. This is a call for a new "material base", a new "order of things" that determines the network of invisible rails along which thoughts travel and collide to create new thoughts and new rails. This is a move from text to territory, to maps of territorialised events. A point of departure for such a research program is a material counterpoint to abstract projects like Husserl's and Heidegger's. It is a graphic (visual) counterpart to the constitutive



*A.-T. Tymieniecka (ed.), Analecta Husserliana LXXVI*, 233–246.

knowledge maps of (linear) progress, a collection of graphic records of "truth" in which material images render visible the underlying scientific, moral and aesthetic concepts. It is a record in which abstract constructs will have become act-objects and their classifying and evaluative impositions will have been neutralized by turning them into signs on a map of the territory in question. That way it might be possible to render visible the bare nonlinear character of concepts, the multiple meanings (presence in many different domains of material life) that previously were held to be autonomous. Paradoxically, it would appear that such programmatic attempts to escape the grand narratives of modernity leave us as the only stable point of orientation for our "journeying" the traces of "Galilean" inscriptions. In the absence of "traditions" we depend on thought processes and signs associated with pseudo-mathematical forms of description of change and order (of symmetry, analogy, induction, translation, superposition, etc.) as the organizing principles with which to identify and share an event.

## FROM COSMOS TO EXHIBITION HALL

Galilean modernity is grounded in the separation of humans from things, in the rejection of the ancient ideal of the undifferentiated experience of life. The Cosmos of Aristotle was an organism, a purposeful unity of gods, things, and humans. Galileans replaced the intuition of ancient Greece with sensory perception. "Pigginess" and "final cause" were replaced by universal variables like force and mass. Reality is what can be measured and quantified. "The greatness of the natural sciences consists in their refusal to be content with an observational empiricism, since for them all descriptions of nature are but methodical procedure for arriving at exact calculations." (Husserl, 1965, p. 151) When Kant set out to reconcile philosophy with Galilean science he divided the world into two realms, phenomenal and noumenal. For Kant knowledge is the knowledge of phenomena, the domain of Pure Reason. What cannot be present in time and space, the beginning of time, God, freedom, belongs to the realm of Ideas. They are thinkable but unknowable. In Kant's *Critique of Practical Reason*, Ideas acquire a positive content via the moral law. However, this law is a mere potentiality. One cannot "know" the moral law. For if one knew how to be moral, then morality would be the object of science! However, because the unknowability of the moral law is known, it can still orient human actions!

The *Critique of Pure Reason* is concerned with what "is". The *Critique of Practical Reason* is concerned with what "ought to be". Any actualization of such theoretical necessity depends on the connection between this necessity and individual freedom. This connection belongs to the realm of reflective judgments, of aesthetics and powers of representation. For Kant the sphere of reflective judgment is also autonomous. The necessary condition that legitimizes this autonomy – and consequently the communication content of the representational faculty of man – amounts to a demand of autonomy for subjects and objects. The world of things – nature – as a neutral referent is Kant's key assumption even when the object in question is a product of sophisticated human preparation and labor.

## THE PERFECT CRIME

The Art defined by the *Critique of Judgment* is the actualization of individual freedom, a way of connecting the necessary transcendental truths about what is and what ought to be with the experience of living. To participate in the project of emancipation and progress underwriting modernity means that art must fulfill its critical role as a guardian of this link. In Greenberg's words, "to be a modernist work is to be a work that takes its own conditions of possibility for its subject matter, that tests a certain number of the conventions of the practice it belongs to by modifying, jettisoning, or destroying them, and that in so doing renders the conventions or conditions thus tested explicit or opaque, revealing them to be nothing but conventions". Yet it was the very critical function of modern art that led it to challenge the autonomy of the material world in which its legitimacy was grounded. "With the onset of modernism painters began to challenge the technical–aesthetic conditions deemed necessary to identify a given thing as a painting" (De Duve, 1998, p. 378). Duchamp appears to have reduced art to its "necessary and sufficient conditions". Artists "... instead of bearing upon qualities contained in those conventions began to bear upon those conventions themselves". This takes us from "this painting is beautiful" to "this is (not) a painting" (ibid, pp. 377–8). Thus Duchamp the modern artist read the binary choice between a painting and not a painting as a convention. Duchamp's objective is simply to mount a challenge to this convention. Duchamp's experiment – whether interpreted in the Greenbergian modernist fashion or in the textualist "postmodern" manner of Michel Foucault – proved that the Kantian pure (autonomous) object was dead. This is the "perfect crime"! (Baudrillard, 1996) Indeed, everything, miniatures, frescoes, pictures, vases, statues, seem to have lost their properties as (autonomous) objects. The crisis of the critical function is also the crisis of the concept of representation.

To act "disinterestedly" is the necessary condition for access to truth and freedom, for access to reliable representations of reality. Since this freedom is freedom only if it remains potentiality, those who are not knowingly pursuing disinterested inquiry must be a majority. For if everyone were a conscious Kantian, there would be no world of events for the Kantian subject to inquire about! The success of Kantian modernity amounts to its irrelevance. Twentieth-century science showed that Newton's laws of motion are not truly universal (e.g., that there are physical processes not correctly described by Newtonian mechanics). No system of scientific knowledge is universal in the sense of the *Critique of Pure Reason*. The grounding condition of modernity (the Kantian autonomy of pure reason; "science", practical reason; "morality", and judgement; "aesthetics") turned out to be a mere approximation. Within its domain of applicability (say, macroscopic reversible events at nonrelativistic speed) there is, of course, nothing "ambiguous" about Newton's laws. Whatever social (linguistic) constructs we put on a set of phenomena, we can always within the chosen rules of measurement produce agreement on the results of measurement. For example, the power and bandwidth of a laser beam used to perform an eye operation may well be replaced by other linguistic constructs, but the operation/cut remains the same and is reproducible; it is reliable ("objective") within certain specified limits. This "objective reality" (the patient can see again) and the world of things associated with it has therefore not disappeared. However, the questions about how this "finite" knowledge reshapes the way people think, e.g., how the actualization of a particular knowledge system affects other knowledge systems (i.e., the perpetual battle over the boundaries and overlaps of meaning and the applicability of competing knowledge systems) lie outside Kantin metaphysics. For they imply that knowledge is not detachable from those (humans) who use it.

When it became apparent that the (utopian) Kantian critique failed the self-imposed criteria of legitimacy, the artist turned to the critique of utopias. Instead of guarding and promoting the conditions that make the critique possible, the avant-garde in a desperate bid for survival criticized the vigilance itself! By pointing to its own history it went out of its way to argue how art had overstepped its intended (Kantian purposiveness notwithstanding) purpose and how it managed to negate social reality in works that no longer anticipated any liberation. Although artistic activity of this kind formally retains its critical function, the freedom of the artist is only adequate enough to show how impure, alienated and illusory such freedoms are. In the condition of advanced capitalism such freedoms are no more than an alibi for justifying bourgeois privilege and consumerism. In the last decades of the 20th century the real "returns" in the form of an injured body; the art lover's pleasure is portrayed as something morbid or at best ambiguous (e.g., Foster, 1994). The assumptions on which the legitimacy of art's critical function came to be based in Kant's Critiques seem almost forgotten, passé, buried under the debris of social and technological "progress". Yet with the fall of the autonomy of the Kantian spheres of enquiry, "art" appears to have been abolished. Or – if one prefers the postmodernist jargon – anything could be a work of art (Danto, 1997, p. 184).

If the basic assumption of Kant's Critiques grounding the autonomy of pure reason was the separation of the subject and the object, of the phenomenal and the noumenal, then any radical critique of the consequences of this position must abolish these divisions as a point of departure. This is the challenge taken on by Phenomenology. "The physical identity of an object is intelligible only through the acts in which the object is present to consciousness since it is precisely the relation to consciousness in these acts that makes it an object." (Lauer, 1965, p. 12) "Phenomenology thus helps the partial sciences ... by clarifying their partiality ... and by showing that what they identify can be seen from perspectives they do not enjoy." (Sokolowski, 2000, p. 209) For Husserl consciousness is tied to intentionality. This is the "law of awareness", i.e., the necessary relation of every conscious act to an object. To reveal and grasp the meaning of the intentionally given in its purity (essences), one has to perform phenomenological reduction. This launches a radical departure from Kantian dualism and Hegelian dialectics, a move that has since informed every serious challenge to sterile historicism and positivism.

Contrary to Kant's Critiques, for Husserl both the form and the content of reality are in phenomena themselves (appearances). The world to be studied is, then, not the world of subjects and objects but the "environing world" or *Umwelt* of the spiritual subject. Indeed, science itself is a product of spirit and cannot be investigated by the type of method whose object is nature! Science is concerned with what things do, with the state of things described by functions. The object is intentionally constituted in acts of consciousness. Yet how is this absolute knowledge communicated to another, how is it brought into contact with or projected upon the contemporary material condition of humanity? In this regard the certainty gained by phenomenological reduction might appear no more useful than the Kantian disinterested critical truth. It is worth noting that other distinguished critics of Kantian modernity such as Heidegger and Derrida fare no better; they too stop short of working out the consequences for technicity of their version of the critique of Western metaphysics, and of Kantian and Hegelian philosophy in particular. Nor do they address with any specificity the ways in which the transition from the Galilean to the contemporary status of "things" might enter their discourse. When Heidegger asks "What is a thing?" (Heidegger, 1976), he turns with his usual brilliance to Van Gogh's shoes! Although the issue of access to and the status of materiality has always been tormenting phenomenology (see this comment on Derrida's *Edmund Husserl's Origin of Geometry*, "This is because the sense to which we have access can always not be incarnated" – Rapaport, 1989, p. 181), nowhere does one find reference to the agenda of manmade heterogeneity or networking. For Heidegger technicity is the mere actualization of metaphysics, a pure negativity or irrelevance. Even Einstein's demolition of Kant's a priori forms of perception (space and time), legitimating intellectual space for truly post-Kantian models of consciousness, earned him "credit" for rediscovering Aristotle (Heidegger, 1992, p. 3E). For Levinas order (law) and otherness are irretrievably separated from material conditions. As in the Old Testament, they are laid down prior to nature. When Derrida (1994) looses time from any form of organization and establishes the undecidability of any such form, he stops. He would not go on to describe time's inscriptions as the very experience of time by humans. The material is just one such organization to which time is not reducible. Yet without the material embodiment of "life" we could not "experience" time.

## THE RETURN OF THE REAL: JOURNEY-AS-ACT OBJECT

For Deleuze and Guattari distinguishing the subject and the object is a "poor approximation" to thought; "thinking is neither a line drawn between subject and object ... thinking takes place in the relationship between the territory and the earth" (Deleuze and Guattari, 1994, p. 85; Bosteels, 1998, p. 145). This is a move from text to territory, without necessarily denying veracity to deconstruction and textuality! The Deleuzian intervention turns to novel "spatial" metaphors ("territory"). Instead of universal variables (that are criticized but still retained by deconstruction), instead of worries about aporias of time and law, they bring out events by examining (mapping) the way they depend on local boundaries and links, on their precursors.

A map is not just a mirror (representation) of "nature" (phenomena). Mapping avoids specifying subjects and objects, and how they interact. It is about pathways, not about cause and effect. The unconscious is a rhizome of "machinic interactions through which we are articulated to the system of force and power surrounding us". "More than ever today nature has become inseparable from culture" (Bosteels, 1998, p. 156). This cartography requires a view of incorporeal events that are neither given in advance as in Newtonian science nor generalized after the observation or application of laws of nature to data. It is neither deductive nor inductive. Experiment is now a world creating event, not a disinterested application of a priori laws! "The map expresses the identity of a journey, and what one journeys through" (ibid., p. 167). It matches with the object when the object is movement. It engenders the territory in question, it territorialises.

What is a "thing" in the age of clones and intelligent programs? Smith (1996, p. 36) proposes to address the problem by turning to a dynamic definition of objectness and intentionality via his concept of registration. "We should not think about what it is to be an object but what it is to act, ... or be treated as an object". Intentionality will be reconstructed not in terms of "meaning" but in terms of "being meaningful". The goal is "to understand how a conception of objects can arise on a substrate of infinitely extensive fields of particularity" (ibid., p. 191). Registration means "to find there, to be a certain way, to carve the world into"; "being a computer is a question of a fit, not of architecture". So an object cannot be an object "on its own". One has to work in order that an object (e.g., a desk) remains the kind of thing it is! Hence physics is just one registration of the world, namely, one in which objects are eliminated at a particular level.

In their last joint work Deleuze and Guattari made a new turn: "are there functions ... of concepts"? They present a challenge to the next generation of thinkers: ground the runaway generation of concepts, ground "concepts" in "functions" (i.e., states of the material world dealt with in, say, physics, design, art and engineering) (Deleuze and Guattari, 1994, p. 162). This is, they proclaim, a task "only scientists" can fulfill! *Phusis*, inquiry into "the coming into radiant being" was, for young Deleuze as for the Stoics, at the heart of philosophy, and particularly at the heart of the question of being (Mullarkey, 1997, p. 446). This is a new turn, for to attach a function to the Deleuzian concept is in effect to "close", to then "deposit in the concrete material context". The subject rises out of the homogenous background via the creative act of event ("object") generation. The thought is "engendered" in the act of registration. But these must remain "philosophical", not social" or Galilean (i.e., quantifying) moves. To attach a function to a concept is to connect the subject with the material specificity of the real, to complete the act of registration making it communicable, recordable, and translatable (e.g., related to other registrations).

The difference is fundamental. In the act of registration the idealized Galilean state of things meets (is projected upon) the (quasi-local) boundary conditions which turn it into an event (bring it into the world). In this language to ground the act of registration (make it amenable for communication) then amounts to specifying a (finite) set of concept-parameters and boundary (spatial and temporal) conditions accounting for the given state of matter and its evolution. The event becomes "actualized" (assigned material and temporal finitude) in the (multiple and contingent act(s) of) registration. The transition from one registration to another can now be identified in terms of a transformation matrix that converts one registration into another. This matrix is a scheme, an operator to take us from one set of "parameters" or "variables" to another. It has a meaning on its own (e.g., as a multidimensional symmetry or order parameter), outside of any "calculations". Now, of course, in addition, the theorist is herself "part" of the boundary markings. And she is herself inscribed by the divisions that the principle of the imposition of boundary conditions demands, by (hidden) coding and (invisible) pathways that make the event registrable today.

Once registration has taken place both the "thing" and the "concept" or "knowledge" of the thing are "intermixed". They contain an extra (partially material and partially virtual) component. It is also this "mixing", this fragmented object-value (perceived as culture) "produced" by registration, that makes the event (quasi-object) a "recorded", communicated event. The "value" of any given description is simply the measure of the success of an individual in performing the "subject-and-object engendering" matching process and maintaining it. If I then say that my experience is traumatic (Foster, 1994, p. 130), I am really saying that the divisions my mind is equipped with are too rigid and limited. My ability to participate in the "journey" is inadequate for the challenge before me! The Nietzschean "everything is permitted" now means that a particular form of registration has been challenged. Neither the modernists De Duve and Foster, nor the post structuralists Deleuze, Guattari, and Smith address this question directly; their work remains programmatic and critical. Yet it has been recognized that the most difficult problem for Deleuzian theory of culture, and material culture in particular, is – as it was for Husserl and Heidegger – that of how to express (record, communicate) "the coming into itself" of a fragmented act-object (Buchanan, 1997, p. 486).

## ART AND TERRITORY: A RESEARCH PROJECT

How is this re-appraisal of human finitude instrumental in opening access for human expression (e.g., artistic, mathematical) to the "real"? How do we "research" the heterogenous space that is territorialised by our journey?

A painting by Raphael is a "representation" of reality, a rendering of a model, be it in the shape of Christ's body. Three centuries later Paul Cezanne sits for days in front of rotting apples. This representational mode of art's access to reality is communicated by assembling paintings in an exhibition hall. A learned critic is called upon to invoke one of the metanarratives to justify the "installation" (classification, valuation). Even the enchantment of the gardens of Villa d'Este in Tivoli may well reach the learned mind in the form of architectural and poetic metaphors depending for their effectiveness on certain "theoretical" constructs instilled into the educated mind by real or virtual tours of (Hegelian) museums and books about museums. If "art" is to open access to the "truths" of the rising and fading of assemblages, of which the artist herself is a part, it needs a generic attempt to escape the divisions in the mind deposited there by two centuries of modernist "paradigms". A base for such a re-appraisal cannot be grounded in another totalising narrative. This calls for a material counterpoint to abstract projects like that of Heidegger, whose chief aim was also to oppose the Kantian distinction between nature and the experience of nature, for providing a graphic alternative to aesthetics generated out of an epistemic system. This calls for a collection of graphic records of "truth" in which material images make visible the scientific, moral and aesthetic concepts associated with it. This calls for a record in which such abstract constructs have been turned into act-objects and their classifying and evaluative impositions neutralized by turning them into dots on a map of the territory in question. In this way it might be possible to render visible the non-linear character of these concepts, the multiple meanings (present in many different domains of material life) that previously were held to be autonomous.

A project conceived under similar programmatic guidelines was planned and partly implemented – albeit for very different reasons – by Walter Benjamin (1999) in the 1930s. Indeed, his aim was not a Heideggerian debunking of technology and the ideologies of Progress. He wanted to recover under the fragmentary appearance of reality the remains of an irreducible material trace of Ur-history, of a pseudo-Marxist "originary" history of humanity that for him was grounded in the Utopia of a classless society. Benjamin thought that the ideology of technoscientific fantasies was not Marx's false consciousness but a source of positive energy, a living proof of the redeeming power of the modern era. He saw the world of things as a neutral background even if no longer part of a shared continuum of narrative history (myth, tale). He saw himself as an archeologist facing the fossil rich layers of (bourgeois) material culture in which to recover the Ur forest of eternal fragments. To "see" these fragments (be they ornaments or roofs) as "eternal" and to believe that their "essence" can be communicated by unconcealing them as always-the-same is to presuppose an impressive degree of collective memory of the tradition in question! Only a few mandarins of the academe of 2000 CE might be blessed with such background knowledge, perhaps the very "Hegelian" knowledge Benjamin wanted so much to undermine. It would seem that Benjamin shares with his contemporaries Adorno, Husserl, and Heidegger the paradox of depending on those very systems of knowledge he wants most to dismantle. His Copernican Revolution can only be effective in the tutored minds produced by the educational establishments wedded to the "age of ideologies". When Benjamin wanted to "assemble large scale constructions out of the smallest and most precisely cut components", "to discover in the analysis of the smallest individual moment the crystal of the total event" (ibid., N2, 6), he was thinking of some pre-modern Ur-objects, fragments and debris of narrative wholes (fossilized into stone blocks, columns, beams, street shops or corners) not of man-made-and-lived nanostructures and genetically engineered bodies living as dynamic nodes of a vast network of virtual messages! His notion of "things" was static, cast in the language of universalistic modernist theories (pseudo-Marxist, Jungian, positivistic, etc.) that he was about to make redundant. Benjamin sees the multiplicity of presence and meaning that contradicts Hegelian and Marxist metanarratives but believes he can make sense of them by redeeming their originary meaning in a classless Utopia. They are reducible to ontologically stable static parts such as men, women, ornaments, mirrors and shop windows even if under different disguises (meaning) depending on contexts (place, time, function).

In fact, in the course of the evolution of his project, Benjamin gradually departs from his initial pseudo-Marxist Expose of 1935. Nor does he please his rabbi friend Scholem. The theory gradually dissolves in the manifold patterns of his graphic records. His Jungian archetypes and witty citations become just one of many "fossilized" act-events gathered or at least anticipated in the various "convolutes" into which the Parisian "data" was to be deposited. Half a century later the traditional bonds that had already been fragmented and scattered in Benjamin's rendering of reality of the 1930s have been fatally weakened by victorious Capital. The contraction of time and space brought about by the rise of postwar electronics turns the eye of an archeologist into the eye of a "multimedia" producer. It is no longer sufficient to update the renaissance cone of perspective with the Lacanian "gaze" (Foster, 1996, p. 139). A new "dynamic" dimension has been attached to the notion of the "real". The result is the astonishing tolerance by humans (including toddlers!) today of apparently arbitrary assemblages of artefact-simulacra in rapid motion. Yet this does not inhibit the educated consumer from making choices – and sacrifices in order to be able to afford such choices! What is it that acts and is recognized as acting behind the chaotic face of the real with sufficient authority to bring humans to the screen or into a gallery and gives them the confidence to take part, to choose? Perhaps under favorable circumstances there rise to prominence amidst the debris of overlapping fragments of the metanarratives (as one of the few remaining ways of access to the order of things) the structure formative mechanisms associated with the pseudo-practices mimicking the Galilean (linear) and the post-Galilean (fractal) machinic mathematisation of experience. The "consumer" is all too familiar with recognizing them as significant (as objectified laws of science-as-God in the material form of artefacts). The trickster who pulls the strings animating the act-object assemblages of today is no longer the Benjaminian shaman-ideologist but a pseudo-mathematician manipulator-inscriber.

Indeed, it was just such pseudo-mathematical inscriptions, i.e., marks of repetition (versus rhythm), limit (versus excess), instability (versus metamorphosis), complexity, the approximate, etc. that Omar Calabrese (1992) chose to characterize the transition from the language of cultural theory grounded in "tradition" to the "sign of the (neo-baroque) time" of today. Is it not possible that the order of visible signs today is recognizable not by the "meaning" granted by a place in shared, be it fragmented, narrative, not by the enchantment inherited from poets, but by a (real or virtual) "driving engine" registered dynamically in a (real or virtual) act-object? It is also this reference to the type of motion or relation, located at the site of experience, that makes the event recordable and communicable. This might constitute the most stable reference – be it in the form of unconscious "inscriptions" – under the overload of overlapping significations whose apparent "meaning" collapses under slightest challenge (e.g., a test of their "positive" veracity). In the catalogue introducing an exhibition of works of fifteen artists called ABRACADABRA (The Tate Gallery, London, September 1999), Nicholas Serota, Catherine Kinsley, Catherine Grenier and others openly declare their distaste for the legacy of the age of avant-garde theories, for the fights and gloom of late modernism. While cherishing the technical advances of the avant-garde "none of the artists wants" – in the words of Jemina Montagu writing in the Exhibition Catalogue (p. 40) – "to pick a fight", "deny or fight reality", be "plunged into despair by the fall of Utopias". For them "art is no longer a place to plant their flag but a territory of exchange", art that needs to be "animated". Consistently with their position, the verbatim of the explanatory or "programmatic" passages is greatly simplified and avoids much of the conceptual gymnastics that is obligatory for those claiming or disclaiming links with ideological castings of aesthetics. Hence, the commentator is not a critic-interpreter but a translator-registrar. In carrying out her job she may have to draw attention to the virtual processes without which the shape of the quasi-object remains hazy or simply invisible. The articles in the catalogue introducing the individual artists are indeed attempts to identify the "local" variables and boundaries employed in the works in question. Instead of Greenbergian-modernist, deconstructive-textual, sociological, etc. "critique", instead of "discourse", they are there simply to assist the viewer in constructing "transmission" matrices that might, say, link the exhibits one to another, translate one territory into another. As for the works themselves, take for example "The Loop"" by Eric Duyckaert, it is an artefact in the shape of a loosened infinity sign several meters long. It is adorned with many different yet simple easily "identifiable" hanging parts (e.g., globes, triangles, boxes, bicycles, toys). In the absence of any "metanarrative" they can only be held together by the "microbonds" invoked when Deleuzian energy is "exerted" along the implied infinity sign. Does the assemblage come to life on the strength of the mind's ability to reach the many different "world-territories" in which bicycles, globes, roads, humans, signs, etc. – the "parts" displayed in the work might have had their place and function? The role of "metanarratives" such as the Hegelian Zeitgeist was not only to hold together but also to constrain the (number of) possible basic units (parts) and "readings" (constructs). This in turn imposed constraints upon artefact makers such as architects to construct their product out of just such a limited number of well-identifiable units and in a limited number of ways. For example, for many centuries "architecture" was a construct consisting of columns, capitals, walls, etc. The paradigmatic shift away from the "metanarratives" of the "age of ideologies" means that another way must be found for "matching" the globes, bicycles, boxes, humans, etc. that form the Loop, for instance, by reference to "usage" (real and virtual) in a modern city. Today city inhabitants from human bodies to rats are, first of all, dominated by techno-science. We are used to recognizing "things" by their function in dialing, shifting, turning, etc. They do that according to a prescription. "Evolutionary mutations" and their confrontations with the "problem of absolute symmetry", the cultural contradictions of harnessing "scientific discourse" in material signs of urban lifeworlds, these are, we are told by the author of "The Loop", the "variables" and "boundary conditions" (domains of applicability) of the Loop. The (sign of) machinic mechanism forms a "registrable skeleton" on which to "hang" the "chaotically attachable" components (garments, roofs, rails, etc.) that give a material meaning to the body of a quasi-object. Most of the decorative baggage one "sees" appears to be interchangeable with other equally powerful parts, colors and shapes. The whole retains a certain solidity and welcome to the visitor to "enter" into an act-object event mainly because of this virtual skeleton of machinic processes or at least a promise of it. It is this "promise", this bare machinic mechanism lurking from behind and remaining after one has stripped the assemblage of its "garment attachments" that is driven by the new trickster engine running on the pseudo-algorithms of our fractal age.

The post-Galilean and pseudo-mathematical seeks order (direction, bonding) by invoking "analogy" on the model of mathematics. It is this "analogy" that figures in all manuals and instructions today. In (linear) mathematics (in the construction of commonly used technology), the relation expressing the quality of two quantitative relations is definitely constitutive so that if three terms of a proportion are given, the fourth can be obtained. Analogy in philosophy or art can only apply to qualitative relations so that when three terms are given, only the relation to the fourth and not the term itself is obtained. In short, analogy yields a "vector" of "force" identifying the process qualitatively but stopping short of a definite closing step. It sends the inquirer into a sequence of searching steps, none of which is truly "successful". Artworks (graphic signs) may be regarded as material records of the failures to implement the expected actualisation of such analogies and the fractionation of experience resulting from such failures. The rise to prominence of "nonlinear" (fractal, heterogenous, post-Galilean) processes completed the mathematisation of nature by territorialising the remaining domains of experience (e.g., nonentropic effects, growth, population dynamics). In this post-Galilean realm of higher complexity, the Galilean expectation (of a linear contact-causal chain) fails. When the inquirer makes a step back to reassess her position she will be driven not into one but several likely answer-seeking sequences. The search "crashes" when the inquirer gives up any hope of finding a stable answer among the overlapping solutions created by the clash between Galilean expectation and the consequences of the post-Galilean multiple character of the interactions (so ably documented for instance in Calabrese's monograph). "Passagen 2000" could be conceived as a book of unfinished maps of journeys-as-act-objects of today. If this is repetition of the same, then the repeats are governed by a Platonic God-Geometer. However, its subjects and objects have been transformed and displaced – not back into a Greek Polis but into the web-like structure of communication networks of today.

*Centre for Research in Knowledge, Science and Society*
*Newcastle University*

## REFERENCES

Baudrillard, J. (1996). *The Perfect Crimes.* London: Verso.

Benjamin, W. (1999). *The Arcades Project.* Trans. Howard Eiland and Kevin McLoughlin. Cambridge: Harvard University Press.

Bogue, D. (1997). "Art and Territory", *The South Atlantic Quarterly* 96: 3, pp. 466–82.

Bosteels, B. (1998). "From Text to Territory", in *Deleuze and Guattari*, ed. Eleanor Kaufman and Kevin J. Heller. Minneapolis: University of Minnesota Press, pp. 145–174.

Buchanan, I. (1998). "Deleuze and Cultural Studies", *The South Atlantic Quarterly* 98: 3, pp. 483–97.

Calabrese, O. (1992). *Neo-Baroque: A Sign of the Times.* Princeton: Princeton University Press.

Danto, A. C. (1997). *After the End of Art.* Princeton: Princeton University Press.

De Duve, T. (1998). *Kant after Duchamp.* Cambridge: MIT Press.

Deleuze, G. and Guattari, F. (1994). *What is Philosophy.* Trans. Hugh Tomlinson and Gragam Burchell. New York: Columbia University Press.

Derrida, J. (1994). *Aporias.* Trans. Thomas Dutoit. Stanford: Stanford University Press.

Foster, H. (1996). *The Return of the Real.* Cambridge: MIT Press.

Heidegger, M. (1976). "What is a Work of Art", reprinted in *Philosophies of Art and Beauty*, ed. Albert Hofstadter and Richard Kuhns. Chicago: Phoenix. (1992). *The Concept of Time.* Trans. William McNeill. Oxford: Blackwell.

Husserl, E. (1965). *Philosophy and the Crisis of European Man.* New York: Harper.

Lauer, Q. (1965). *Phenomenology and the Crisis of Philosophy.* New York: Harper.

Mullarkey, J. (1997). "Deleuze and Materialism. One or Many Matters?," *South Atlantic Quarterly* 96: 3, pp. 439–63.

Rapaport, H. (1989). *Heidegger and Derrida.* Lincoln: University of Nebraska.

Smith, B. C. (1996). *On the Origin of Object.* Cambridge: MIT Press.

Sokolowski, R. (2000). *Introduction to Phenomenology.* Cambridge: Cambridge University Press.

JOSÉ LUIS BARRIOS LARA

# LOS BORDES IMAGINARIOS DEL ASCO Y EL MORBO: UNA FENOMENOLOGÍA DEL TIEMPO EN LAS FRONTERAS DE LA ANIMALIDAD EN EL CINE DE PIER PAOLO PASOLINI Y DAVID CRONENBERG

## INTRODUCCIÓN

Si pudiéramos sintetizar en tres palabras la reflexión filosófica del siglo XX, sin duda ellas serían: cuerpo, otro y tiempo. Los problemas de género, de multiculturalismos, de deconstrucción, de globalización, neotribalismo, y tanto otros, son variables muy importantes de estos conceptos. Pensar la cultura y la sociedad contemporánea desde ellos, significa asumir un descentramiento del sujeto y de la conciencia como génesis de organización del mundo. Descentramiento que no tiene que ver con las lecturas posmodernas del fin de la historia y de la muerte del arte. Sencillamente tiene que ver con la posibilidad de asumir el problema de la diferencia como el problema de nuestra época.

En este contexto, los estudios de la fenomenología y la hermenéutica francesa ocupan un lugar imprescindible. Merleau-Ponty, Emmanuel Lévinas, Jean Louis Chrétien, Phillipe Némo, Paul Ricoeur y otros han dedicado buena parte de sus investigaciones al problema de la diferencia. Problema que por lo demás tiene una clara referencia a la corporeidad como dato fenomenológico irreductible de toda originalidad insustituible del sujeto y del otro. Si algo caracteriza el pensamiento fenomenológico francés, es la importancia que tiene la sensibilidad, el sujeto encarado, en la construcción del sentido del mundo. En esta tradición fenomenológico-hermenéutica se ubica, en primer lugar, este trabajo.

Pero también se contextualiza en la estética de lo grotesco del arte del siglo XX. Estética que está definida y si no cuando menos orientada, desde la concepción que relaciona lo grotesco con el erotismo y la subversión. Me refiero particularmente a las estéticas de la transgresión que remontándose al Marqués de Sade, nos llegan a través de los poetas malditos y sobre todo de la obra de Georges Bataille.

Atendiendo pues a la fenomenología y la estética descrita, este trabajo abordará la construcción del valor de lo grotesco de estas estéticas en las variaciones emotivas y sensibles del asco y el morbo.

Para hacerlo me centro en el fenómeno artístico del cine. La invención de este lenguaje, no sólo para la historia del arte, sino para la cultura en su



*A.-T. Tymieniecka (ed.), Analecta Husserliana LXXVI*, 247–262.

totalidad, significó un cambio en el paradigma narrativo de la humanidad. El cine introduce el tiempo objetivo como condición de construcción del sentido en la textualidad. Tiene un enorme poder de convocatoria y funciona como un mediador social incuestionable en el mundo actual. En un sentido más preciso, el cine cambio el modo de construir la trama en la cultura: introdujo el poder del "como si fuera verdad" en el arte, lo que supone un sentido de la vivencia estética inédito en la historia (Gilles Deleuze, *L'image,* 165).

En síntesis, este trabajo quiere elaborar las reducciones fenomenológicas del asco y el morbo a través del análisis de dos películas – *Saló y los ciento veinte días de Sodoma* de Pier Paolo Pasolini y *Crash* de David Cronenberg – y mostrar como estas dos vivencias responden a dimensiones muy específicas del sentido del cuerpo, del otro y del tiempo de la cultura contemporánea.

Una precisión más: los argumentos y análisis aquí expuestos son parte de mi investigación con la que obtendré el doctorado en Historia del Arte en la Universidad Nacional Autónoma de Mexico, por lo que en buena medida son apenas esbozos del trabajo que estoy realizando.

## 1. LA PREMISA: EL PROBLEMA ESTÉTICO FENOMENOLÓGICO DEL ASCO Y EL MORBO

En la tradición del pensamiento estético es prácticamente inexistente la reflexión sobre el asco y el morbo. Lo poco que existe lo contextualiza dentro de la estética de lo feo y de lo cómico y como una variable de lo grotesco. Esto se debe a dos cosas. Primero, la producción artística que pudiera ponderar estos dos valores los relaciona más bien con sentidos edificantes del arte y con los usos político-sátiros de subgéneros de la comedia; segundo, la emancipación, real o supuesta, de estos valores funciona en el contexto de lo que se ha dado por llamar posmodernidad y que yo prefiero simplemente nombrar como cultura contemporánea. En cualquier caso, algo se pone en evidencia a la hora de reflexionar al respecto: los usos del asco y el morbo, más allá de los contextos de lo sublime o lo cómico, es son fenómens que se desarrollan en el del arte contemporáneo. Si bien podemos reconocer importantes antecedentes en el culto a la reliquias en el medioevo o en buena parte de las manifestaciones de la pintura religiosa del barroco-donde están asignados al infierno, –, e inclusive en la literatura del siglo XVII de Quevedo; o siendo aún más radicales, en la literatura francesa del siglo XVIII con las obras del Marqués de Sade; estos "valores" se asumen como condición "legítima" del arte hasta el siglo XX.

Los antecedentes estéticos de esta ponencia se remontan a los trabajos que de fenomenología he realizado. En ellos he abordado el modo en que la corporeidad funciona como un elemento fenomenológico trascendental en la construcción de los valores de lo bello y lo sublime.[1] Parte importante de lo que ahí se afirma es la función que el cuerpo tiene, como analogado vital, en la construcción del sentido de las artes visuales y cinéticas. El cuerpo articula modos fundamentales de intencionalidad motriz que determinan en buena medida los usos estéticos y simbólicos del arte. Son básicamente cuatro las articulaciones fenomenológico-vitales del cuerpo con las que se intenciona el mundo: lo vertical-horizontal, lo interior-exterior, el arriba-abajo, y un lado y el otro. Intencionalidades motrices que a la vez se relacionan estructuralmente con el movimiento y la organicidad corporal.[2] Estas estructuras posicionales y situacionales del cuerpo son, desde la perspectiva fenomenológico-hermenéutica aquí supuesta, el equivalente sensible-vital de las formas puras de la sensibilidad de la teoría del conocimiento kantiana.

Sin duda estas estructuras fenomenológico-trascendentales están en una relación indisoluble; sin embargo, hay ciertas constantes que dominan a la hora de construir valores estéticos determinados. Por ejemplo, en la construcción del valor estético de lo bello, tanto la configuración de la obra como la refiguración por parte del espectador, se fundan en estructuras vitales tales como lo vertical, el movimiento y la acción corporales. Esto, en otras palabras, significa que lo bello tiene que ver con una posición y una situación corporal en que entra en juego el dato fenomenológico del mundo como lo representable a travé de lo dominable, medible o lo a-la-mano, como lo diría Heidegger.

Desde esta perspectiva es fácil adivinar la estructura fenomenológico trascendental a partir de la cual se configura el sentido estético de lo grotesco en sus determinaciones específicas del asco y el morbo. En el morbo y el asco, como formas de lo grotesco, entran en juego aquellas mediaciones simbólicas y pragmático vitales que ponen en peligro la interioridad orgánica y la integridad motriz de la corporeidad. En otra palabras, abren la dimensión vital inmediata del horror-fascinación, a través de la disolución o fragmentación de la unidad motriz u orgánica, lo que significa el acontecimiento metafísico de la muerte en "carne propia".

Esto explica, desde mi planteamiento, las tipologías básicas de las mediaciones simbólicas del miedo y el horror como reacciones propias del asco, pero al mismo tiempo dan razón de ser la atracción que es lo que corresponde al morbo: 1. Las simbolizaciones que nacen de la interioridad orgánica y que se vinculan corrupción/ destrucción/ fragmentación y, 2. Las simbolizaciones de la alteridad del otro como lo monstruoso y lo contagioso.

La desintegración corporal, como vivencia fenomenológica del tiempo mortal, se relaciona con las formas de mutación, exhibición y transgresión de los límites:

> El foco de todo simbolismo de la contaminación es el cuerpo, asimismo el último problema al que induce la perspectiva de la contaminación es la desintegración corporal. El simbolismo corporal adquiere unas connotaciones altamente emotivas, todo lo que sea un desperdicio corporal es sinónimo de peligro. Todo aquello que hace referencia a los límites del cuerpo, que atraviesa sus fronteras (cualesquiera de sus orificios), que signifique restos corporales (de piel, uñas, pelo ...), que brote de él (esputos, sangre, leche, semen, excrementos ...), tiene el calificativo de altamente peligroso, de impuro. Siendo la contaminación más peligrosa la que 'se produce cuando algo que ha emergido del cuerpo vuelve a entrar en él, (José Miguel G. Cortés, *Orden y...*37).

Por su parte, la alteridad se percibe como aquello que pone en crisis el orden de la sociedad y la vida; la otredad como lo feo, lo monstruoso lo siniestro:

> ... la sociedad tiene miedo de todo aquello que parece extraño y raro, de lo que se escapa de la norma. Existe una profunda tendencia a parangonar lo feo y/o lo distinto con lo anormal y lo monstruoso. El sujeto ante lo informe, desordenado y caótico percibe un peligro que se cierne sobre su integridad, que pone en duda su seguridad y no puede soportarlo. Por ello necesita separar de su lado todo aquello que es diferente (35).

Estas nociones preliminares me servirán como andamiaje de las aproximaciones que haré al cine de Pier Paolo Pasolini y de David Cronenberg. El nivel de lectura que realizaré de sus obras se da en tres registros fundamentales: en la estructura formal-cinematográfica, en sus mediaciones simbólicas y en la fenomenología de la temporalidad que se muestran como dimensión última del asco y el morbo.[3]

## 2. EL ANÁLISIS. MIRAR Y SENTIR: LAS CONSTRUCCIONES DEL ASCO Y EL MORBO EN EL CINE DE PASOLINI Y CRONENBERG

A diferencia de los otros sistemas artísticos, el cine no construye su sentido a través de la ficción, sino a través de la falsedad.[4] En este lenguaje los acontecimientos y los hechos se viven como si fueran verdad. La imagen movimiento, introduce en el lenguaje del arte la posibilidad de integrar la totalidad de lo real a través del movimiento de la cámara, de la presencia de personajes y de la narración de una historia. En este contexto, el sustrato óntico del cine tiene que ver con la objetividad de la toma. Donde objetividad significa, sobre todo, la relación entre tiempo-movimiento y la referencialidad de la toma. La toma es entonces la unidad de sentido mínima de la cinematografía. Dependerá del tipo de toma el modo en que se genera la estructura básica de la intencionalidad en el cine.

### *1ER. CASO: SALÓ O LOS CIENTO/VEINTE DÍAS DE SODOMA, PASOLINI*

El cine de Pasolini se explica por medio de una estética que él mismo llama "cine de poesía". Contextualizado en la tradición de la semiótica italiana, a este director le interesó crear un lenguaje donde la imagen fuera en sí misma la máxima concentración de sentido. En el im-signo, como él define la imagen cinematográfica; "los arquetipos lingüísticos (...) son las imágenes de la memoria y del sueño o sea imágenes de comunicación con nosotros mismos" (Pier Paolo Pasolini 19). Éste no se concibe como mimetismo y naturalismo. La toma sobrepasa la referencia objetiva, en ella siempre está en juego el punto de vista del artista, la toma es ante todo la posición del que mira. Se trata, en palabras del mismo Pasolini, de la "subjetividad libre indirecta" donde la imagen analoga el sentido en términos de cuerpo y no de lenguaje. En Pasolini el cine no es lingüístico sino estilístico.[5] La intención se construye por el modo en que la imagen cinematográfica articula un punto de vista no narrativo, sino evocativo. La imagen cinematográfica es movimiento hacia la interioridad del sujeto. Nacida de lo mirado, la condición prejudicativa de la imagen cinematográfica, abre una dimensión donde el límite entre el exterior de lo mirado y el interior del que mira, se disuelve en el tiempo mismo de la evocación. Es sorprendente el modo en que Pasolini construye el sentido de sus imágenes a travé de la toma fija y a media distancia. Este recurso permite, según Pasolini, introducir al espectador en el sentido del sueño, pero además impide que el tiempo cinematográfico se identifique con la acción o la trama. El cine es pues, una construcción perceptual donde importa más la densidad y concentración del sentido que el deslizamiento de un posible significado a travé del discurso lingüístico. El cine como lo piensa Ingarden, facilita a través de la ampliación, la alteración, la sucesión y la fusión, la aparición de "... *eventos en su desarrollo total concreto* temporalmente extendidos" (Roman Ingarden 378). En el caso de Pasolini los eventos cinematográficos se relacionan con la analogía que se establece entre el sueño y la imagen cinematográfica: "... el cine es, de momento, un lenguaje artístico no filosófico. Puede ser parábola, nunca expresión conceptual directa. He aquí por consiguiente, un tercer modo de afirmar la prevalente artisticidad del cine, su violencia expresiva, su corporeidad oníríca: o sea, su fundamental metaforicidad" (Pasolini 18). El lenguaje cinematográfico es poético, se construye por la toma a distancia, el encuadre y la suma de instantes; lo que en otras palabras significa que no es discursivo. Esto explica por qué en el cine de este director, las imágenes son juegos hipnóticos, donde la acción o el suceso acontecen por la distensión interna de la toma. Recordemos por un momento la secuencia del círculo de la mierda en *Saló o los 120 días de Sodoma*, que sirve de preámbulo o prólogo a todo este círculo: en ella, la cámara siempre está a distancia. Esto permite dos cosas: hace soportable la secuencia y al mismo la acción se da en tiempo real. Esta escena es paradigmática de la poética de Pasolini. Así pues, la construcción de las tomas a partir de la estética del distanciamiento, estructura una intencionalidad donde el tiempo tiene que ver con la distensión de la acción a través de una cámara casi fija y donde la acción acontece a cierta distancia del espectador. Esto en otras palabras supone un aletargamiento del espectador, lo que en buena medida hace evidente la noción de lo onírico con la que Pasolini explica el recurso visual de su poética cinematográfica. En ella todos los elementos irracionales u oníricos están filmados en primer cuadro, lo que significa traerlos a la conciencia del espectador. El poder hipnótico de la toma permite que se liberen los niveles inconscientes y generalmente reprimidos por el espectador. De ahí que en el im-signo: "... los arquetipos son las imágenes de la memoria y el sueño son sean imágenes de comunicación con nosotros mismos" (19). La construcción del sentido descansa en el poder que tiene la imagen cinematográfica de sumergirnos en un estado intermedio de percepción entre la conciencia y el sueño, entre el yo y la alteración. Esto facilita la creación o irrupción de un espacio libre de las pulsiones primarias de la violencia, el erotismo y la perversión. En otras palabras, el recurso mismo de la imagen como subjetividad libre indirecta significa abrir un espacio incontrolable, donde se ponen en juego las formas primarias de la subjetividad y donde, de alguna manera, se entra en el juego de la fascinación-repulsión, propia de estos impulsos fundamentales.

Las mediaciones simbólicas de Pasolini se explican por una constante: el uso político de la sexualidad y sus perversiones. Uso que en el caso de *Saló o los ciento veinte días de Sodoma* está definido por tres niveles de referencialidad: el nivel de lo inmediato, construido sobre las expresiones del erotismo, la escatologia y la muerte; el nivel de lo mediato literario, edificado por el cruce sintagmático/paradigmático con las obras de Dante y el Marques de Sade, en lo narrativo del film y en la obra de Georges Bataille, en lo conceptual y estético; y en el nivel mediato también sintagmático/paradigmático de la iconografía en las tipologías de los personajes y las topologías de los ambientes y espacios. La resignificación visual de retratos de personajes históricos del renacimiento y, la resignificación topológica del ambiente a través de las referencias visuales al arte de la vanguardia son muestra clara de esto último.

La distensión de la temporalidad a través del recurso cinematográfico de la toma a distancia y semifija, unido a las mediaciones simbólicas y vitales que el director construye, hacen que la concretización temporal/vital del sentido, se resuelva en términos de estética de la crueldad. En ella el director apuesta por la distensión emotiva del tiempo del dolor y el placer a través de las formas vitales de la perversión. En Pasolini, la perversión tiene que ver con la posibilidad metafísica de la aniquilación del otro como gozo de sí mismo. Acaso por ello, todas sus mediaciones producen un umbral de la conciencia en el receptor que lo ancla en las fantasías. En este sentido, la crueldad tiene una ambivalencia vital: pone al descubierto el grado de atracción que esto ejerce en el espectador y al mismo tiempo rompe y transgrede el orden moral y político del propio receptor. El asco entonces, aquí, presenta una ambivalencia. El carácter temporal que abre, muestra al mismo tiempo el rechazo moral del espectador hacia la crueldad y la atracción por la pulsión de destrucción. En este sentido se da una ambivalencia vital del sentimiento del asco y el morbo: atrae a condición de objetivar al otro en lo imiediato del placer del yo. El sentido ideológico y político que esto pueda tener en la obra de Pasolini desborda el límite de este trabajo.

Las secuencias en las que se basan estos análisis, dejan ver la complejidad en la construcción de los elementos a los que se hizo referencia. La distancia de las tomas, unida a la mediación expresiva que hace de la coprofagia o de la tortura y de la mediación semántica de paradigmas culturales y artísticos, tal como se observa en la ambientación renacentista de la sala donde se reúnen los personajes para oír las narraciones; o, en la significación que adquieren las pinturas de la vanguardia europea de la primera mitad del siglo XX, en el salón desde donde se observa la tortura final del círculo de la sangre, o finalmente, el cruce sintagmático que hace con el retrato de *Federico de Montefeltro* pintado por Piero della Francesca hacia 1465, muestran el modo en que el director introduce el acoso y la tortura como un dato visual-fenomenológico y los sentidos ideológicos y discursivos que adquieren estos valores en su cinematografia.

### *2do. Caso: Crash. Extraños placeres, David Cronenberg*

A diferencia de Pasolini, la cinematografía de David Cronenberg seconstruye a través del cine de montaje.[6] Éste se caracteriza por la esquematización del tiempo por medio de tomas fragmentarias y preestablecidas de la acción. Se trata de la construcción de la trama a través de instantes privilegiados y previamente determinados por el guión y la dirección. Desde luego esto en el cine de Cronenberg se explica por la relación que él mismo entabla con la tradición de los *serial killers* norteamericanos, con el cine negro y con cierta estética de la ironía de la que echa mano. En su lenguaje cinematográfico existe una estrecha relación entre el manejo de las tomas de acción y la construcción de la trama. Lo que en otras palabras significa que hay un esquema preestablecido de la acción, a partir del cual se introducen el resto de los elementos: los personajes, las situaciones, los ambientes, etc. A este recurso habría que añadir el manejo que hace Cronenberg de la cámara. De movimientos más rápidos y en desplazamientos interiores y exteriores, la toma se introduce en la acción al grado de superponer los puntos de vista del director, los personajes y del espectador mismo. Esto desde luego redunda en una construcción de la acción más dinámica, cuando menos en lo que se refiere al sentido de la imagen-movimiento como elemento estructural del tiempo como representación. En este sentido, el cine de este director no aporta mayores elementos a la estética propiamente cinematográfica; sin embargo, la importancia de estos recursos se explica en relación al modo en que logran involucrar al espectador en la violencia propia de su obra. Los acercamientos, los cambios de punto de vista y la rapidez de la acción permiten, a diferencia del cine de Pasolini, que el espectador quede atrapado en cierto estado de embriaguez visual. Además, la estética de Cronenberg se orienta hacia el lado subjetivo y psicológico del asco, el morbo y la violencia.

En este sentido, las mediaciones simbólicas de las que echa mano este director entran en relación con un contexto más regional. Regionalismo psicológico, cultural e histórico. En Cronenberg la mediación simbólica descansa en dos elementos: la relación que establece con los imaginarios construidos por la sociedad posindustrial norteamericana y la mediación simbólica de la disolución posmoderna de la identidad del sujeto, a cambio de una estructura paranoide del individuo perseguido y acosado por la máquina.

La mediación simbólica de los imaginarios posindustriales, se lee a través del paradigma de los espacios marginales de las ciudades. En Cronenberg los espacios fronterizos de la ciudad funcionan como bordes simbólicos de lo horroroso y lo fascinante y lugares del deseo y el erotismo. Las acciones en sus películas,

> ... tienen lugar en un barrio, en la part industrial, en part portuario, desierto y lleno de charcos donde la noche es interminable. Un alucinante descenso a los infiernos plagado de referencias siniestras y cercano a la locura, una atmósfera opresiva y sórdida donde los diálogos poco abundantes y un ambiente sonoro ininterrumpido crean una situación de absurda irrealidad (Cortés 187).

Esto además se vuelve más interesante cuando el director de *Crash* introduce el uso del automóvil como condición indispensable de la perversión erótica de la vida urbana. El maquinismo automovilístico funciona como mediación del movimiento y el peligro y es el detonador del erotismo y la seducción. Su lenguaje cinematográfico integra el sentido del movimiento y la velocidad como una variable determinante de la configuración de lo urbano en la cultura contemporánea. A esto habrá que añadir. Los usos semióticos propios de los materiales industriales: sonidos metálicos, chatarra, desperdicio no degradable que crean las atmósferas de esta película.

La fascinación que siente Cronenberg por lo industrial y los espacios vitales que el desarrollo tecnológico produce en la sociedad contemporánea, lo llevan a configurar un sentido del deseo del cuerpo femenino maquínico y con ello a construir una mediación simbólica donde el asco y el morbo son producto de esta ambivalencia corporal:

> En la imagen corporal que nos propone Cronenberg en sus películas, no nos reconocemos (o nos da miedo reconocernos): La fusión de lo animal y lo humano, la mutación genética, la ambivalencia de sus fronteras, y la debilidad de su existencia. Sus *monstruos* no proceden del exterior no son producto de la magia por el contrario, están dentro de nosotros, son nuestros propios cuerpos (190).

Los avances de la ingeniería y la informática médica han abierto un nuevo imaginario corporal en nuestra sociedad, el que tiene que ver con el sentido interior del cuerpo y con la reconstrucción de la organicidad y la vitalidad a través de la máquina. Esto Cronenberg lo lleva al extremo de hacer de las máquinas las formas mismas del deseo y la perversión: ya sea ce espacio erótico y erotizable del coche donde se realizan buena parte de los actos sexuales en esta película, hasta la maquinización del cuerpo en uno de los personajes femeninos, cuyos miembros son una gran prótesis mecánica. El maquinismo en este director significa la exhibición del estado interno de destrucción/corrupción/infección del deseo y la sexualidad.

> El elemento central de la películas de Cronenberg también es la representación del cuerpo como lugar profundamente vulnerable y en contínuo proceso de degeneración, que subraya su decadencia en su cine el cuerpo se convierte en materia viscosa, en parásito vírico, en malformaciones genéticas que mezclan excrecencias y el deseo sexual, en mutaciones fisiológicas que muestran la complacencia del director por la poética de lao pútrido y nauseabundo ... Ello traerá consigo la anulación del narcisismo, pues la imagen unificadora del cuerpo se desmorona en ese proceso de transformación inaprensible (...). El cuerpo se degrada y empieza a apestar, es la putrefacción cadavérica que acecha en el interior y avanza sin descanso. Cronenberg observa el cuerpo como algo siniestro, algo de lo que se desconocen sus procesos y se teme su fragilidad (188).

Así, el maquinismo es el recurso simbólico a través del cual introduce la forma del deseo como peligro y fascinación ante la muerte. La máquina no es una recuperación de la vida, sino una exhibición del estado interno de corrupción del cuerpo. En este sentido, existe una diferencia fundamental con Pasolini. En el italiano la ambivalencia del asco y el morbo funciona como una subversión ideológica donde el espectador queda atrapado en la equivocidad de la estética del torturador/torturado, lo cual tiene que ver con un uso ético del asco a través de la exhibición de la propia perversión de la que el espectador es víctima. En cambio en Cronenberg, la construcción del asco y el morbo funcionan como apertura de la obscenidad y la promiscuidad del deseo que se objetiva en las formas de lo maquínico. Es decir, en este director el asco es un desdoblamiento de la propia estructura de perversión del sujeto moderno. Cronenberg disuelve la identidad del sujeto a través de la estética del cuerpo mutilado y expuesto. Esto no en más que la otra cara de la conciencia moralista de las clases medias norteamericanas y el desenmascaramiento del discurso sexual de poder de la cultura occidental, donde el deseo y el erotismo son el lugar de lo monstruoso. Se trata de una mediación simbólica del horror que pone en evidencia el doble discurso de la cultura de la higiene, tan querida de la sociedad contemporánea.

En Cronenberg, el asco que producen sus películas se contextualiza en el acoso de las fantasías y las perversiones de una sociedad temerosa de todo lo que trasciende los límites simbólicos de su pertenencia e identidad sexual, racial y nacional. En este sentido, la exhibición del erotismo a través de la sangre y las cicatrices tiene la función de poner en evidencia las formas de perversión que una sociedad no asume como principio. Así pues, en Cronenberg hay una doble construcción corporal del sentido del asco: la que tiene que ver con la perversión y el delirio de un sujeto en estado de fragmentación y el modo en que este estado se signa y se asigna al otro como el extraño, lo monstruoso y lo perverso.

## 3. LA FENOMENOLOGÍA DE LA TEMPORALIDAD DEL ASCO Y EL MORBO; EL CUERPO, EL TIEMPO Y EL OTRO

En la primera parte de este trabajo hablé de las estructuras trascendental-fenomenológicas del cuerpo. También armé que de éstas, la que le correspondía a la estética de lo grotesco era la relación interior-exterior del cuerpo. La interioridad orgánica introduce las funciones y los órganos como un *elan vital* sobre los que se construyen, en su nivel trascendental fenoménico, las vivencias estéticas del asco y el morbo.

Ahora bien, el orden estético fenomenológico del asco y el morbo habrá que ubicarlo, en principio y por cierto derecho de tradición, dentro de la axiología de lo grotesco y dentro del género de lo cómico. Ubicación que de alguna manera he venido sugiriendo a lo largo de este trabajo.

A propósito de lo cómico, Aristóteles afirma en su poética:

> La comedia es como hemos dicho, imitación de hombre inferiores, pero no en toda la extensión del vicio, sino que lo risible es parte de lo feo. Pues lo risible es un defecto y una fealdad que no causa dolor ni ruina; así, sin ir más lejos, la máscara cómica es algo feo y contrahecho sin dolor (Aristóteles &5, 1449a).

Desde la perspectiva fenomenológico-hermenéutica supuesta a lo largo este trabajo, la idea aristotélica de lo risible como lo "defectuoso sin dolor" se convierte en una orientación importante para llevar a cabo el análisis de esta última parte del trabajo.[7] De la negación del dolor y su relación con la risa cómica (reírse con y la alegría de la vida) puede desprenderse una primera consideración en torno a la definición de lo grotesco, el asco y el morbo. Lo feo y lo monstruoso en la estética de lo cómico se relaciona con cierto sentido de solidaridad vital implícito en la risa; en lo grotesco, en cambio, se relacionan con la ambivalencia del placer y el dolor ante la reducción a la animalidad del cuerpo y el peligro de fragmentación del sujeto. Un primer dato vital se pone en juego en lo grotesco: la posibilidad de ser dañado, de ahí que la risa grotesca tenga que ver con la crueldad como forma vital de supervivencia del sujeto. Inclusive cuando Aristóteles se refiere a la relación que la comedia tiene con los cantos fálicos como génesis de este género (&5, 1449b), supone cierta vinculación de lo cómico y lo grotesco con formas de la conducta animal. En este contexto, el asco y el morbo son dos manifestaciones vitales de estas conductas primarias, son un dato fenomenológico emocional que nos permite explorar el funcionamiento y sentido temporal de esta vivencia. Las conductas primarias tienen que ver con la liberación de las pulsiones sexuales y escatológicas en su estado animal, de ahí se desprenden la construcciones semántica e iconográfica, cuyas mediaciones simbólicas se explican por la liberación de gestos "irracionales" y por la liberación de estados orgánicos de la materia. La ambivalencia del asco y el morbo es un modo de identificación donde el sujeto receptor reconoce estados propios de perversión o abyección en la objetivación del otro y en la excitación de sí mismo. "Lo auténticamente monstruoso es descubrir la bestia en el seno del ser humano y, con ello, destruir toda la seguridad en la identidad del hombre" (Cortés 165).

El asco y el morbo, si bien no necesariamente se relacionan con la risa, está muy cerca de lo risible ridículo. Como lo afirma Jauss: "La propiedad estético-afectiva de lo cómico sería, pues, algo así como un filtro capaz de convertir la simple negatividad y la suficiencia ética de la risa en algo positivo." (Hans Robert Jauss 202) Mientras que lo ridículo se vincula con lo cotidiano y la exhibición de la torpeza ante situaciones sociales de estandarización de los individuos, la risa cómica tiene que ver con la distancia que el espectador guarda con el personaje o la situación. Lo ridículo involucra la crueldad donde el espectador es juez o víctima de la situación. En este sentido, lo grotesco está más cerca de lo ridículo que de lo cómico, con la variable que en lo grotesco se involucran no sólo transgresiones de la vida social, sino transgresiones de las estructuras vitales y orgánicas básicas. Entre la broma y la crueldad, el elemento que está en juego es o, la alegría de vivir o, la exhibición del daño al otro o a uno mismo. La burla establece una relación directa con lo grotesco y tendría una forma específica en la crueldad y por ello sus salidas a través del asco, la tortura y el morbo.

Si bien lo grotesco tradicionalmente se relaciona con lo risible, sin embargo puede devenir en formas trágicas. En este sentido, lo grotesco es un valor intermedio entre la compasión trágica y la simpatia cómica. Para Jauss, lo ridículo conserva su fuerza negativa en tanto forma de conservación del orden social establecido (165). Desde esta perspectiva, una de las derivaciones hacia lo grotesco y su diferencia fundamental con lo ridículo, se define por la ambivalencia de sentimiento. Es decir, la identificación no funciona como restitución del orden a través de la risa de exclusión[8], sino con la identificación de las perversiones del receptor.

Así por ejemplo, la organización social de seres monstruosos o marginales en la comedia funcionan como un *sympathos*, donde los personajes y las situaciones provocan la "alegría de la vida", el sentido de la expectativa de la acción como deseo. En cambio, en la construcción grotesca estos seres monstruosos exhiben la tendencia destructiva y corruptora de la organicidad vital y social, de ahí que se lleva a cabo la subversión de las normas y el fenómeno de atracción rechazo en la recepción de este valor:

> ... podríamos entender la noción de informe como un término que, en su doble concreción de araña y escupitajo, apela tanto al desbordamiento del pensamiento racional (liquidación del sentido), como a la perdida de la identidad (animalidad) tanto a la extenuación de los valores (ambivalencia) como a la degeneración y licuescencia (realidades indefinidas) (...) "Partiendo de la normalidad de lo humano, lo animal o del vegetal, la forma puede evolucionar de dos modos opuestos: o bien se exhorbita, se hiperboliza hacia la combinatoria monstruosa, o bien se disuelve, se altera, se deshace para alcanzar lo informe. (Cortés 165)

Estas aproximaciones a lo grotesco, y su especificidad en el asco y el morbo, dibujan las condiciones fundamentales de su estructura temporal. Si como lo afirma Ricoeur, en última instancia, lo que está en juego en la recepción de una trama es una vivencia temporal; la de lo grotesco es una refiguración del tiempo como vivencia emotiva de la degradación de la vida o de la vulnerabilidad ante lo otro.

Lo grotesco, en su ambivalencia de asqueroso y morboso, no temporaliza la voluntad de poder sino la voluntad de destrucción y de muerte. De ahí que los modos en que se simboliza lo grotesco se relacionan con estructuras subversivas del orden personal, social y biológico, es por ello que las mediaciones simbólicas de lo grotesco son configuraciones que expresan los aspectos corruptibles y transgresores de las sensaciones y emociones corporales y psicológicas primarias del ser humano. Esto explica también porque las mediaciones simbólicas ponen en crisis los órdenes normativos de la sociedad y la cultura. Espacios marginales, lenguaje prosaico, gestos animales, transgresiones sexuales y escatológicas, son los modos en que se semantiza la temporalidad en los signos culturales que utiliza esta estética de lo grotesco.[9]

En cualquier caso lo que interesa mostrar, en última instancia, es el carácter temporal que se revela en el cine de Pasolini y Cronenberg a través del asco y el morbo. Desde luego hay que tener en cuenta un aspecto fundamental: el sentido de la interioridad orgánico-corporal. Ya sea por la exhibición médica del cuerpo o tan sólo por las funciones orgánicas que muestran los desechos y los excesos del cuerpo; la organicidad vital tiene una primera relación con el asco en cuanto que evidencia el tránsito de la vida a la muerte a través de la descomposición. En ella se pone en juego el sentido del tiempo como corrupción, fragmentación y destrucción. La interioridad orgánico-vital significa la apertura del sentido metafísico de la espera como temor y disolución. De ahí que Bataille considere que el asco en el fondo es el temor mismo:

> Esas materias movientes, fétidas y tibias, cuyo aspecto es horroroso, en las que la vida fermenta, esas materias en las que bullen los huevos, los gérmenes y los gusanos están en el origen de esas reacciones que llamamos *náusea, repugnancia, asco.* Más allá del aniquilamiento por venir, que dejarla sentir todo su peso sobre el ser que soy, que espera ser aún, cuyo sentido propio, más que ser es esperar ser (como si no fuera la *presencia* que soy, sino el porvenir que espero, que sin embargo no soy), la muerte anunciará mi retorno a la purulencia de la vida. Así puedo presentir – y vivir en la espera de- esa purulencia multiplicada que por anticipado celebra en mi el triunfo de la náusea (Georges Bataille 80).

Temor que significa la irrupción del tiempo como finitud echa de came y hueso. En este sentido la reducción fenomenológica de la temporalidad del asco y el morbo, se refiere a la posibilidad de la muerte como angustia ante la vivencia humana de la temporalidad.

> La angustia es al aguijón del mal. Enfermedad, mal de carne viva, senescente, corruptible, perecimiento y podredura: estas serían las modalidades de la angustia misma, por ellas y en ellas, el morir es en alguna manera vivido, y la verdad de esta muerte resulta inolvidable, irrecusable, irremisible; en la imposibilidad de disimularse uno mismo su propio morirse estriba la indisimulación misma y quizá el desvelamiento y la verdad por excelencia, lo de por sí abierto, el insomnio originario del ser: roedura de la identidad humana que no es un espíritu inviolable abrumado por un cuerpo perecedero, sino la *encarnación*, con toda la gravedad de una identidad que en sí misma se altera (Emmanuel Lévinas 155).

La estructura temporal del asco y el morbo son un colapso en tiempo. El instante del asco y el morbo demuestran el límite de una espera: el de la vida que se anuda entre el deseo y el temor.

¿Qué se pone pues en juego a través de la cinematografia de estos dos directores? Sin duda una construcción del sentido del asco a través del juego analógico del "cuerpo". Éste contempla y al mismo tiempo se introduce en la acción a través de la toma. El cine es un lenguaje que acorta, de manera casi absoluta, la distancia entre la narración y el tiempo. La estructura óntica que define el lenguaje cinematográfico, en última instancia tiene que ver con una fenomenología del cuerpo en movimiento. El como si fuera verdad, es entonces la construcción del sentido a través de la intencionalidad corporal que el cine asume como condición perceptual trascendental de su lenguaje. En el cine las fronteras entre el interior y el exterior de la experiencia estética quedan prácticamente disueltos. La motricidad corporal hipostasiada en el "ojo" del director, permite hacer del movimiento el elemento mínimo de sentido en el cine. Pienso que sólo así se puede explicar por qué el espectador se involucra a tal grado en la vivencia estética de este arte. Acaso también por ello, se explica por qué el asco y el morbo tienen que ver con la coincidencia de la mirada con la acción, de la acción con la emoción del espectador y de la emoción del espectador con el tiempo como vivencia cuasi-verdadera de un evento en su desarrollo concreto. Visto así, la primera condición de construcción del asco y el morbo se define por la manera en que los usos cinematográficos permiten la construcción del sentido a través de un soporte pre-judicativo y expresivo del tiempo en términos, no de ficción, sino de falsedad. En el cine la diferencia entre verdad y falsedad, entre ficción y realidad es tan endeble que permite pensar que este arte ha abierto un campo aún por explorar: el del arte como tiempo puro.

*Universidad Iberoamericana*

## NOTAS

[1] Al respecto pueden verse mis trabajos *El arte y lo sagrado. Una aproximación fenomenológico-hermenéutica a lo sublime.* El primer capítulo de esta investigación está por aparecer en la Memorias del Coloquio Internacional de Historia del Arte del Instituto de Investigaciones Estéticas de la UNAM. También *Tiempo Narrado. La obra pictórica de Roberto Rébora.*

[2] Estas estructuras intencionales del cuerpo se relacionan con las investigaciones y análisis filosóficos que Merleau-Ponty, Paul Ricoeur, y Emmanuel Lévinas han trabajado. También tienen una relación directa con la fenomenología y la antropología de lo imaginario de Bachelard y Durand.

[3] Los tres registros de lectura están tomados de Paul Ricoeur y hacen referencia a los tres niveles, que conforman Mímesis I: la semántica de la acción, la simbólica vital y la fenomenología de la temporalidad (Paul Ricoeur 113–168). Sin embargo, en tanto que el objeto directo de estudio es el lenguaje cinematográfico, reinterpreto la semántica de la acción a través del concepto de imagen movimiento de Deleuze. Esto significa que la fenomenología de la acción se asume como dato sensible y perceptivo inmediato de la imagen movimiento. En el cine la cámara es una prolongación *in-mediata* del movimiento corporal.

[4] Para Deleuze los poderes de lo falso en el cine significan varias cosas. 1. Desde el punto de vista descriptivo; la descripción orgánica y la descripción cristalina, 2. Desde el punto de vista de la narración: la narración verídica y la narración falsificante. 3. El tiempo y el poder de lo falco y 4. Desde el personaje falsario: su multiplicidad y su poder de metamorfosis. De éstos, el más importante para el filósofo francés es el del tiempo y el poder de lo falso. En sus palabras, los poderes de lo falso son "… la reponse de Borges à Leibniz: la ligne droit comme force du temps, comme labyrinthe du temps, est aussi la ligne qui bifiirque et ne cesse de bifurquer, passant par des *présentes incompossibles*, revenante sur de *passés non-nécessairment vrais* (…). Ce n'est pas du tout <<chacun vérite>>", une variabilité concemant le contenu. C'est une pusisance du faux que remplace et détrône la forme du vrai, parce qu'elle pose la simultanéité de présentes incompossibles, ou la coexistence de passés non-nécessairement vrais. (Gilles Deleuze, *L'image* … 171)

[5] El concepto pasoliniano de estilo se relaciona con el de Merleau-Ponty. El estilo es sobre todo un esquema interior del cuerpo con el que el artista se relaciona con el mundo.

[6] El cine de montaje es construido, Lo que en otras palabras significa que la narración se basa en la esquematización del acción y la historia y, sobre todo, en la ilusión de la acción a través de la fragmentación del tiempo en instantes privilegiados. La fórmula que propone Deleuze es la siguiente: cortes immóviles + tiempo abstracto (Gilles Deleuze, *La imagen – movimiento* … 269).

[7] El desarrollo de esta investigación se ha orientado bacia el pensamiento aristotélico, sobre todo el que se refiere a la generación y la corrupción, al problema de la materia prima y al problema de movimiento en la metafísica.

[8] La idea de la risa de exclusión responde a las formas de la sátira que sirven para legitimar el discurso de poder. Significa el modo, en que a través de la burla de lo que es peligroso para una sociedad, el sistema social reprime la alteridad y la somete a sus normas propias (Jauss 213–214).

[9] Los modos de simbolización vital y de semantización de la acción, en el contexto de la cultura contemporánea, definen de manera específica el sentido de lo grotesco. La primera definición se refiere al procesos de secularización de nuestra época. En ésta, lo grotesco y sus variaciones, el asco y el morbo, no se interpretan como sublimaciones, sino como condiciones de realidad. La segunda involucra la noción de crisis de identidad del sujeto: ésta libera formas radicales de la alteridad psicológica, sexual, cultural, social y política que adquieren significados subversivos para los sistemas monologistas de la cultura. Unido a lo anterior, y como determinaciones de la secularización y crisis del sujeto, el imaginario simbólico de lo grotesco en la cultura contemporánea, se relaciona con el desarrollo de los sistemas culturales de nuestras sociedades. En

particular importan; el desarrollo de la ciencia y los subsecuentes descubrimientos de microorganismos que producen formas terroríficas de la enfermedad, las formas de la marginalidad económica que se manifiesta tanto en las sociedades, en la configuración de espacios simbólicos "underground" y los habitantes imaginarios de este espacio los punks, yonkies, los skin head) y, los espacios marginales de las sociedades económicamente pobres; los modos imaginarios de lo monstruoso y lo siniestro producidos a través de los medios masivos de comunicación, finalmente los modos imaginarios de lo grotesco que los sistemas políticos discurren como estrategias de legitimación del poder.

## BIBLIOGRAFIA

Aristóteles, *Poética.* Madrid: Grédos, 1974.

Barrios Lara, José Luis. *El arte y lo sagrado. Una fenomenología-hermenéutica sobre lo sublime.* Tesis. Mexico: 1997.

– *El tiempo narrado. La obra pictórica de Roberto Rébora.* Mexico: Oakl/FONCA, 2000.

Bataille, Georges. *El erotismo.* México: Tusquets, 1997.

Cortés, José Miguel G. *Orden y caos. Un estudio sobre lo monstruoso en el arte.* Barcelona: Anagrama, 1970.

Deleuze, Gilles. *La imagen-movimiento. Estudios sobre el cine.* TI. Barcelona: Paidós, 1984.

– *L'image-temps. Cinéma.* TII. París: Editions de Minuit, 1985.

Ingarden, Roman. *La obra de arte literaria.* México: Taurus/Universidad Iberoamericana, 1998.

Jauss, Hans Robert. *Experiencia estética y hermenéutica literaria.* Madrid: Taurus, 1986.

Lévinas, Emmanuel. "Trascendencia y mal", en Phillipe, Némo. *Job y el exceso del mal.* Madrid: Caparrós, 1995.

Pasolini, Pier Paolo. *Cine de Poesía.* Barcelona: Cuadernos Anagrama, 1970.

Ricoeur, Paul. *Tiempo y narración.* TI. México: Siglo XXI, 1995.

# SECTION IV

LEONARDO SCARFÒ

# ON THE NECESSARY FORM OF PHILOSOPHY IN THE VITAL DETERMINATION OF EVERY BEGINNING THEREOF

There is a musical work that we should listen to every time that the exasperation or the tiredness that are inevitably associated with every search, no matter how serene it may be, threaten to disenamour us of philosophy.

The "Goldberg Variations," a rare example of perfect accord among structure, sense and content, of genius and method (profound harmony – indistinguishability – of expressive freedom and formal necessity), esthetically represents the most profound characteristics of metaphysical thought, the ones that are most difficult to express and to recognize.

May the reader, therefore, permit me to use this seemingly extravagant metaphor.

As "*Vierter Tell bet Clavier Übung*" (the third, surely not by chance, is a "*Dreifaltigkeitsmesse*" that, in keeping with the sacrality of the number that gives it form and name, seeks to express the perfection of creation, the vastness and the clear firmness of the divine), these variations, steeped in the mystique of the number "4", the number of creation, but also of the human, of space, of the physical elements and of time itself, stand in contrast with "3" as creation with demiurge, as activity with intelligence and necessity, imperfect with perfect, incomplete with eternal.

Here we find sublimity side by side with irony, grandeur with modesty; the "*Variationen*" close by reopening, end by commencing all over again (they are "*Übungen*" and cannot be anything but such).

They return to the starting point after a surprising articulation of variations without ever having ceased to be, at every point and moment, one and the same thing. The fine poetry of the principle of the variation (on a theme) makes it possible to give a form to the logically impossible, to the perceptively obscure, to the apparent absurdity of every feeling, to time itself; in it – and through it – what remains identical is the changing or, if you prefer, what changes is the manner of being identical.

Each of these passages brings a philosopher to my mind: one rigid and obscure, another sadly beautiful, one arid and involved, another obsessive and vehement, yet another serene and elusive.... What distinguishes them is – always and invariably – also what they have in common.

In the same way philosophy is similar to that initial (and final) "Aria"; it is the guide and the profound trace or path that, even though you can follow it in many different ways, remains originarily identical. In Plato as in Kant, in



*A.-T. Tymieniecka (ed.), Analecta Husserliana LXXVI*, 265–279.

Aristotle as in Husserl, we have essential and vertiginous evidence: "Knowledge commences with experience but does not derive from it." Living this evidence and exploring its implications, its radically problematical and even equivocal nature, feeling its pervasiveness and imminence, all this is philosophy.

Not only does knowledge commence in somebody who lives and therefore thinks, takes form in a subject to the extent to which he or she gathers (or, if you prefer, "rediscovers" or "remembers") its form, a subject who, constituting this knowledge, is also constituted by it (*Ausbildung*).

But knowledge cannot derive even from this, it is not here that it has its origin; if it were so, there would exist no obstacles to knowledge, it would be as natural as breathing and the limits of things would also be the same as those of the mind that thinks them and this mind would be one thing and these things, thoughts.

Between the experience of the subject and knowledge in general there does in fact exist an ineludable but not by any means obvious link, a necessary but hidden link. Indeed, if knowledge, always presupposing its evidence, were to find its occasion in something presumptively possible without losing any part of its necessity in crossing this accidental nature of its giving itself – rather, if it is valid solely by virtue of this apparent contingency and nonessentiality that undoubtedly represents its limit and therefore also its possibility – it is clear that this possibility of the person who knows would have to carry within it exactly the same form, a form that is neither occasional nor casual, as the form of the thing that thereby comes to be known.

Without which, indeed, there would be neither knowledge nor ignorance. It is by virtue of this inevitableness of the subject, of the subjective giving itself of the objectively known (true), that every radical philosophy – straining towards the origins – notwithstanding its many thematic and methodic "variations," cannot but constitute itself at first as anything other than the science of subjectivity (monadology).

Philosophy as the science of "knowing subjectivity" has its object in the clarification of the principle of experience; this is the point from which it must necessarily start and to which it must return time and time again. Feeling strains to seek its own beginning, its necessary (determinant) form, starting from its own occasional commencement; it can do so, because while this initially hidden beginning reveals itself, it determines – first of all – this selfsame and seemingly nonessential form from which one must necessarily start.

But recognizing one's own life and the authentic living of the subject in the knowledge and the intentional activity that inaugurates and directs it may seem like a paradox; knowledge as "activity," thematized into this radical form, appears to us as the most obvious operation, but – at one and the same time – also as the one that is least comprehensible for each individual consciousness.

Let us consider the pure phenomenological structure of consciousness; it can be recognized as identical within the unforeseeable flow of one's own lived experiences, in the flood-like opacity – the inexhaustible variety – of one's own eventual correlates. This recognition is given as continuous experiencing of uninterrupted modifications in the continuous modification of one's own lived experiences (*Vormeinung, Mitmeinung, Täuschung, Erinnerung*...) and cannot be separated from grasping (gathering) the unity and the uniqueness – the permanence and complete and secure subsistence – of this flow (the horizon of experience as unbroken and inevitable experience of the horizon in its indeterminate aperture), starting from determinate and finite (successive) modifications and always returning to them. This feeling the succession (of perceiving in perceiving) contemporaneous with the constitution by successive moments of the feeling, which thus becomes a feeling of the succession within the succession of the feeling, prevents consciousness from becoming or thinking of itself as an "object." The so-called "self-consciousness" is in fact evidently given in each feeling, it is innate in it, as it were, and is neither its condition nor its consequence. A "reflecting" consciousness is already fully self-conscious; it feels the feeling. This feeling the feeling is not "object" but always no more than *of* objects.

Indeed, "perception" cannot but be the thought of perception. Reducing consciousness to just one thing among other things (i.e., reducing perception to an object) means thinking of it as something fictional, or as a function or abstraction. On the other hand, the objectivated subject lends itself to being analyzed by variously orientated specialist disciplines; it is the object of logic inasmuch as it is knowing (reasoning), of biology inasmuch it is living, and of psychology inasmuch as it is moved emotionally by certain feelings or is affected – or afflicted – by a character.

Nevertheless, a consciousness cannot be reduced either to any one of these abstract forms nor to any of the theories connected therewith; the latter are obviously derived from it. Not vice versa.

Unlike any one particular knowledge, philosophy should thus give itself as the "*Idee einer Wissenschaft von den Bedingungen der Möglichkeit von Theorie überhaupt*."[1] The characters of this mode of knowledge will be altogether particular and will place it "hierarchically" higher than (above) any special theory or technique.

If I understand the "subject" in the most obvious and natural sense, i.e., as an individual, as a single person, philosophy becomes to all intents and purposes impossible. If subjectivity "is subjective," philosophy is impossible, because in that case every one of its beginnings is also an originary beginning, though the very fact that one affirms the absoluteness of subjectivity makes it wholly impossible for such an originary beginning (principle) to exist. One can readily pass beyond this argument by limiting oneself to not denying the evidence by virtue of which a consciousness is recognized as a subject that in general – constitutively – knows and which therefore "by chance" does not "know" something or something else, but in which, just as in every other individually determined cognitive act, it always finds in knowing the typical activity within which every individual learning and reflecting finds its own relative place and a sense by virtue of which also every individual subject, precisely through its limitation, irreducibility and changeability, the intimate solitude – the imperfection – that are peculiar to it, has to operate by starting from its being necessarily "consciousness in general," being universal. The object in general, on the other hand, seems an empty matter, a substrate that is as inert as it is indifferent – merely passive and static, but also irreducible, stolid – just as the subject that "in general" knows this object seems a rigid and mute fiction.

"Subject" and "object," nevertheless, transcend themselves to the extent to which they "are said" in an abstract manner, become disjointed in the reiterated attempt of giving to feeling the ordered logical structure of a proposition and in feeling the absurdity of this statement and this attempting. The contemporaneity of subject and object, their unity in perception – and in thinking the perception (the reflecting) – originarily oppose themselves to the nonessential act that endeavours to suspend its insuperable character of contemporaneity in the form of subject-attributes-predicate. The principles – as also the "logical" expressions – are in permanent contradiction to the temporality from which they derive.

The first and most delicate passage of the phenomenological method consists of the modification of the natural attitude into the naturally unnatural one of the "epoché"; philosophy teaches, first and foremost, the need for unlearning.

It is in the profound oscillation between the idea of an "objective" science independent of experience (logic or mathematics, for example) and the powerful intuition of the evidence of subjectivity operating as objectively constituent that the occasion for phenomenological reflection imposes itself. Indeed, for phenomenology it becomes clear that it is impossible to reduce knowledge to subjectivity and subjectively learnt objectivity to relativity; it has to be precisely in the inevitability of objectivity giving itself subjectively – of the hidden but inexorable operating of an "objectively subjective" – that one has to recognize *both* the objectivity of this subjective constitution *and* the objectivity of what nevertheless has come to be known subjectively.

The Husserl of the first few pages of the *Sechste Untersuchung*, precisely because he still moves within the prejudice of an ultimate and analytical definition of the principles of knowledge, finds himself in the continuous and inevitable embarrassment of not being able to demonstrate the unitary (identical) character of "*Intention*" and "*Erfüllung*" (in knowing the true as also in knowing the immediately false), and this precisely because, distinguishing them abstractly, he is obliged to assume the idea of "*adequatio*" as evidently valid and, making this assumption, he has to consider as really separate moments that in intuition are given as inseparable. Analytical and incorrect, because not originary, is the distinction between (acts of) "*statische*" and "*dynamische Erfüllung*,"[2] whereas the former – thematized (or reflected) as *Intention* – should really represent the (logical) condition for the latter; the temporality that distinguishes them – since that of the former is an invariant logical space and that of the latter a linear and punctual succession of lived experiences causally arranged within that space – does not confirm either the possibility that acts of static and dynamic "Erfüllung" (realization) can give themselves separately (i.e., be distinguished) or the necessity that "outside time" there exist permanent and invariable contents (logical truths) that do not derive from some act of realization or, and this really amounts to the same thing, an intention, independently of a mind that thinks them, of a consciousness that lives them.

As an aside: The perfect unity of feeling as the identity of consciousness and correlate or of intention and realization (which are a metaphor thereof) cannot be "said" by any language that postulates its opposition, that induces its (in)difference. *This syntactic equivocation gives rise to the tendency to identify contradiction with error.*

The sole language capable of rendering account of "pure" feeling – with which every kind of abstraction can try to grapple without avail – lies beyond having to be logical, makes methodical and ironic use of paradox itself, affirms it continuously to deny its presumedly contradictory nature, and lies even beyond the very "principle of contradiction," affirming its radical absurdity and recognizing its character of negator of time and, consequently, also of itself. There exists a pure language capable of grasping the continually recomposed discard, the revelation of what is hidden, the uninterrupted loss and gain of feeling in feeling without abandoning to the extreme edges of this desert garden just two, distant things. Eyes different from one's own way of looking. This is poetry.

Here we are concerned with a very dangerous equivocation: An analysis of language based on the assumption of the principle of the logical nature of knowledge (that "*Wissenschaftslehre*" is founded on logic) is a concommittant cause and an argument seemingly favourable to the legitimation of the difference between "theoretical" and "normative" (or practical) just as, among others, Husserl himself proposes, at the beginning of his "*Logische Untersuchungen*." This difference has mortal effects on philosophy, which cannot but encounter this death at its very beginning and therefore – as it were – will eventually have to free itself of it for no other reason than to be able to commence. The real phenomenological "turning point," so it would seem to me, consists of this movement that seeks to clarify (*Aufklären*) not the need for "uniting" and "harmonizing" the theoretical and the practical, but the originary (constitutive) impossibility of separating them.

In this sense, then, the *there is no "turning point*." This sense has been at work uninterruptedly from the "*Sechste Untersuchung*" right through to the *Krisis*. It becomes explicit in the *Krisis* as the inevitability of a task and the recognition of a horizon that is as necessary for thinking as it is for living.

Within the terminology and the problematical spirit of the *Sechste Untersuchung*, for example, one may admit that, in the relationship of the "*dynamische* Erfüllung,"[3] the cognitive act places in relationship with each other two moments that are substantially and genetically separate, namely, the moment of the "thought" and the "intuitive" moment, even though that act remains temporally separate from them; whereas in the "*statische Erfüllung*" – understood as "*bleibendes Ereignis*"[4] – of this temporal process, they (intuition and the intuited) would coincide in a (true) *third object*. One cannot arrive at admitting the (nonlogical) evidence of the "compresence" of the two clearly different temporalities given, as it were, one within the other, of the two different modes of "being separate" for "intention" and "realization." Analytically, I find them united (compresent) when I think them (to think them as) successive and discover them successive only on condition that I think them identical (simultaneous).

What Husserl neglects here is the obscure temporal nexus between "*Bedeutungsintention*" and the eventual (successive) "*Bedeutungs*-erfüllung"; on the one hand, this nexus would correspond to a causal nexus formed on the linear chronology of objective time, but on the other – if for no other reason than thinking a similar causality as a "physical" translation of the logical form of "premise-and-consequence" – it would have to correspond to an atopic temporality – one without either duration or real succession in which the terms would be synchronous and, therefore, no longer be terms.

This "second" temporality, which would be related to "causal" time as the latter is to space, is logic (logical) space. How "logic space" is related to "usual" space thus becomes an enigma, though it does so in the surprising evidence of the (technical and technological) efficacy of all their indirect relationships.

The "time" is necessarily the same and, for this reason, necessarily has *two forms*.

It is on this evidence that even the terms of "transcendent" and "immanent," as opposed or contraposed or even irreducible, find themselves faced with a necessary revision, find themselves faced with the insuperable evidence of the *immanence of the transcendent*, which is given – lived – time and time again (anew) as the *transcendence of the immanent*.

Here there comes to be lost their equivocal sense of irreducible opposition, of impossible relationship; in the nonessential terms of "what gives significance" and "what receives significance," on the other hand, the nexus between them resolves itself in signs, is merely and emptily allegorical, without time; whereas in transcendental-phenomenological terms the nexus, intact and clear in and with its essentially temporal character, constitutes itself as a "*Theorie der Transzendenz auf dem Boden der Immanenz*."[5] This latter is the necessary form of the first philosophy (*Erste Philosophie*).

In fact, experiencing an immediacy "happens" in time, but what is experienced by means of it is not temporal in the same manner as the feeling that lives it, for we feel immediacy as the result of a reflection, as determined, and we think of this reflection as being successive to a generic "having felt"; phenomenologically treated, the difference between "feeling" and "having felt" is an *abstraction* that is nevertheless real precisely on account of the fact that (inevitably) it has to be reflectively thematized. Evidently, the immediacy of the experience (experiencing the immediacy) is not successive to anything at all, just as the perceived rendered "thematic" (in retaining it, in preaching it, in remembering it, ...) is undoubtedly always "here and now," but the perception, precisely because it is always "here and now," *does not have a place*. Nor does it have a time.

In "*Logische Untersuchungen*," however, it seems that the "*statische Erfüllung*" may be something like the result (or, indeed, the thematization) of the "*dynamische Erfüllung*" whenever the latter – without prejudice to the fact that a "*letzte Erfüllung*" is the ideal of perfect knowledge[6] – does really have to presuppose the former, which in its turn presupposes it (the latter).

Were I to be convinced of knowing a certain object in my intuitions just as I think it[7] (of immediately intuiting, as it were, its significance), I could describe this manner of knowing in the following terms: Knowledge is true when intuition and thought are *of the same object*. In this way I have not only separated the "known" from the thought, turning even the latter into an object, but also intuition from the known object, turning the latter into a thought.

*The same error in method that had induced me to think of knowledge as the activity of a description – and therefore of "Wissenschaftslehre" as a description of knowledge – is the one that at this point, becoming unexpectedly operative, makes me infer from the intuition and the thought of "one and the same object" that this intuition and this thought, under the form of a true enunciation, are themselves "the same object."*

I can no longer understand it, seeing that the very possibility of this coincidence had to be excluded first and foremost to postulate an "intention" and a "realization" that were temporally separate and a cognitive act (that mediates them) in its turn temporally separate from their relationship.

On the other hand, only in the identity of thought and intuition (of intention and realization) would there be a perfect adequation, true knowledge in the form of an identity without residues, an objective unity.

If I, continuing to err, but – as it were – doing so "methodically," liken this knowledge to an activity, it is clear that the very identity thereby constituted will not be something like an index or a scheme (pattern) external to the "person who knows" – a comparative term – but *this very identity will give itself as an "Erlebnis"; the truth will be a lived experience and the adequation that opens to it will be an act.*

The identity (of "*Intention*" and "*Erfüllung*") cannot be the result of a (reciprocally determinant) cross confrontation of intention and object, object and significance, significance and realization, because such a confrontation would presuppose all these elements (or moments) as already given (distinct) *outside the relationship* from which they can never be separated. Within the horizon of common sense, as also within the scientific horizon, "things" are given, but within the transcendental horizon they are given to the extent to which they are constituted (*made*); a givenness stands in need of being judged and a judgement is true or false, whereas a "constitution," as it were, has to be lived and seen; it is evidently insuperably imminent.

The analytical criterion of "truth" is a logical "squaring" of the originary, phenomenological criterion of evidence.

Grasped in their problematic concept (which, however, is not really a concept), "evidence" and "truth" reveal an obscure nexus and together with it an even more obscure *difference*. But a necessary one.

Still in the terms of the *Sechste Untersuchung*, let us therefore admit that the adequate identity obtained dynamically as a result can be called a true knowledge, understood as lived experience (*Erlebnis*) – which can certainly have a true or false "content" (in a *Täuschung*, for example) – though it is never itself either true or false, but simply lived experience; *it is therefore evidence*. All this is in agreement with Husserl's conception as set out in the *Prolegomena*, where he considers evidence to be "the lived experience of truth."[8]

All this is also ambiguous because, if truth is also a (certain type of?) *Erlebnis or a particular Erlebnis* that would necessarily accompany true representations, the evidence – for its own part – would greatly resemble a "reflection" but bear little or no resemblance to something immediate.

Some time was to pass before Husserl clarified this equivocation with his analysis of the difference between "*immanente und transzendente Erlebnisse*";[9] lived experience of a lived experience cannot but itself be a lived experience. In his *Prolegomena zur reinen Logik* we are still moving within a conception of knowledge in which the motive of the "*adequatio*" prevails over the irreducible temporality of the *Erlebnis*: "Evidence" is said to be the lived experience of the agreement between intention and what is given, as a fact, *where the "truth" is the idea of this agreement and this agreement is also an act.*

And yet, just as a "purely" considered *Erlebnis* is neither true nor false, in just the same way an act cannot of itself be either true or false.

Up to what point, therefore, is it correct to consider the activity of knowing – knowledge (the activity of a subject orientated on its own object) – as a *tertium* placed between these two poles of "*Intention*" and "*Erfüllung*"? It is surprising that in *Logische Untersuchungen* "truth" still appears as the preliminary condition for "evidence"; all said and done, this esthetic paradox derives directly from the pretence of founding "*Wissenschaftslehre*" on logic, this in conformity with the idea of knowledge as *adequatio*; if anything, it is the evidence that is the necessary and – in any case – *not* sufficient "condition" of truth.

This means that logic in general cannot be considered as originary knowledge or theory of knowledge; it cannot be "*Erste Philosophie*."

*In fact, the "false" is as evident as the "true"; rather, it appears as evidently true, because otherwise it would not be truly false.*

Having thus critically excluded the idea of knowledge as "*adequatio*," we are left with the general form (i.e., without deciding, as it were, whether it is something or nothing) of a problematic relationship in which subject and object in the cognitive act necessarily coincide in the intuition and can be separated only abstractly by virtue of a methodical reflection that, time and time again, has its beginning – returning to it – in the immediacy of all experiencing. "*Bewußtsein*" and "*Korrelat*," "noesis" and "noema," are but one thing in the uninterrupted life of feeling that is the feeling of uninterrupted life. But they are *not* a thing.

On the other hand, the perfect transcendental horizon (*Horizont*) by virtue of which – within which – this principle remains always valid is not immediately evident on the multistratified, prejudicial terrain of objective time that, notwithstanding its own obtuse character of an automatism (a running along by inertia), in hiding it, also reveals it. Indeed, it is within the narrow horizon of objective time that one finds the occasion for grasping the "transcendental"; it is precisely this latter that – by virtue of a passively reiterated abstraction of the evidence of the indescribable identity of "consciousness" and "correlate" – gives schematically linear and consecutive form to the former. One may certainly object that the "transcendental," here and wherever else it may be evoked, is essentially a fiction (*Fiktum*); nevertheless, it is altogether essential also for the concept (and the experience) of "normal" (objective) time understood "ingenuously" as real.

Phenomenologically, it is necessary to found so-called "reality" on a "fiction" precisely in order to allow one to glimpse the fictitious (abstract) character of the presumedly real and thus also the real character of the fiction.

The necessary fiction that founds the real is therefore transcendental; the character of fiction of transcendental time, given its evident (apodictic) necessity, therefore makes it "more real" than normal time, this in the same way in which the experience of art or experience as art are more profound and more authentic than merely undergoing (being subjected to or exposed to) it. There exists and is continually operative – with similar modalities – a consciousness of (and from) this "time"; the difficulty of conceiving this consciousness as likewise transcendental consists of the insuperable (and at the limit constitutive) fact by virtue of which I can neither seek the definition and the attributes of this consciousness as if they belonged to an autonomous "*substantia*" subsisting for its own sake, nor contrapose it to something individual (or particular) in the same way in which "genera" are contraposed to "species." It is within me and is not I; nevertheless, I am I because it is within me. I can find (and lose) this "transcendental consciousness" only by starting – commencing – from my individual consciousness, though the latter is less authentic, less originary than the former; "transcendental consciousness" is also the "*Fiktum*" that founds "individual consciousness." It is therefore misleading to consider the transcendental (ideal, abstract, constituted,...) and the real (concrete, empirical, given,...) as irreducibles; opposing the transcendental to the immanent is possible only by negatively deriving one from the other. This apparent derivation is wholly illusory, first and foremost because when they really give themselves – when they are lived and experienced – the "transcendental" and the "immanent" never give themselves separately; the problem consists precisely of the fact that they always and invariably give themselves at one and the same time. Extracting them clumsily one from the other, placing them (annulling them) outside themselves means depriving each one of them of its essential character of being itself within the other and the other within itself. The transcendental is not elsewhere in the same manner in which the immanent is not here. It is surprising – but undeniable – how every single (unique) person who knows and all single (unique) knowledge (each known thing), no matter how "private," irreducible, fleeting and perhaps incommunicable they may be, can give themselves and be said to be such only inasmuch as (relative being) they have in common this form – undoubtedly variously filled and infinitely variable as far as the possible modifications are concerned – *of uniqueness*. The necessary form of uniqueness is such as to make it be identical from time to time, giving itself continuously as different in the finite determination of its open but constitutive indeterminacy.

The form of this uniqueness is that of a real multiplicity within the unity of a horizon. This is the point; this "uniqueness" – this being necessarily unique of every beginning – is not in its turn unique with respect to (different from) the manner of what from time to time is said to be unique or, better: It is not something from time to time unique (which indeed would be something "different"); its "uniqueness" is necessarily identical and is expressed in lived experiences of differences that continually render a certain lived experience unique (singular).

It is therefore a "pure" (eidetic) form; it remains the same to give itself in an infinitely open series of uniqueness.

The uniqueness of an individual is not essential as compared with that of the form of the uniqueness; the former is a multiplicity of unity, the latter is the unique principle of multiplicity. Thus to say that knowledge is "subjective" can only mean that it "commences" in a subject, but not that it is relative to that subject and that for another it either is or could be "different."

In this connection Husserl has written a beautiful passage regarding the presumed skepticism of the Sophists, about its non-existence; for them, too:

zum ersten Mal wird das real Weltall und wird in späterer Folge die Allheit möglicher Objectivität überhaupt "transzendental" betrachtet, als Gegenstand möglicher Erkenntnis, möglichen Bewuβtseins überhaupt. Es wird betrachtet in Beziehung auf die Subjektivität, für die es bewuβtseinsmäβsig soll dasein können, und rein in dieser Beziehung: d.i., auch die Subjektivität wird rein als solche transzendentale Funktionen übend betrachtet, und ihr Bewuβtsein, die transzendentale Funktion selbst, als dasjenige, in dem oder wodurch, alle erdenklichen Objekte als solche für ein Bewuβtseinssubjekt jedweden Gehalt und Sinn erhalten, den sie für dieses Subjekt sollen haben können.[10]

It is precisely this insuperable manifestation in the immanence of the transcendental – of the subjective character of all nonetheless objective knowledge – in the form of uniqueness in general that represents the problem of philosophy; it does not look to uniqueness in a merely obvious sense, as irreducible difference; philosophy does not remain bound to empirical and obviously inexhaustible multiplicities (even "feeling" this inexhaustibility is not something "empirical" or "subjective"); it looks to uniqueness as a typical mode of "unity" (singularity, individuality, givenness, factualness,...); within the open horizon of feeling it sees the "identical" or the "similar" in what gives itself time and time again in some different manner. *This giving itself differently is identical.* A science of subjectivity – intuitively but rigorously delineated by Husserl already in his "*Erste Philosophie*" – is both possible and necessary, because (and to the extent to which) it comprises not only the objects of all the particular sciences but also all the sciences inasmuch as they are objects. Going "backwards," philosophy frees itself of its objectivism by continuously reflecting about itself and about this selfsame reflecting (*Selbstbesinnung*). Philosophy is born and takes form, it grows and lives in an individual and has a beginning in time; given its inevitability, whether it begins at a certain time or at some other, in this individual or in some other, it is nevertheless something to which it remains wholly indifferent.

Just like artists, philosophers – and their works – are all contemporaries. The subject in which this philosophy gains ground and slowly and inexorably reveals itself (takes shape by yielding itself) is as indispensable to it as the nature of that subject is of no importance: The end of philosophy is simply that of commencing. How much truth there is in it depends on the extent to which it belongs to a subject (person who knows), that is to say, the extent to which this subject, being similar to it and being recalled by it – recognizing it – belongs to it. To be such, this "truth" must from time to time give itself – cannot but give itself – through a particular subject, because it is only in relation to this "truth" that the particular subject can be said to be "finite" and "relative."

The logic of the "whole and the parts" is obviously inadequate for this essential and irreducible relationship.

In knowing, in every "part" (individual or moment) there appears – gives itself – the "whole" (the idea) and this whole appears there as part as well as part of the "one" of which it is part.

It is not by chance that the crucial themes of philosophy come to the fore in "*Erste Philosophie*" in their ineludable connection: the problem of evidence understood as "*unbedingt gültige Notwendigkeit*,"[11] and the fundamental starting point as point of arrival and completion (perfection). The pure structure of feeling (*Fühlen*), *time*, the nexus between "evidence" and "feeling" (and its possible distortions) – present knowledge – as constituted and orientated by a "natural" – and fully esthetic – "*Tendenz zur Harmonie*."[12]

The fact that these aspects reciprocally confer form upon each other, their being in a necessary connection, may or may not be seen. Nevertheless, they are not immediately clear, for each tends to hide the other; one arrives there after having passed through their ambiguity, after having risen to the top and then descended again to the (presumed) obviousnesses that occult common sense as the traditional scientific image of the world. "Pure" philosophy, as the recomposed and extremely rich (open) residue of the radical critique of all prejudicially (not originarily) founded "*Wissenschaftslehre*," exists in the permanent relationship between transcendental consciousness and its eidetic correlates within an "*in sich unendliche und dabei völlig geschlossene transzendentale Erfahrung*,"[13] in transcendental time. The open place for this philosophy is "*in der Alleinheit eines endlosen Lebenszusammenhang*,"[14] in the unity of an infinity and in the infinity of every unity (as of every object).

Transcendental phenomenology (or idealism), as the preliminary phase of a pure philosophy, is the experience of "*Transzendenz auf dem Boden der Immanenz*."[15] No subject that begins to philosophize can do so without the form of this relationship, because it is within it and only within it that he can possibly recognize himself as a subject; the task of philosophy (of the subject who knows) is inexhaustible not because, for example, it requires an infinitely long time or goes beyond the limitations of one and all; rather it is precisely by virtue of this teleologically necessary unbalancing of philosophy that, moment by moment, philosophizing is (becomes) perfect, just as the beauty of every splendidly beautiful thing is *beauty*. But the task is and remains inexhaustible, because the end here is making a start.

Philosophy in its necessary form of teleology is not an instrument (or technology). It is theoretical, and its end is not different from itself. Better, the nature of its being is that it has to be (itself); this "*Wert als Zweck*" character of philosophy defines its ethical nature in an unequivocal manner. Philosophy is characterized by an intrinsic and natural ethicalness that is neither symbolic nor extrinsic, neither moral nor exterior; nothing conventional (no fruit of convention) can give itself in it, and not even anything original (peculiar to one only), because its very structure renders it inflexible and constant. For philosophy, therefore – no matter how much it may authentically tend to become what it has to become and how continuous and uninterrupted it may be in this movement towards itself, into itself – "theoretical" and "normative" are but one thing; it is theoretical to the extent to which it is normative and normative (practical) to the extent to which it is theoretical (pure).

The apparently obvious and seemingly "ultimate" distinction between "theoretical" (to be) and "normative" (to have to be) is the first one that has to be unlearnt and forgotten. Rigidly postulating, on the one hand, a logic (or a mathematics or a geometry or any other formal deductive system) and on the other various technologies, this in the illusion of being able to found (or of having founded) the latter upon the former – whereas in actual fact one has absurdly brought the former (as means) into line with (made them commensurate with) the latter – represents a mortal danger for philosophy; it turns it into a particular discipline and turns the philosopher into a species or into an imitation of the specialist.

*University of Siena Munich*
*Ludwig Maximilian University*

Translated by Herbert Garrett.

NOTES

[1] E. Husserl, *Logische Untersuchungen*, Husserliana, Vol. XVII/I (The Hague: Martinus Nijhoff Publishers, 1975), p. 248 (21–23).

[2] Cf. E. Husserl, *Logische Untersuchungen*, Husserliana, Vol. XIX/2 (The Hague, Boston, Lancaster: Martinus Nijhoff Publishers, 1984), p. 567.

[3] Ibid.

[4] Ibid.

[5] E. Husserl, *Erste Philosophie, Gesammelte Schriften* (Collected Works), Vol. 6 (Hamburg: Felix Meiner Verlag, 1992), First Part, p. 15.

[6] E. Husserl, *Logische Untersuchungen*, Husserliana, Vol. XIX/2 (The Hague, Boston, Lancaster: Martinus Nijhoff Publishers, 1984), p. 540.

[7] Ibid. Cf. §8.

[8] E. Husserl, *Logische Untersuchungen*, Husserliana, Vol. XVIII/1 (The Hague: Martinus Nijhoff Publishers, 1975), p. 190.

[9] E. Husserl, *Ideen zu einer reinen Phänomenologie, Gesammelte Schriften* (Collected Works), Vol. 6 (Hamburg: Felix Meiner Verlag, 1992), First Part, p. 60 (1–11).

[10] E. Husserl, *Erste Philosophie, Gesammelte Schriften* (Collected Works) Vol. 6 (Hamburg: Felix Meiner Verlag, 1992), First Part, p. 60 (1–11).

[11] Ibid., Second Part, cf. pp. 35, 36, 49, 50, 68.

[12] Ibid., p. 46.

[13] Ibid., p. 170.

[14] Ibid., Second Part, p. 153.

[15] Ibid., First Part, p. 153.

TADEUSZ CZARNIK

# IS FREEDOM A CONDITION OF RESPONSIBILITY? AN ANALYSIS BASED ON ROMAN INGARDEN'S NOTION OF FREEDOM

Roman Ingarden claims that every event and process has its cause. This does not mean any radical determinism.[1] For him, there are relatively isolated systems in the world. The relatively isolated system is a system which is isolated temporarily and only in some respect and to a certain degree, but which is not isolated in other respects.[2] The system can be isolated by an isolator or by neutralization. The isolator is any material which takes over any influence.[3] Neutralization is the nullification of the efficacy of any process or event vis-à-vis others; for instance, the nonreactivity between chemical compounds which need a catalyst. According to Ingarden, the world is a big system made up of many relatively isolated lower level systems. The relatively isolated system is no oasis of freedom. Every event in an isolated system has its cause,[4] but not every influence is its cause. This means that Ingarden is a moderate determinist.

Freedom is found in the world, but for Ingarden, freedom is not any simple absence of a cause. Freedom is any ability (power) to generate *my own* decisions, *my own* acts.[5] Freedom is an activity, which has its cause in *myself.* My activity must go out from *myself.* What does "my" mean?

For Ingarden, man is a system of relatively isolated systems. The main systems are: the body, the psyche and the ego.[6] The body is a relatively isolated system of other such systems like organs, nerves, and so forth. The psyche is a system of psychic systems. In this way influences from the external world are not always the causes of internal events. My decisions can be independent from another system, but of course they have their causes. These causes are within my own system. They can be in my body, for example, great hunger; in my psyche, love; and in my ego, for instance, moral decision. The ego influences the psyche, but also the psyche influences the ego. The same is true between the ego and the body, and also between the body and the psyche. We can see the differences between them in that moment when one system goes against another. For example, the body goes against the ego when we are tired but want to do something more to help someone. The ego is against the psyche when moral principle is in conflict with a desire. In these situations my psyche is more mine than is my body and my ego is most of all my own.[7] This means that I am the most free when my decision comes from my own ego.[8]



*A.-T. Tymieniecka (ed.), Analecta Husserliana LXXVI,* 281–287.

Now, we would like to recall four different situations in which responsibility emerges, as they are distinguished by Roman Ingarden: 1) somebody is responsible for something; 2) somebody assumes responsibility for something; 3) somebody is called to be responsible for something; 4) somebody acts responsibly.[9] In our opinion the term "responsibility" comes from the term "response" (answer). The same is true in other languages: in French, in Polish. For example, in German, *Verantwortung* – responsibility derives from *Antwort* – response. "To be responsible" means to "answer" for something. For us, this answering is the essence of responsibility, but it must exist amid further circumstances. Let us consider two situations – first, one where I have done something wrong and, second, one where I have done something good. In the first instance I am responsible for this act, in the second one we do not say that I am responsible for my good act. This is because in the second situation my action did not entail any bad effect. It is not necessary to call somebody to account for good acts. What then is the purpose of calling somebody responsible? We thank someone for something good done, but we do not speak of that person's being responsible for that action. This is because in consequence of our good act there is a state which should exist. All the above means is that we call somebody to answer if he does something wrong. Broadly speaking, we do not want bad situations and we prefer good situations. Then if something does go wrong, we usually react (respond) to that. In the beginning there are two answers in which responsibility emerges. The first response is a reaction in the shape of discontent with something wrong. The second is an attempt to eliminate this bad situation. Later, we wonder if a person who has done something wrong, is the real (true) cause of this bad situation. (We say about this person that he/she is the cause, if there be any cause within him/her which began an action culminating in the bad situation.)

These first two responses are only first steps toward responsibility. But they are the next which are closer to essence of responsibility as we shall see later. A response means that there is somebody who asks and somebody who answers, i.e., there are two conscious subjects. Otherwise the question and the response would make no sense either regarding content or regarding purpose. What is responsibility for? It eliminates wrong situations, otherwise why would one ask such questions? Eventually, we could react only emotionally, but in that case our emotional reaction would not need any response. To call somebody to respond aims at preventing something wrong, preventing misery. This means that first there must be an evil (something wrong), and later we can talk about responsibility. Responsibility, therefore, is not a condition of morality (morals), but rather an instrument by which to eliminate or minimize evil. Responsibility is a moral coercion which makes us good, which forces us to respect its precepts.

We have said that taking responsibility is to prevent the recurrence of a bad situation or a bad state. This means some new circumstances. The most important is that there are causal connections which secure our result in the future. We do not want any bad state in the present to continue. A new situation must be possible. This means we must know the pertinent causal laws and so make plans and apply them. If there were no causal connections, we could not change our world through moral rules and any concious acts, and the result would be that we could not be rational.

As we said previously the inquiry concerning the cause of the bad situation follows the first spontaneous reaction. We want to know what or who has been the cause of the bad effect. We inquire into it because we are going to keep this situation from recurring. Therefore, we ask if somebody in particular has been the cause of the bad situation. If so, that party can answer, but he does not have to answer. This is the first moment when the initiator of the state of affairs is called on to respond and to be responsible. If this other, who has been the cause of a bad effect, does not want to undertake a response (which is not yet to assume responsibility), that does not mean that the process of taking responsibility is finished, because we ourselves may call him to be responsible for his bad influence, i.e., we acknowledge that we or other people may yet do something to prevent the next such predicament. We can reprimand him, but we can also punish him. That all means, "to call somebody to be responsible for something." Of course regarding this man who is responsible for something, it means that we know for sure that he is the cause of a bad effect and may also assume responsibility for something. He can consider if a decision of his has been the cause of the bad effect, and he can acknowledge (accept) that he can be reprimanded or punished. In this way he responds to (answers to) us. A person may also act (be active) with consciousness of which things cause which bad effects, and he can be very careful. He can also be ready to accept a reprimand or punishment if he has done something wrong. That is called: "acting responsibly." Of course, there are many levels of responsibility, depending on the magnitude of effects, on personal consciousness, and so forth.

But if there are causal connections in the world, we can ask whether responsibility can be assumed without there being freedom in the world, whether responsibility requires freedom. Usually people and philosophers claim that responsibility is not possible without freedom. Let us consider this problem. If responsibility is the process of our response to something bad – our recognizing that we personally cause the situation and our working to prevent similar bad events, that means that it is not necessary to be free. Every act of response of an agent and of a sufferer (victim) can have and has a cause. We do not need any freedom at any moment of this process. Our call to be responsible is the cause which changes the decisions of others. This is a special conscious reaction.

Of course sometimes people claim that freedom is necessary to responsibility, but this is only an imperfect recognition of the personal causation of bad situations. When somebody deliberately makes a bad decision without pressure he is said to be responsible for that decision. Even if he were under pressure and his every decision had an impetus (because it had), he would be responsible for that decision. Responsibility depends on whether our response (reaction) can yet change that decision or can change his next similar decision and the outcome. This is because one of the main aims of responsibility is to prevent evil, now and in the future. This means that the limit of responsibility is in the case, where none of our (or others') calls to be responsible can or could change the decision of the agent. If the pressure is so strong that it is not possible by our calling him to be responsible, to change his decision, that means that he cannot assume responsibility and that he is not responsible for his decision. Of course the pressure of responsibility has many levels and they depend on the given situation. The punishments, like values, are fixed over a long time of "moral evolution." The punishments are proportional to the values of bad acts. In this way the pressure of morality cannot be infinite and is proportional to the fault. All this was described above in an individual case, but usually all this takes place among people through the influence of values. Morality is impersonal like language, but it has its foundation in people. Responsibility is founded on morality, because it is a process which is in relation to evil. But there is also another relation: morality needs responsibility, because otherwise people could ignore values and morality. Morality must have a weapon, otherwise it is unprotected from our egoism. Responsibility is this weapon.

*The Institute of Philosophy*
*The Jagiellonian University*

NOTES

[1] "I should like, however, to underscore the dissimilarity of the conception of the 'freedom' (i.e., of the ownness of the decisions and deeds) of a human being developed here from that of the customary treatment of the problem of freedom and its possibility. For the most part, one

subscribes either to the view which completely denies man's freedom, since it is excluded by a radically conceived determinism of the real world, or, at the opposite extreme, to the view that a human being is endowed with total, unconditioned feedom. Cases which lie, so to speak, in between these conceptions are not considered at all; nor is what is primarily at issue here taken into consideration, namely, that decisions and deeds are one's *own*." Roman Ingarden, *Man and Value*, trans. A. Szylewicz (Munich, Vienna: Philosophia Verlag, 1983), pp. 61–62 and "To begin with, if the real world were to form the kind of causal system which emerges from the conception of Laplace, and in the last analysis from all of modern natural science and philosophy, then, as has been frequently ascertained, there would be no free decision of the will possible in this world, nor any free human (and animal) action. For this conception includes the assertion that *all* events in the real world (thus, also volitional decisions) together form one *single* system of causal relations, in which, as effects, they are uniquely and necessarily determined by their causes. I call this conception 'radical determinism.' If one admits that every decision of the will comprises an event in the world, then it is impossible to consider it as not having a cause. Therefore a positive solution of the problem of freedom can be expected only when 'freedom' is not identified with having no cause, but rather when it is conceived as the agent's 'independence' from external factors, and when it is at the same time demonstrated that radical determinism is untenable." (*Ibidem*, p. 101.)

[2] "Closed systems were in fact regarded as an ideal limiting case, which could never be fully realized in actuality except, perhaps, in the case of the whole real, material world itself. 'Open' systems in contrast were frequently treated as if they were 'open' *from all sides*, i.e. nowhere 'shielded,' 'delimited,' 'isolated,' so that they would then have to decompose into mere events, which would extend in all directions and vanish into the infinite manifold of the world-process. These 'open' systems were therefore once again something like a merely conceptually formed ideal. If however an 'open' system is to be able to sustain itself effectively within the real world for a time, as something identically the same, then it should not be universally open but must, at least in some respects, be bounded off from the surrounding world and partially isolated or, better, shielded from it." (*Ibidem*, p. 86.)

[3] "On the one hand it protects the system from certain strictly defined kinds of influence to be found outside the system, and always in response only to a certain degree of intensity of influences and to a determinate kind of impress exerted by them; on the other hand, however, it permits certain special kinds of outside influences to encroach into the interior of the system, and it allows certain processes taking place in its interior to pass outside...." (*Ibidem*, p. 90.)

[4] "For freedom of action or, as we also ordinarily say due to a shift in the problem context, the decision of a 'free will,' is interpreted in the sense of 'lacking a cause.' This is considered incompatible with the deteminacy which, in the real world, reigns universally. This is also at bottom Kant's positon, who in fact introduces into the world of things in themselves a certain 'causality inherent in freedom' and distinguishes it from the 'causality of nature' (in the phenomenal world). In some remarkable way, however, Kant allows this 'causality inherent in freedom' to break into the seamless web of causal interconnections in the phenomenal world and to initiate in it new causal chains. This chain is itself not supposed to have any cause in nature, yet it is supposed to apear in it and to affect its course. In the world of appearances therefore 'freedom of the will' is tantamount to 'lacking a cause.' How Kant can speak about a causality inherent in freedom in the world of things in themselves – in which, after all, no categories whatever have a place – and how he can suppose that it can penetrate into the uninterrupted, causally determined manifold of events in the phenomenal world are matters concening the Kantian philosophy itself which we need not delve into here. We should not forget

however that the Kantian determinism in nature tacitly presuppposes a causal ('natural') order of the world in the sense of Laplace. In this context two things need to be undertaken: 1) not to interpret freedom in the sense of absence of cause, but to set in its place the concept of one's own deliberate decision and of one's own action; 2) to consider whether Laplace's conception cannot be replaced by some other conception of the causal structure of the real world. I shall not go into the latter until later.

"Nicolai Hartmann, who does not accept Kant's transcendental idealism and his two-world theory, has already emphasized in a noteworthy manner that it is impossible to demand that the free volitional decisions have no cause. This is so first of all because such would be impossible in a world which is causally determined through and through, and secondly because it could not then be rational and suited to the real situation which calls for it. Hence it would be unmotivated, and could be made neither by an I nor by a non-I. It would not, accordingly, be a decision of the given person, and the action stemming from it would not be that person's own deed. The person could not at all be held responsible for it. If we still wish to maintain that the acting human being is, and can be, responsible for his own, and only for his own deeds, then we have to admit at the same time that these deeds have, in their way, to be causally induced." (*Ibidem*, pp. 59/60.)

[5] "A person who is to bear responsibility for his deed must, as we have already established, be *free* in his decisions and deeds. According to our earlier considerations this means nothing other than that the given conduct is his *own* deed. But this in turn means that the deed follows from the person's initiative and at the moment of its being undertaken and in the course of its performance it is at least *independent* of any factual matters which do not include the person in his immediate environment, factual matters which could in principle exert an influence on the person's decision and on the implementation of the deed. This presupposes, on the one hand, a definite formal structure of the person, and on the other, a particular structure of the real world in which the person lives and acts." (*Ibidem*, pp. 84/85.)

[6] "Every human being is a corporeal and psychic being whose personal 'I' is engaged in a special way whenever he is engaged responsibly." (*Ibidem*, p. 78) "Still, the understanding, that a human being as person is such a very complicated, partially isolated, system of a higher order, hierarchically built up out of many lower systems, has heretofore not dawned on the consciousness of anthropology and its subdisciplines of psychology, anatomy, and human physiology." (*Ibidem*, p. 87.)

[7] "In order to be 'independent' of the surrounding world in his decisions and in the actions issuing from them, the person must, above all, contain a centre of action, which enables him to take initiative and at the same time to have defence mechanisms which prevent his being disturbed in his acting. But he must also be sensitive to outside intrusions, insofar as his responsibility springs from a determinate form of his living together with the surrounding reality, and particularly with other people. The person must therefore be open and receptive in his behaviour and in his so-being, and at the same time protected and insensitive in other respects." (*Ibidem*, p. 85.)

[8] "A deliberate decision and action can pass for a given person's 'own' only when they spring forth directly from the I-centre of that person, have their authentic origin in it, and when this I-centre commands and directs the execution of the action emanating from it, therefore the action can be accepted as one's 'own' not when the I-centre merely has some personal interest at stake, but when it holds within itself the decisive impact over the total course of the evolving action. This can come about in two different ways. In the first case the I accepts only what takes place within its own psyche, in the sphere of its personal being (of its being thus and so [*Sosein*]), or what encroaches into this sphere from without and which the I accepts only out of necessity, as it were, because it cannot do otherwise. Here, however, approval is granted to some mode of

behaviour, without the I actually assimilating it or making it its own. In the second case, however, the I draws the decision out of himself, out of his own deliberation which is unswayed by extraneous reasons, and proceeds to engage in the endeavour of acting. Complete unconditionality by external motives and reasons thereby comprises the optimum situation in which a volitional decision is made, and in which an action undertaken by a personal I is performed in the strict sense as the given person's own action or deed." (*Ibidem*, pp. 60–61.)

[9] Vide *ibidem*, p. 53: "To begin with, we need to distinguish four different situations in which the phenomenon of responsibility emerges: Someone

1. *bears* responsibility for something or, differently put, is responsible for something,
2. *assumes* responsibility for something,
3. is *called* to account for something,
4. *acts* responsibly.

"The distinctiveness of the first three situations is demonstrated in the first place by their factual independence of each other, although determinate interconnections of sense undoubtedly obtain between them. One can be responsible but neither be called to account, nor assume responsibility ("take it upon oneself," as Nicolai Hartmann puts it). And conversely, one can be called to account for something without being in fact responsible for it. One can also factually assume responsibility for something, without being actually responsible for it. Given that someone is responsible for something, he should both assume responsibility and be called to account for it. If one does not assume responsibility, or even evades doing so, despite the fact that one is responsible for something, then one is (also) responsible for behaving in such an [irresponsible] manner. But even the very assumption of responsibility for something that one is not responsible for appears to fall under the proviso of responsibility. Still, an essential interconnection of sense appears to obtain between these factual states of affairs, irrespective of their mutual independence in fact."

J. J. VENTER

# ECONOMISM: THE DEBATE ABOUT THE UNIVERSALITY CLAIMS OF ORTHODOX ECONOMICS

## 1. INTRODUCTION

Present-day economic theory is the product on the one hand of abstract thinking and on the other of a long interaction with economic reality. This is why we find the strange paradox that while economic theory is accused of being totally urealistic in its assumptions, in the outcomes it functions normatively in affirming the economic system as it is, as good. For the unrealistic assumptions have been formulated in such a way that their outcomes are in support of the economist society as we know it. Serge Latouche characterizes this society as follows:

> The economy is the religion of our time. This is attested to by many analyses and recognized by certain economists themselves. Modem society – many have noted it – did not chase away the idols, myths and dogmas; it only succeeded in replacing them with others. There is even an abundance of candidates for the divinity: Reason, Progress, Science, Technology – to name only the most credible. In any case the devotion to Progress, the dogma of Development, the cult of Technology, the appreciation as if sacred of Material Well-Being, up to the sacrosanct Human Rights and the untouchable Democracy, are at bottom directly or indirectly linked to the economy via utilitarianism. The calculus of pleasures and pains, of duties and rights, of costs and benefits, inhabits our projective imagination and nibbles away at the major part of our practices. (Latouche (a), p. 10)

We may recognize something of our own environment in Latouche's words – the way our lives are directed by the divinity of Material Welfare too. At the same time we know that problems like unemployment, poverty, the world market coming to our doorstep with international competition, are forcing us to submit in some way to the demands of this religion of our time. Almost all the social contexts which previously focused on non-economic activities, such as the university, the high schools, the sports and art institutions, have been refocusing and are assuming functions and structures of profitability. We know that the economy is a real part of our lives, and we have to function in it. Under the pressure of the International Monetary Fund (IMF), the state is adjusting to late capitalism also. We cannot just pretend that this environment does not exist.

The tendency towards encompassing the totality in economic terms is not limited to our practice, but has been supported for a long time by orthodox economic theorists. One of the grandfathers of economic theorizing, the



*A.-T. Tymieniecka (ed.), Analecta Husserliana LXXVI*, 289–320.

eighteenth-century thinker, Adam Smith, already extended the idea of a free market based upon self-interest to all of society, believing that we do not need benevolence as a foundation of society, since the "exchange of good offices" can do the job (cf. A. Smith, p. 86; J. J. Venter, (c), p. 49). Whereas Adam Smith metaphorised the idea of exchange in non-economic areas of social life, it seems as if recent and very influential Monetarist economists simply took it literally. Karl Brunner on the one hand explicitly rejects a multidisciplinary approach to economic problems (including the Keynesian "sociological perceptions of non-market situations"); on the other hand he wants to apply the basic principles of Monetarism to other social disciplines, as if economics provided the only valid social scientific approach:

> We reject, on the other hand, an escape into sociology, which offers no relevant analytic framework. We maintain that socio-political institutions are the proper subject of economic analysis. This entails an entirely different view of the political institutions and their operation. The sociological view typically supports a goodwill theory of government and yields conclusions favouring a large and essentially unlimited government. An application of economic analysis, in contrast, alerts us to the fact that politicians and bureaucrats are entrepreneurs in the political market. They pursue their own interests and try to find optimal strategies attending to their interests. And what is optimal for them is hardly ever optimal for the "public interest" (K. Brunner in Klamer, p. 186),

Heterodox economists and social scientists are attacking the imperialism of orthodox economists' claim to universal truth. At stake is the claim of orthodox economists that theirs is a positive science, based upon the calculability and measurability of the objective facts of individual rational decisions. This claim is expanded to cover all of social reality – society itself is supposed to be a market and all forms of social intercourse are supposed to be transactions – at base, says a heterodox economist, is the *grand récit* of the two (individual) savages who meet and find a bond in the action of bartering (cf. G. Bérthoud, pp. 61ff). This narrative expresses on the one hand the individualism and on the other the social totalitarianism of exchange as supposed in capitalist theory – both typically Western.

The beliefs underlying the absolutizations (of reason, progress, science and technology), in particular the idea that the economy is based on universal and atemporal principles (which underlies the pretence of economics as positive, predictive science), and the hypothesis of the universal utilitarian rationality of individual decisions (which makes economics as an axiomatised calculus possible), as well as the supposed enmity between humankind and nature (which motivates exploitation), are all questioned by heterodox economists. Orthodox economics pretends to be and is accepted as the most "scientific" of all the human sciences, but, claim the heterodox, by the same token it has become the least "human" of all the human sciences. These latter favour a holistic approach in which economics is "dethroned" and returned to its rightful status as one science among others in the context of the body of social sciences, and at the same time capitalism is "dethroned" and relegated to the position of a modern, "provincial", Western belief.

It is my intention in this paper to show, through an analysis of the truth claims by orthodox economists and the counterclaims of heterodox economists, that the so-called universal economic laws are actually norms (and that economics unavoidably is a normative discipline), and that we need a meaningful repositioning of the economic system as well as scholarly economics through adopting the principle of the simultaneous realization of both non-economic and economic norms. It is only in this way that we can take account of the variety of facets of human life and avoid the stereotypical approaches of calculi of human behaviour (like those of Ricardo, Malthus, Edgeworth, the Neo-Classicals and the Monetarists).

Given the collapse of Marxism and the context of Neo-Liberalism, critiques of capitalism are fairly unpopular at the moment. This in itself makes the criticism of the heterodox economists noteworthy. The accusation that we are faced with a religion of some sorts is so strong, that we need to give consideration to both sides in order to evaluate the criticism – since this religion gives direction to the social context in which we live.

## 2. THE BASIS OF ORTHODOX CAPITALIST THINKING

P. Combenale (pp. 165ff) summarizes the main points of orthodox economic theory as follows:

1. Moving away from a metaphysical thinking in which one asks "why" (for example "why exchange of goods?" or "why maximizing profit?") and "what" (for example "what is the essence of economic behaviour?") to a more positivist approach in which one asks "how" ("how to explain the formation of prices" or: "how to maximize profit?") and in which one studies relations among phenomena, in order to find regularities (as the natural scientists). Economics is a human science, but its practitioners practice it like a natural science; this is why Latouche ((b), p. 15) calls it the least humane of all the social sciences.
2. The obligation to distinguish between positive (descriptive; saying how things are) and normative (saying what ought to be) enunciations, and to

stick to the ideal of the objectivity supposedly characteristic of science. This is why economists have difficulty in answering the kinds of questions journalists pose to them, for example, how do we tackle our unemployment problem or move out of underdevelopment, for this forces them out of the descriptive and into the normative mode. (Factually, the economic sphere is a normative sphere, and finally the regularities we find must serve as indications of norms or the transgression of norms – but this presupposes that we must have some general idea of the good in advance. The economists have found such an idea in "utility".)

3. The necessity of finding for itself a particular study object, by cutting out of the wealth of social reality that which is called "the economy". (Louis Dumont says that there is nothing resembling an economy in external reality up to the moment when we construct such an object – which is, I think, going too far. But it is true that in the process of establishing economics as a science, phenomena with certain characteristics are isolated from all others as a field for study.) This area then should be regulated by endogenous laws which one can study, all other things remaining equal. Ideally this area should be regulated by itself; otherwise there won't be a field for the economists if technique, politics and culture regulate all its phenomena. This autoregulation is found in the market which is supposed to co-ordinate the plans and actions of separate individuals; if the market cannot be assumed to do the job, then one would have to ascribe the co-ordination (if one assumes it to be there) to an institution like the state or the church, or to the social dynamics of values, which will in turn imply that there really is no field of study for the economist.
4. The ideal of bringing to light universal and atemporal principles. Like the laws of physics, the laws of economics are supposed to be applicable everywhere and all the time. The one thing that, according to economists, is a universal invariant upon which a model can be constructed, is scarcity. Among all human beings of all cultures and persuasions this one determinant remains: their unlimited wants have to fit in with limited resources so that human beings are always faced with choices – apple tart or porridge, going to the mountains or going to the beach, butter or cannons. (One could say that there is a relationship between scarcity and wants: the assumption of the universality of scarcity is a correlative of the assumption of unlimited wants, which rationalizes the creation of wants by advertising. In a Stoic or Epicurean community, with an ethic

of limited wants, such rationalizing would have been considered immoral.)

5. The hypothesis of the rationality of human behaviour, which otherwise would be erratic and unpredictable. One cannot develop a calculus of utility (increasing pleasure and avoiding pain) without some criterion, in this case that of rational decision (maximizing advantage and minimizing costs). This defines the human being as an economic being – a being who supposedly always chooses rationally to increase his/her pleasure and decrease pain. (Ironically, one of our pains is that of labour, so we work in order not to work.) These choices are supposed to be made by individuals, who are treated as if they are asocial, isolated, well-informed and self-sufficient; the social system is not taken into account – thus the political organization and the value system adhered to by the community does not enter into this perspective. More precisely, I would say that rationality is understood in terms of a supposedly universal value of utility.

These ideals and demands create their own problems. To understand some of the problems, we can study what heterodox economists say about the orthodox (Neo-Classical) point of view.

The implied view of science is that of the *hypothetico-deductive method*. In other words, with the help of certain assumptions the human being is stereotyped as being always rational (self-interested) in his/her behaviour and always fully informed about his/her interests and needs, the products available to satisfy those needs, the alternatives to this product, and the prices of all relevant products. On the basis of these assumptions economic theory is set up as a calculus of decision making. Given the universally stereotypical behaviour of human beings and the universality of the scarcity principle, one can deduce and predict future behaviour under certain conditions. (Von Hayek calls this an empty [a priori] logic of options, devoid of any information about real situations of competition.) This implies that the socio-historical side of human behaviour, its changes, conflicts, group dynamics, is not taken into account. That this latter is important, one can see from the work of Marx (the contradictory movement of capitalist history), Schumpeter (innovation as the basis of capitalism rather than the market), and Keynes (the radical uncertainty of future equilibrium) – even if one does not necessarily agree with them. (In brief: the assumption, for example, of full information eliminates from discussion the fact that the economic agent may not know what products are offered elsewhere or may come into the market in the future.)

The dominant paradigm in economics is the Neo-Classical one. Neo-Classical economics and its near kin, Monetarist economics, both tend to *avoid talking about institutions*, even economic ones. They limit their research to supply, demand, and prices of money and goods, and treat institutions as black boxes which only mediate supply and demand (and therefore prices) (see further Venter, (c), p. 52; Machlup, p. 9). For the understanding of the laws of economics one does not need to know what happens inside these black boxes. Rather only input and output are of importance. Heterodox economists (see Combenale, p. 168) ask about the contextualization of economic activities. They want to study individuals in their socio-economic context such as the relationships between capitalist and salaried employee, entrepreneur and banker, entrepreneur and lender, creditor and debtor, looking at agents as differentiated by their function: enterprises by production, banks by financing, households by consuming, et cetera. They also want to look at power relationships and their possible hierarchies such as at how households are dependent upon enterprises for income and products, enterprises upon banks for finance, and so on.

This already indicates that there is a wider variety of factors than only supply and demand as such to take into account. The idea of a *self-regulating* market is construed in terms of the one-sided assumptions of the stereotyped "homo oeconomicus". Keynes, for example, in his attempt to take expectations into account, could no more successfully construe a general equilibrium; von Hayek thinks that general equilibrium theory does not take into account how competition really works, and that it presents no more than an a priori logic of choices, options, which show themselves logically on the basis of certain assumptions (cf. Venter, (b), pp. 241ff). We also know from the very fact and problem of overspending, both by individuals and institutions, and from the depletion of resources – even resources which are supposed to have no economic value because they are supposed to be in abundance, like water – that the market is controlled by decisions which are often not well informed or responsible, and does not automatically solve problems by re-establishing equilibrium. Combenale (p. 169) calls the idea of a self-regulating market a "faith" or "belief", since the stability of general equilibrium has not been proven.

The heterodox economists blame the Neo-Classical economists for *neglecting money* in their theory. Neo-Classical economists tend to focus on exchange as a process in which real goods are swapped. The heterodox economists want to study the role of money as a social factor, since, as they say, money is endogenous to the process of exchange. In fact they go much further, stating that the unity of the market society formed on the basis of the division of labour is based on the circulation of money and that if the latter stops, the system of exchange will collapse. And it is endogenous, since it is created by the banking system in the form of credit, so it should be studied as part of the economic process. The heterodox economists want to study money even as a social bond, since it functions under the authority of the state (and therefore expresses political power), and is therefore also a vehicle for the importation of exogenous disturbances (for example, state intervention). They view money as a factor which "depersonalizes" and even "decommunalises" (Combenale, p. 169) the objects which circulate (in other words money makes it possible to eliminate person to person bartering while it enables us to sell and buy to total strangers even without direct contact). Money makes the "science of economics" possible, for it makes it possible to evaluate goods in terms of a common denominator, and therefore makes them calculable, and they say, it is the only possible social bond in a decentralized economy. Focusing their attack on the Neo-Classical economists, the heterodox economists do forget that another influential current of economic thinking, the Monetarists, did focus on money and even took its political manipulation into account – but only again to defend the self-regulation of the market, viewing all of society as one big market where interests are exchanged (see further Venter, (c), pp. 51ff). Heterodox economists, to my view, are too vague about what they mean by a "social bond" – although they may be correct that the economic links in society have grown out of proportion in their importance, these are even today not the only links that exist.

The Neo-Classicals tend to provide a *static analysis* of the economy – in terms of mathematical equations where both sides of the equation remain in balance. This tends to provide a fixed system with predictive power: what is added or subtracted on one side has its effects on the other side of the equation. Heterodox economists feel more at home in the neighbourhood of Keynes' view of radical incertitude. Keynes held that in a monetary production economy the variation of expectations about the future can influence the actual volume of employment – in other words if producers suddenly feel uncomfortable about the future demand for their product, they may produce less and retrench some workers. A number of characteristics, which clearly delineate the heterodox economists from the orthodox ones, are important here: "... the monetary production economy stands over against the economy of exchange of real goods of the Neo-Classicals; the accent is placed on incertitude; the level of employment, therefore that of production, and the volume of resources in general, are variable; full employment is not automatic" (Combenale,

p. 170). Keynes drew two consequences from this radical uncertainty: firstly, that the importance of money follows essentially from the fact that it constitutes a link between the present and the future; secondly, the economy evades chaos only by certain conventions, for example, the present level of the interest rate conventionally depends to a large extent on its future value according to estimations of the dominant opinion.

The orthodox views of *time and rationality* need revision too. In orthodox theory *time* is reversible: if some exogenous disturbance occurs and the economy is thrown out of balance, then orthodox economists can model a return to equilibrium. Heterodox economists want to work with time as experienced in real life, where one event follows another, and a process is not reversible to the previous position. This demands a second look at the concept of rationality. In Neo-Classical theory *rationality* means that the individual takes care of his/her own interests by calculating pleasures and pains (utility) and settling at maximum advantage. Heterodox economists (Combenale, p. 171) believe that rationality in this context is procedural – problems are solved one by one (trying for the least evil possible rather than the optimal solution immediately), so that one finds avenues along the way which allow for specialization or benefiting from economies of scale; a basis for relative prices develops which helps to expand the market, et cetera.

## 3. MILTON FRIEDMAN AS ORTHODOX ECONOMIST

Most of the basic ideas about orthodox economics outlined above find expression in both Neo-Classical and Monetarist economics (even though these currents differ about certain matters). Milton Friedman, Nobel Prize winning Monetarist economist, may serve as a case in point.

Friedman sides with Pragmatism in his defense of the idea that theories are instruments that are useful for the prediction of empirical outcomes. Therefore, he says, the assumptions of a theory do not have to be realistic; they only have to predict consequences correctly. Friedman therefore believes that any theory is acceptable as long as it predicts correctly. (As we shall indicate later, heterodox economists contend that the world construct of orthodox economics does not have sufficient linkage with the world in which we live economically, neither does it succeed in predicting major changes in the economic world.) Pragmatism believes itself to be free of any of the grand metaphysical keywords such as "God", "reason", "being", and to be functioning only as a method; it is not supposed to imply any idea about the nature of reality or of humankind or of knowledge. Pragmatism is one of the irrationalist descendants of Positivism in its rejection of metaphysics.

Friedman, following this line of thought, distinguishes with Keynes between a "positive" and a "normative" economics (and adds a third: economics as a practical "art"). Though recognizing "normative" economics as valid knowledge, he focuses his own attention on "positive" economics:

Positive economics is in principle independent of any particular ethical position or normative judgments. As Keynes says, it deals with "what is", not with "what ought to be". Its task is to provide a system of generalizations that can be used to make correct predictions about the consequences of any change in circumstances. Its performance is to be judged by the precision, scope, and conformity with experience of the predictions it yields. In short, positive economics is, or can be, an "objective" science in precisely the same sense as the physical sciences. Of course, the fact that economics deals with the interrelations of human beings, and that the investigator is himself part of the subject matter being investigated in a more intimate sense than in the physical sciences, raises special difficulties in achieving objectivity at the same time that it provides the social scientist with a class of data not available to the physical scientist. But neither the one nor the other is, in my view, a fundamental distinction between the two groups of sciences. (Friedman (a), pp. 4–5)

Friedman distinctly recognizes that there are "oughts" regarding economic life; but he separates them from his preferred positive science (and as we shall see later, makes them somehow dependent upon positive science). Friedman exemplifies orthodox economics in placing positive economics in the vicinity of the physical sciences (and therefore also in a non-metaphysical tradition). But his distinction and separation between "what is" and "what ought to be" is still an ontological distinction, which in fact poses the question of the ontological status of "what ought to be". (Does not the "what ought to be" have some kind of existence?) Friedman's view of economics as a science and his practice of this science could not clear itself of all metaphysics, as we can clearly see from his view of the relationship between theory and reality:

A fundamental hypothesis of science is that appearances are deceptive and that there is a way of looking at or interpreting or organizing evidence that will reveal superficially disconnected and diverse phenomena to be manifestations of a more fundamental and relatively simple structure. And the test of this hypothesis, as of any other, is its fruits – a test that science has so far met with dramatic success. If a class of "economic phenomena" appears varied and complex, it is, we must suppose, because we have no adequate theory to explain them. Known facts cannot be set on one side; a theory to apply closely to reality on the other. A theory is the way we perceive "facts", and we cannot perceive "facts" without a theory. Any assertion that economic phenomena *are* varied and complex denies the tentative state of knowledge that alone makes scientific activity meaningful. ... The confusion between descriptive accuracy and analytical relevance has led not only to criticism of economic theory on largely irrelevant grounds but also to misunderstanding economic theory and misdirection of efforts to repair the defects. "Ideal types" in the abstract model developed by economic theorists have been regarded as strictly descriptive categories intended to correspond directly and fully to entities in the real world independently of the purpose for which the model is being used. The obvious discrepancies have led to necessarily unsuccessful attempts to construct theories on the basis of categories intended to be fully descriptive. (Friedman, (a), pp. 33–34)

On the one hand Friedman tries to sustain a physicalist form of "objectivity", but on the other hand he attacks those who criticize theories for "unrealistic assumptions", precisely because they want the theory itself to be objectively realistic, instead of testing it against its fruitfulness in predicting subjectively intended outcomes. On the one hand he wants a "positive" economics, in which we strictly separate that which *is* from that *which ought to be* (i.e., the kind of economic policies we have to follow); on the other hand, once he works in "positive" economics, he relinquishes the description of "what is", stating the impossibility of a "realistic" descriptive science, since "facts" are not separate from "theories" – these latter being the way in which we see "facts". The criterion for seeing well is the realization of one's aims. It is not clear whether the simplified structures which are crystallized in theories are supposed to have a certain measure of objectivity. Implicitly, in his rejection of theories which leave phenomena varied and complex, he claims universality for the "relatively simple structure"; but since they are structures crystallized in terms of subjective aims, this claim is relativised with regard to those who share the same aims.

Clearly Friedman, under the heading of "methodology", is not simply spelling out procedures for finding new knowledge in his discipline – he is interested in the problem of describing reality accurately (which, he concludes, is not possible). In fact he is busy rationalizing a philosophy of power. His whole view of the "real" is concentrated into those elements which are predictable given certain assumptions. Some of these assumptions are "universal" (generalizations), such as that economic agents act "rationally" (i.e., in their own interest) and that they are fully informed – orthodox assumptions, as we have indicated above. Assumptions of this kind co-operate with circumstantial assumptions in order to predict certain specific empirical outcomes. If these outcomes are realized, then measures of control follow from them. For example, if the mentioned general assumptions (together with the hypothesis that a specific rate of increase of the money supply will cause a certain increase in the inflation rate) are realized in a concrete situation, then we also know that we can control inflation by controlling the money supply. It seems that Friedman, attempting to develop a pragmatist methodology of economics which avoids a foundation in values and grand ontological keywords, re-admits at least the ontological problems of the relationship between "what is" and "what ought to be", as well as of that between subject and object, with subjective power of control as the hidden key-element in his meager ontology.

But one can have more questions about the universality of Friedman's approach. Since the predictability of inflation does not in itself imply that inflation is bad, it makes no sense to keep the inflation rate down, or in fact to do research on the inflation rate, if there is no certainty that inflation is bad. In fact the attempt to predict inflation can only make sense if the assumption already exists that it is something which can and ought to be controlled. But this means that Friedman's positive economics presupposes values without which the whole undertaking does not make sense. Friedman himself believes that control measures can be deduced from positive economics, but not the other way round:

> Positive economics is in principle independent of any particular ethical position or normative judgements. As Keynes says, it deals with "what is", not "what ought to be". ... Normative economics and the art of economics, on the other hand, cannot be independent of positive economics. Any policy conclusion necessarily rests on a prediction about the consequences of doing one thing rather than another; a prediction that must be based – implicitly or explicitly – on positive economics. There is not, of course, a one-to-one relation between policy conclusions and the conclusions of positive economics; if there were, there would be no separate normative science. Two individuals may agree on the consequences of a particular piece of legislation. One may regard it as desirable on balance and so favor the legislation; the other as undesirable and so oppose the legislation. (Friedman, (a), pp. 4–5)

Even though Friedman contends that positive economics is in principle devoid of any particular ethical position, there are some "oughts" hidden in Friedman's "is". He does not provide a clear defense for his contention that normative and practical economics are necessarily dependent upon positive economics. During thousands of years, trade existed among kingdoms, and governments made decisions about taxes and therefore influenced the money supply and trade, without much positive analysis of economic "causes" and "effects". In fact, the influence of economists in this area is of a very recent date. For Friedman positive economics must play the role of uncovering the consequences to be expected from certain actions, in order to come to correct policies. Since normative economics is one-sidedly supposed to depend upon positive economics, this indicates a belief in the power of science to guide human life, and that positive economics can provide value-free guidelines. Are these not in themselves normative beliefs which direct Friedman's approach to economics? Friedman does not seem conscious of the fact that he had already founded "positive" economics on a normative basis when he gave meaning to it in terms of the ideal of control.

Friedman does not account for the origin of the value system which dominates normative economics. Thus he does not really help us to determine which supposed consequences are to be taken into consideration as the basis for policy decisions. Suppose two parties in parliament are in agreement that a specific piece of legislation will leave the poor five per cent worse off in the first year following the implementation of the legislation. There may still be differences about further consequences: one party may believe that this will cause labour productivity to rise, so that in the longer run the poor will be much better off, while the other one may belief that the poor will land in an unstoppable downward slide. And there may even be a debate about the good or bad of each of these possibilities. Friedman himself puts his trust in the homogeneity of the value system of the West to eliminate differences of opinion which may become nasty – this he writes in 1953, less than a decade after a world war which had a nasty ideological origin. Friedman trusts the progress of positive economics to make more and more precise predictions, so as to overcome most of the differences finally.

Furthermore, he talks in one breath about "policy", "legislation", and "norm", siding apparently with the Western tradition that there is an encompassing society which finds expression in the state, which then, through policy and legislation, expresses the values of society in words. This creates a problem, for the direct association of society with the state is already totalitarian. Values are also formulated and inculcated by a variety of non-state structures (even in spite of totalitarian situations). Thus Friedman's assumptions about society are debatable.

Neither does Friedman give account of the nature of economic generalizations which positive economics uses as the basis for its predictions. He does not see any fundamental difference between economics and the physical sciences; therefore we must accept that the general statements of economics express something like "laws". He uses the term "structure" for the simple predictive explanations of economic phenomena, but they are products of subjective aims. It ought to be explained just how they are generalized into universal applicability.

Furthermore, when we reach the highly abstract level of a quantitative model, like Friedman's model of the money supply, then economic generalizations and their implications can look very much like the laws of physics. But what are the implications of this for human freedom? Without duly analyzing the relationship between economic laws and freedom, Friedman becomes a defender of freedom in his positive economics. Has not thereby the normative idea that freedom is good come in through the back door into Friedman's formation of economic theory, as a directive basic idea?

Friedman does not give us an analysis of freedom in its relationship to "law" or "structure". He uses the market view of society, however, and thus economic freedom of choice becomes his guiding model for freedom. He distinguishes between economic markets and so-called political markets – the political market is actually only an inefficient system to co-ordinate individual self-interest, and a system for co-ordinating self-interest is nothing but a market. The political market robs us of our freedom in the measure to which we put our trust in it. The contrast made by some between the economic market as directed to self-interest, and the political market as directed to the public interest, is a myth, according to Friedman. The latter actually only serves the self-interest of the bureaucrat – to be human is to serve your own interest; humankind is nothing but *homo oeconomicus*. He furthermore believes it to be a myth that political "transactions" take place according to the system of "one man one vote". It is rather more a question of the aggregation of weighted group interests into a package for which one can only say "yes" or "no". In contrast the economic market is much more sophisticated and democratic – it provides much more individual freedom and fairness through the means of proportional voting in terms of one man, one dollar, one article (Friedman, (b) pp. 6ff).

From his almost "physicalist" point of view, Friedman believes in a mechanical connection between markets, which transmits political disturbances into the economic market. If the government, on the insistence of the electorate, spends too much, then this is balanced by a hidden tax increase which we call "inflation". Such an increase in government spending also has the consequence of an increase in security measures – government will want to control the spending of the extra money – and this will imply a decrease in freedom and effectivity and an increase in collectivism. This causal relationship also works positively in the opposite direction: any increase in capitalist free enterprise, Friedman believes, is accompanied by an increase in political freedom; in fact he views the free enterprise system as a necessary condition for a free political system. Even social welfare in its broadest sense (i.e., the establishment of non-profit organizations, like universities, libraries, hospitals), flourishes, Friedman believes, when a laissez-faire policy applies (Friedman, (b), pp. 25ff).

Friedman's belief that social welfare activities do much better under a laissez-faire policy than under a system of high social spending by the state gives an indication of the extent of his trust in the free market system. (He does not seem to remember the history of the circumstances of the workers under early capitalism.) His model can be summarized in two regularities: (1) a proportionally negative relationship between government spending and effectivity/freedom; (2) a positive relationship between the extent of market freedom and effectivity/freedom. The non-profit sector needs no further economic analysis – it functions analogically according to these two relationships. One could call this a calculus of freedom.

With the help of these regularities, Friedman defends freedom – positive economics is thus used for the protection of a value in which Friedman believed in advance. At the same time he neglects freedom, for the question of whether, under the guidance of free ethical decisions, a more effective use of resources is possible is not seriously investigated. For example, when volunteers (such as nuns) work in a government subsidized hospital, resources may be used more effectively than at a private hospital, since the volunteers work without remuneration. And if we can accept the proposition that nuns can work responsibly with government money, why should we presuppose that every single government official works only for her/his own interest. Surely the final balance for civil servants may not show the same effectivity as that of nuns, for civil servants work for salaries and nuns do not. But Friedman does not allow for varieties of economic behaviour for he does not allow the question of responsibility to enter his analysis.

Friedman's closed approach does not even allow for the study of varieties of economic behaviour, for he needs to stereotype economic agency in advance (in terms of self-interested behaviour) in order to be able to treat economics as a science of the same nature as a physical science. Therefore, as the heterodox economists will surely note, important economic variables are excluded from his analysis. And we can ask whether it is not his reductionist assumptions about control, man as self-interested being, the market as the model of society, which has led him into stereotypical approaches.

In his rejection of "realistic" assumptions, Friedman loses perspective on the coherence of economic regularities with other and different patterns through which human beings give a specific form to their lives. Outside the economic sphere the only area he refers to is the physical sphere; he clearly avoids reference to the psychic factors which may play a role. By reducing the whole of society to good and bad markets, he, for instance, loses perspective on the suggestive power of his own theory. The theory suggests that the market approach is the right way of living – in this he is supported by the Neo-Classical economists. When senior economists and politicians (for example, the IMF and the Reagan-Thatcher economic policy initiatives), take their cue from theories such as these, a whole society can be turned round to this line of thinking. That the state has to ask questions about fairness and justice – which do not coincide with ethical questions – Friedman is unable to see. His theory suggests practically that a competition of everyone against everyone is actually the only way of living. And this suggestion is growing day by day in the minds of contemporary people – caring for others is decreasing.

## 4. THE PARADOXES OF ECONOMICS AND THE ECONOMY

Latouche ((b), pp. 15ff) draws attention to nine paradoxes of the economy and (orthodox) economics in interaction (both of which in French are indicated by the very same word, "économie", which accentuates the paradoxes and misinterpretations).

### *4.1. The Mysterious Familiarity Paradox*

Although everybody (Latouche, (b), pp. 17–18) is necessarily familiar with the economy, it remains mysterious and incomprehensible for the majority. The majority of people do not understand the pages in the newspapers devoted to economic and financial matters, even though none of us can live outside the economy. Everybody participates in the economy. We are like gears in a machine which determines our place in society in terms of our job, income, and consumption. Life has been reduced to its economic aspect; therefore, everybody is obsessively busy with economic problems. All of us live with our attention focused upon economic indicators, be it the price of butter or the tax rate. Thus everybody is also an amateur economist, each with his/her own model of demand and supply. Each fits to his/her own situation ideas derived from the press or from rumour, ideas that are rather more like magical incantations than explanatory or coherent probing.

### *4.2. The Theory-Practice Paradox*

Economics is theory, yet it is also practice. In French this is even more paradoxical since, as was mentioned above, the same word is used for both the science of economics and the economic praxis. The "economic" indicates both a concrete domain and a theoretical sphere; the first is generally known and the second is an area for specialists. But the two are inextricably linked –

woven into one. This is not the case with the "hard" sciences, says Latouche ((b), pp. 18–19), in which we do not have such a fusion between nature and the study of nature. According to Latouche, in economics the interdependence of subject and object is total. The economist invents the economy, just as the economy invents the economist. For Latouche this is linked to the development of an autonomous sphere with its own terminology and institutions to execute it. The invention of money (as we know it) somewhere before the fifth century before Christ, made some economic reflection possible (as we find it for the first time in Plato), since it made certain market relations possible. The development of capitalism in the Renaissance gave economic reflection a push forward. The bourgeois authors, the Mercantilists, viewed the whole of society as one large market and transferred market methods to the understanding of society: calculability and accounts, working in the direction of strengthening the state by increasing the income of the prince. The Physiocrats shifted the question of the state's wealth and power away from the nobility to the productive sector (agriculture, according to them), modeling the material sphere of society as a living body (with "blood" circulating in veins and arteries) in which wealth is created and consumed. The development of economic life as an independent sphere and reflection upon it is historically something almost exclusively Western. (A more expanded analysis of the history of the economic system and reflection upon it can be found in D. Clerc, pp. 41ff). (The heterodox tend to forget that although the role of the economy has grown into something autonomous in the West, it is not the only sphere that has been differentiated in civil society – there is also the state, the churches, the educational institutions, and many more.)

### *4.3. The (In)Human Science Paradox*

"Political economy", as the discipline of economics has traditionally been called, is both the most scientific of all human sciences, and at the same time the least human of all the social disciplines. Economics boasts of being the most scientific of all the social sciences. If scientificity is defined in terms of logical rigor, then given the mathematisation of its base, it is superior to all the other social disciplines. Price theory has assumed an axiomatic form with the work of G. Debreu. Economic models are ensembles of mathematical equations. From Ricardo to Walras economics has taken the form of a social physics. The laws and mechanisms of economics are considered equivalent to Newtonian universal gravitation, self-interest taking the place of forces of attraction (cf. further Venter, (c) pp. 48ff). If one defines "scientific" as that which the learned consider to be such, then economics still the gets the prize among the moral and political sciences, for its scientificity is vouched for by major thinkers from different disciplines (including F. de Saussure, C. Levi-Strauss, K. Popper and T. Kuhn). But this is all a grand bluff according to Latouche ((b), pp. 20–21) and Caillé ((a), pp. 52–60), for in constructing itself as a rigorous and coherent scientific system, it has precisely abolished moral and political questions. Pretending to realize the greatest welfare for the greatest number, economics has not only transformed human beings into hats, as Marx accused it of doing, but both human beings and hats into equations, and thus has become an inhuman science. "The Ricardian naturalism, mechanism, and determinism made it possible to reduce human activity and history to a vast machinery by postulating a machine-man: the *homo oeconomicus*. An economic society made out of an ensemble of subjects/individuals which function as calculating automata brings about an impressive clockwork. The economists needed only to put these gears into equations". (Latouche, (b), p. 21) But morals and habits have been eliminated from this self-regulating system – the whole social bond, in fact the "human".

### *4.4. The Arrogance Paradox*

Economists have become indispensable experts even though their fame is inversely proportionate to their ability to provide correct diagnoses and satisfactory solutions. Economics is in an imperialistic position with regard to other so-called moral and political disciplines; it is supposed to be the only one of them with a real claim to be scientific. Economists are recognized and consulted, even often requested to lead the way in saving the state from bankruptcy. From the beginning of Modernity up to the First World War, it was not the scientists but the financiers who gave the lead, not with scientific knowledge, but with know-how, and it was expected of them to reset the state finances. But since then what has been expected of economists is not budgetary promises but rather the relaunching of economic activity, prosperity, even the development of a country. This expectation is founded on the pretence of economics and its practitioners: "Initiated into the mysteries of the conjuncture and growth, the high priest of economics declares himself to be able both to give diagnoses of the situation and to predict the evolution of events and provide the appropriate remedies" (Latouche, (b), p. 22). Verification often takes place only in the very short term where the diagnosis is visibly true, the prediction trivial and the solution evident. The pretence fails in important circumstances – the great crises and the crashes of the stock markets have not been foreseen, as is the case with promised ends to unemployment and renewed growth, which never come at the predicted time, say the heterodox. Important planning, such as vast development projects, in which the advice focused on the means to be used, has failed to a large extent – the reasons provided for the failure are usually such as do not threaten the competence of the expert. The general nature of the failure could cause doubt about the relevance of economic science, but this does not happen – the religion remains above attack. The result of this paradox is that other social disciplines such as sociology, psychology and political science have better records in terms of successful analyses of situations, predictions and solutions.

### *4.5. The Paradox of the Failed Calculus*

In spite of its obsessive attempts to evaluate everything, economics ignores whole areas of material reality, be it of nature or of domestic life. Economics reduces the human being into a calculating machine and all of reality into numbers. The Nobel laureate economist Maurice Allais characterized this as: "one assists in a new scholastic totalitarianism, founded upon concepts which are abstract, a priori and detached from all reality, in this species of *mathematical charlatanry* which Keynes already denounced in his *Treatise on Probability*" (quoted in Latouche, (b), p. 23). Nevertheless economists evade whole sections of that area which touches on the satisfaction of our needs, notably two aspects: nature and domestic life. Traditionally nature does not come into the field of economics, since, as Say said, nature has no price; it is not produced and is inexhaustible. Solow, a Nobel laureate, wrote recently that natural resources can easily be replaced by other factors so that there is no real problem regarding the exhaustibility of natural resources. The environment is treated, under the pressure of facts, still very indirectly in economics. That which is not produced by an enterprise, for the market, and does not command the intervention of money, escapes the economists, for example, domestic life, the black market and the informal economy. The very conception of the field of economics is in discussion here, for the rules for inclusion are arbitrary: bringing up a dog is considered a non-economic activity (in the West), but the Chinese eat dogs.

### *4.6. The Ethnocentrism Paradox*

In spite of its pretence of universality, economics and the economy are possibly no more than a very provincial praxis and theory – that of the West.

The economy presents itself as a practice common and natural to all human beings. For this reason the science of economics claims to be universally valid and trans-historical. As we have seen above, the economists appear together with the economic (money and the market). (The heterodox seem to ignore that the economic is an aspect of all human life, even where the economy does not function as an autonomous unit.) The emergence of the economy/economics in modern times is not surprising, according to the heterodox, since the project of Modernity pretended to construe social life on the basis of rationality, independent of tradition and transcendence. According to the Enlightenment thinkers the economy is nothing but the realization of reason. Rationality expresses itself inextricably in technology and the economy – it is all about increasing efficiency, economizing the means for maximum output "according to the norm of 'always more'" (Latouche, (b), p. 25). This rationality became an end in itself. Economic science is obsessed with this calculative rationality. The world triumph of Modernity has, by imperialism, first militarily and politically, and later culturally, established the economy as a practice and as world projective imaginary. But the one-sidedness of economic rationality clashes with the diversity of national traditions (even in multinational companies), which implies that to be efficient, one has to be irrational, and the question comes up: is it really rational to be rational? People in the Southern countries very often react irrationally from an economic point of view, but is not this also a reasonable position? Non-Western societies produce their collective life with a care which is evidently efficient, but on the basis of qualitative ends – this is not a calculable efficiency. Their decisions do not follow an economic model, but are based on prudent judgments, based upon experience and custom.

### *4.7. The Sexism Paradox*

The economy wants to be neutral, yet the struggle for power and money indicates that it has a strong dose of male chauvinism in it. Distinguished economists are persons who show not much joy and humour and who are very seldom of the female sex, say the heretics. If we have to believe the distinguished economists, economic activity is perfectly normal and healthy; it is transparent and guided by positive motives – the satisfaction of needs. Their problem is that the moralists and romantics have for centuries propagated the idea that money is a corrupter which infects the soul, and that the rich – the investor and saver – have been singled out as corrupting agents. Recently this accusation came from the indebted third world against the IMF and the World Bank. And although the entrepreneur is praised in some of their literature, heterodox economists see him as driven by the will to power – the striving for efficiency for efficiency's sake is called "somewhat maniacal" by Latouche ((b), p. 26), even if it does go under the noble pretext of searching for the greatest happiness for the greatest number. However, change is possible, for one is not born *homo oeconomicus*, one becomes it through an apprenticeship in the context of the market economy.

### *4.8. The (A)Morality Paradox*

Economics and the economy have a tactful relationship with morals, which they pretend to do without by replacing them. The age-old characterization of economics as part of the moral and political sciences, or even older, of moral philosophy, implies a focus on morals, thus social life, which has an indissoluble bond with norms and value judgements. But in his search for the "common good" (i.e., a value), Adam Smith was a proponent of competition based upon self-interest. This is a foundational paradox: the realization of the aim of social morality, the common good, makes such morality redundant and even contradictory – at least in the economic area, which for some encompasses all of social life. Generosity, self-sacrifice, altruism, and the main reason for an ascetic life, frugality, and contemplation, have become useless, even noxious, asocial and have been driven back into the private sphere. But one has to admit that theft, fraud, and violence have been prohibited. Business ethics is based upon freedom of contract, the autonomy of the person, the equality of the participants and mutual respect – a very cold and rigorous morality but still a moral code! Historically the economy can boast of having softened the rude and bellicose morals of the Middle Ages, but the "kind commerce" of Montesquieu does not need altruistic dispositions to soften the striving for profit. The limiting condition is that it is in everybody's interest to show respect for the minimum rules of the game: nobody has an interest in defrauding, stealing, violating, and killing. This is the lay science of a purely civil society: the economy eliminates morality by accomplishing it. But libertines aside, no economist or businessman sustains such a radical position: they are asking the state to moralize economic life and itself in its relationship with economic life, since false invoicing, dirty money, fraud, industrial espionage, and even contract murders have become the order of the day in the business world of today. They do not hope to base these codes of conduct simply upon the interests of the economic agents alone, but rather on a moral disposition which finds its ultimate source precisely in the mystery of the social bond, totally upstream from economic life. So, paradoxically, the supposedly universal code of the market has to be defended by the state, which is supposed to not interfere, imposing a moral code, which, if we follow Friedman, is itself supposed to be purified by the good functioning of the market.

### *4.9. The (Trans)Historicality Paradox*

Orthodox economics pretends to be trans-historical (according to the heterodox), yet it is founded in economic life, which takes place in history. With the value problem eliminated, economics can claim that it does not have a link with contingent realities which take place in space and time. The science of economics presents itself as a system of abstract laws, just like physics, in other words, as universal and trans-historical. Nevertheless it remains the science of economic life, and economic life unfolds exactly in time and space. However abstract the axiomatics of general equilibrium theory may be, it still pretends to provide a minimum of clarification of the world in which we live. (Or as Friedman would say: the assumptions may be unrealistic, but the predictions have to be successful.) Marxists find a relevant role for historical study in economics, but the dominant economists do not reciprocate this. This is a surprising situation, since there is a history of the economy, which economists cannot ignore: prices have a history and money fluctuates in time. There are crises which disturb the nice equilibria of economic theory, and the economy unfolds in cyclical fluctuations – equilibrium may even never be reached. The construction of an abstract model which is immovable talks to us about a world which is not ours, and the heart of standard economics is such a model. There are the softening of hypotheses, tactical compromises, and in this way economics sustains its position. But this only makes the question of an alternative approach more acute.

## 5. CONCLUSION

### *5.1. Normativity*

Latouche quotes John Stuart Mill as saying that there are a vast number of social phenomena which are caused by the desire for wealth, determined by the psychological law that we always want more rather than less. These social phenomena are those related to industrial production and productive human actions as well as the distribution of the products of industrial action. Abstracting these phenomena (arbitrarily, according to Latouche) for social reality, makes it possible to have, says Mill, a *predictive science* called political economy (see Latouche, (b), pp. 30–31). Economics and the economy in the West have not yet deviated from the idea that more is better than less – it is still a widespread belief that economic growth – meaning growth in production measured by the GNP – is the solution to problems of poverty, the environment, and unemployment. Social costs of increasing production are by and large considered to be the responsibility of government and too expensive anyway. (Cf. Goudzwaard and De Lange, pp. 68ff.)

The point which interests Latouche most is the implicit utilitarian view of humankind, which remains unquestioned, since we all accept that our own interests must play a role. "Self-interest" has been defined in terms of "utility", which has been defined in terms of unlimited "needs". Since these needs are unlimited, they cannot be defined clearly. We only have to watch advertisements to realise that many needs are created in the market itself – this is one of the historical dimensions of the market which is forgotten in orthodox static economic theory. We tend to interpret "needs", "utility" and "self-interest" as simply "natural" in terms of our own struggle to survive and to manage our economic problems. But a major part of the world economy is controlled by people who collect money and power for its own sake, by creating needs which previously did not exist, for an example, the luxury "SUV" vehicles which we see in shopping centre parking lots these days. Another example is this: According to a recent article in *Time*, a new tendency in the advertising world is to produce pleasant fragrances for shops, and this induces people to buy. But there are questions about this. Paul Fitzgerald of the anti-consumerism pressure group, Enough, is quoted as saying: "Why is all this necessary? ... One-fifth of the world's population already consumes over 80% of the natural resources. This technology is just going to add to the problems". And even George Dodd, who is involved in this research, says: "Having done it I am not convinced about the ethics. You are being assaulted by a chemical and not getting a choice" (Butler et al., p. 34). The mass consumption of resources is much more evident in developed countries than in countries where people are struggling for personal survival. The assumption that needs are unlimited creates the modernization of needs even in the most underdeveloped contexts, and so in fact "utility" does not have a content. This also finds expression in the accumulation of money by some – seemingly to serve no purpose.

Since the Idealist philosopher, Kant, and the positivists, it has been said that science can only contribute to the knowledge of natural laws, and not of norms. Norms cannot be deduced from facts, and science starts from facts. Only those disciplines which can give us natural laws in the form of mathematical formulas are scientific according to this view. According to Kant, we do formulate our own rational norms in practical knowledge, but not in scientific theory. Economics also went this way; with the help of the assumptions mentioned above, it succeeded in formulating laws of equilibrium, in a supposedly neutral way. In other words, it attempted to exclude value statements and a view of life-and-world. It did not succeed in this, as we have shown above, since it implied a view of the human being as a selfish utilitarian. Latouche also shows that it has a view of reality – the world – implied in its view of nature as our common enemy and as unlimited (to which we shall pay more attention later). In our discussion of Friedman above, it was shown that the norm of the promotion of freedom is lurking behind his economic theory.

The attempt to avoid norms in theoretical economics is reductionist, and there will always be hidden norms in the aims of economics. This is already clear when we study personal or national finances – the equilibrium is simply absent, even in the long term. For "efficiency", "utility" and "rationality" are normative terms in themselves. And when "utility" is defined as that which satisfies needs, then it is taken as a given (datum) that needs are unlimited (in order not to make any normative pronouncements about needs). The relationship between needs and aims is not spoken about for this again would compel the economist to use normative language. But in this manner it normatively takes for granted that only those needs, which are registered in the market, do count (for they are the only needs that are calculable) – thus the needs of the poor, the unemployed, and the environment disappear from the calculations. It was not always like this:

> Economic science has always said that it wants to orient itself to people and their needs.... But if we accept this classical economic maxim, then clearly we cannot mean only some people and their needs, even if they number several hundred million. Nor can we mean satisfying the material desires of several hundred million people without paying attention to the basic subsistence needs of future generations. We must, in other words, locate an economic paradigm that incorporates the needs of *all*. Western society cannot accept the reductionist battering by economic science of its original premise. Instead, we must frame a renewed interpretation of the old economic maxim, rooted in the desire to meet the needs of the other. (Goudzwaard and De Lange, p. 71)

I am not one of those who believe that utility is inferior in comparison to so-called "higher" ideals such "the good", "the true", "the beautiful", "justice", "love", et cetera. Utility is very often associated with the "ordinary", or with self-interest, or as a means to wealth, or as referring to the means for the satisfaction of needs. These limited ideas of utility do not save their adherents from normative pronouncements – rather the criteria for good or evil means are then limited to the measure in which the means serve the "ordinary", or self-interest, or wealth, or the satisfaction of needs. This is not neutral; it implies siding with an apparently utilitarian morality which can acquire different hidden (but limited) meanings. But another approach to utility is possible: "Useful are all those matters that are efficient as means for the attainment of valuable/worthwhile goals. When the aims fulfill the criteria, an efficient application of means is still an indispensable requirement for the realization of the aims" (T. P. van der Kooy, (b), p. 28).

We have noted above that it does make sense for economics to ask the question of needs and their satisfaction. Goudzwaard has been arguing in several books that the West has more than enough (even though there is still poverty and unemployment) and, as was said above, that we have to attend to the needs of all, especially those in the Third World. Van der Kooy argues that we should not limit utility to "satisfaction of needs" as such, since the latter is a very personal affair, and very difficult to calculate. For example, we have satisfaction of needs when the most urgent needs have been fulfilled. But this should not be identified with the need of the consumer for products and services, since the worker has a need to work; a producer for a suitable product to produce, as well as skills, opportunity, capital, and buyers of the product (Van der Kooy, (b), p. 26). To identify utility with satisfaction of needs would then deliver us to the arbitrary whims of the individual. Van der Kooy wants needs satisfaction to serve welfare in the widest sense of the word (i.e., not limited to "wealth"). Thus he wants to subject utility, as the means to welfare, to the norm of human dignity. Satisfaction of needs only means welfare when the needs of which the fulfillment is desired, and the objectives strived for, accord with human dignity ((b), p. 30). The criterion for human dignity – positively stated – lies in the question of whether an activity serves *community*, *justice* and *love* for the other ((b), p. 18). On the negative side this means that people who do not have the basic means are not living a dignified life; the same with those who do work the meaning of which escapes them, who have no joy in their work; but human dignity is not reducible to these negative criteria ((b), p. 16).

What we need, therefore, is a different understanding of human life than that summed up by the slogan that more is better. The heterodox economist stimulates us to at least think again and to revisit the dominant economic paradigm. For this paradigm fails us in different respects:

Because it [neo-classical economics] operates in terms of the market, it misses entirely the large shards of poverty that the market is unable to register; because it approaches scarcity solely in terms of prices, it cannot assess the economic value of the ecological problem; and because it views labour solely as a paid production factor, it bypasses the problem of the quantity and quality of work. Neo-classical economics was not designed to help solve these problems. It seeks to understand and support only that which relates to production, consumption, income, and money in the market economy. (Goudzwaard and De Lange, p. 61)

### *5.2. Justice*

In discussing the question of justice, we realize that economics should pay attention to the importance of the simultaneous realization of norms with regard to economic activity.

In the traditional Neo-Classical economy and in Monetarism, justice is a product of the self-regulating market mechanism, given freedom of choice. The task of the state is more or less reduced to the protection of freedom of choice, which implies that it should not intervene in the individual's decisions, except where the freedom of choice itself is threatened. In fact, all of society is viewed as a market in which interests are played out against one another and which balances automatically into a harmony of interests. Latouche ((b), p. 34) refers to Adam Smith's famous example of self-interest: that we address ourselves for our needs of bread and meat to the selfishness of the baker and butcher, and not to their benevolence. The supposed harmony of such interests has been expressed almost ironically by Elie Halévy in a metaphor saying that it is in the interest of the wolves that the sheep be fat and many, which amounts to saying that sheep and wolves have a common interest in the welfare of the sheep, which latter, to realize their interests, have to address themselves to the selfishness of the wolves.

The argument for the harmony of interests is in itself very Modern – it constructs nature as the common enemy of all humankind, thus all human beings can be victorious in the struggle for life and there will be enough for everybody if we extract from nature its resources. In terms of the "trickle down" effect from the richer to the poorer, every nation can be developed by the growth of the economy. The IMF wants the poor countries to solve their problems by exporting more, but they do not answer the question: if every country produces more than enough, who is going to import (Latouche, (b), pp. 34–35). The recent banana war between Europe and the United States is a case in point. The Americans were pressuring the Europeans (by boycotting some of their products) to buy more bananas from them, while the Europeans wanted to reserve a part of their import for bananas from the Third World countries. Even more serious are the ideas contained in the secret report of Lawrence Summers, a World Bank expert, on the harmony of interests between North and South regarding the question of pollution:

> It is satisfactory to export waste and to establish polluting industries in the least advanced countries. In conformity with the economic calculus, this conclusion corresponds to the interest, correctly understood, of everybody. In the first place the cost of cleaning is much lower in these countries, if one takes salaries into account. Secondly, the costs of pollution in these countries are also less serious, both because the degree of pollution is less there, and because in case of an accidental catastrophe, the costs due to the loss of human lives is there much lower. This cost is calculated by taking the actual value of the series of income in proportion to the expectancy of active life. As the duration of life is less and the salary much lower, this brings the value of life of an ordinary Indian to a hundredth of that of a British citizen! Thirdly, the demand for a clean environment growing with the level of life is much stronger in the North than in the South. This massive exportation of pollution from the North to the South will create employment and stimulate the development of the poor countries. All taken into account, one would rather prefer to live polluted than to die of malnutrition! (Latouche, (b), p. 35)

What is justice in this context then? Van der Kooy notes that we cannot say exactly what justice is ((b), p. 88), but he does say that it implies the recognition of the claim of one's fellow human being to a humanly dignified existence ((b), p. 87). He says that capitalism in its naked form (as practiced in the 19th century) has removed the dignity not only of the worker, but also of the farmer, the middle class, the entrepreneurs, for it reduced consumption goods, land, labour power, all to marketable goods, thus delivering human livelihood to the vicissitudes of the market. As was said above, he associated justice with human dignity, saying that every human being has the right and the duty to co-operate according to own ability and skill in the efficient expansion of culture, in community with others and in recognition of the right of the other to co-operate in the same task according to his/her ability and skills ((b), p. 88). The diversity of humanity is expressed in this, and it is also a requirement of justice that the individual retain a sphere of freedom to organize his/her life according to the demands of human dignity. There is a negative side to justice in the sense of limiting, through legitimate authority, the abuse of freedom, and fixing rights and duties. But Van der Kooy also sees a positive sense of justice which implies the recognition by each human being, as a criterion in striving for progress and welfare, of the claims of one's fellow humans to a humanly dignified existence ((b), p. 93). This implies, for example, that in selecting aims and in handling means the interests of others are taken into account in advance. For example, for the entrepreneur it implies that in his calculation of costs he will also take into account the sacrifice – not measurable in money terms – of the dislodging of people from their social environment, and in the calculation of income he will take into account the – not calculable in money terms – joy (or misery) in labour which may be provided by working in this business. It is possible to make up a "social balance" next to the money balances of a business, and it must not be considered unreasonable or impossible in advance ((b), pp. 95–6).

Justice is not the automatic product of the market. We must not hide behind competition when it comes to prices, for it is the businessman who sets the prices within the limits of certain conditions. Perfect competition is a fiction, for the number of participants in the markets is not infinite; there are many markets where supply is in the hands of one or a few suppliers under conditions which simply make it impossible for others to enter that market; very often there are personal relationships between suppliers and clients; not all participants have complete information on supply and demand. This implies that there is more at stake here than only unrealistic assumptions intended for specific predictions: the free bartering society – Friedman's ideal to which the predictions are supposed to apply – does not exist either (Van der Kooy, (b), p. 67). Even in the commercial situation, therefore, it is important to remain conscious of the humanity of the other.

The enormous spreading of welfare in the West, since the middle of the nineteenth century, was not, as the free marketeers (like Friedman) contend, the spontaneous product of the market. In part – probably for the major part – it was the product of pressure by the economically weak, as well as the state. It was also in some cases the voluntary contribution of Christians and humanists recognising the claim to self-development as dignified existence (Van der Kooy (b), pp. 44ff). Although the voluntary approach is the most efficient means to realise human dignity, there is a role for the state with regard to social justice. The state is the organ of public justice, for which it carries authority. Free marketeers tend to deny the state this task with regard to economic life. This will mean that the weaker will be left without protection – and the weaker may differ from circumstance to circumstance ((b), p. 48). The task of the state remains to intervene in situations where public justice is denied to its citizens (Van der Kooy, (a), pp. 113ff).

### *5.3. Subject and Object: The Economic Aspect of Reality*

As we have seen above, the heterodox economists view the economic phenomena as arbitrarily cut out from the interwoven social relationships. On the other hand they are unable to relinquish the economic sphere altogether. Alain Caillé, for example, argues that the market economy is not universal, since in primitive societies exchange is done on the basis of gift and countergift, and not on the basis of price equivalence. In fact he calls both the so-called necessity of work because of scarcity and the necessity of the market method of exchange fictions (falsehoods) with which Modernity represents itself, and yet when he proposes his solutions he does not attempt to move out of these fictions but rather makes proposals for the limitation of the role of the economy in our lives (Caillé, (b), pp. 187ff).

The orthodox economists are therefore correct when they assert that economic behaviour is "natural" and a necessity. The heterodox are correct in their view that the market economy has not always existed, and that economic behaviour is interwoven into all human behaviour. The point is that to state the interwovenness of economic behaviour does not imply that we cannot distinguish such behaviour as having an own identity. It is surely true that all human beings attempt to control resources needed for their survival and that they use means such as tools and skills in order to save on energy expenditure in the control of such resources. It is also true that in original societies the sharing of necessities of life takes place, and that generosity (in giving and receiving) does not express a market relationship. Market relationships take place outside of the social bond of the community (one can buy and sell from total strangers). But the very fact that the original societies maintain a distinction between necessities of life and luxuries, and between ordinary consumption and festivities, indicates that human beings are conscious of a differentiation of relationships in which one lives.

The heterodox are correct in stating that in Modernity the economic sphere has been disclosed as something of its own, and actually hypostatised into something autonomous. We must recognize the validity of distinguishing an economic aspect to life, which has its own norms, in the sense that taking care of our household in an efficient and effective way is healthy. This economic aspect is coloured by the context in which it functions. Two villages in a rural area in Africa will after some time be connected by a footpath. The footpath will follow the easiest route possible, which expresses a sense of efficiency and effectivity, a sense of economy, but one integrated into the social life of the communities. In the West this aspect has been differentiated into something standing on its own, and unfortunately in some cases it has become an end in itself which submits to no norms. Here is another example of the economic aspect: A local church, even though it is a community of believers, still has an economic aspect – it is supposed to remain within its financial means in its planning of the congregation's activities. But whereas a business enterprise will ask after its productivity very soon, the church will do mission work for a very long time and be thankful for only one convert. The business enterprise will want profits for its shareholders, but the church will hold a bazaar and not provide any profits for those who have contributed things to sell, for the selling at a church bazaar differs from that of a business. (Bazaar prices may be much lower than the market value, or members may be prepared to pay a much higher price when money is being collected for a specific project.)

What is important further is that although the business enterprise will function according to economic norms first and foremost, it is not autonomous with regard to other norms. For life is connected and unified by the interwovenness of norms and laws, thus the norms of protecting the environment, justice to its workers, still do apply. One example: if the profit of the firm grows, and there is inflation, the firm will have to do justice to its workers and compensate them for the loss in earnings through inflation, and not give all the profits to the shareholders.

### *5.4. Rationality*

We have already noted that the heterodox economists question the obsession with calculative instrumental rationality in orthodox economic theory. According to orthodox economic theory the economic agent calculates which approach is going to serve self-interest maximally, that is, to render the maximum output/consumption for the least input/financial outlay. Latouche ((b), p. 25) argues that efficiency can also have a cultural meaning. One could make up an example: a footpath in a traditional community may follow the route of the least resistance. This not a calculated act: people just naturally follow a certain route. But it is purposeful in some sense: children are able to use it; women carry water along the footpath. For market purposes it does not function well at all, since for such purposes a fairly straight road is needed which can carry high volumes at great speed. Another example may be that of a modern mother who will still buy apples even when the price is very high, in order to give her children a balanced diet. Thus for Latouche it is a question of whether efficiency in organising collective life can be calculated. Latouche's own example is that of a villager who has enough in milk from his two cows for his whole family, and will not buy a third cow in order to sell the milk for a profit. In the West the outcome of the maximization thesis is continuous growth of production and consumption. But if we take these cultural actions into account, then we are faced with a paradox:

"The economy, is it universal?" The one-sidedness of economic rationality collides with the different national traditions, even at the level of the subsidiaries of multinational firms. The principle of efficiency meets with an insurmountable paradox: to be efficient, one has to act the part of the irrational!. But more fundamentally economics stumbles against the antinomy of rationality: is it truly rational to be rational? (Latouche, (b), p. 25)

The calculative rationality of orthodox economics was also questioned some decades ago by the otherwise orthodox von Hayek. He rejected the reduction of economics to a natural science, saying that it is impossible to understand social actions from a physicalist perspective – one needs a set of teleological categories as a hermeneutical key to interpret these actions and their products. He denied that economic agents are calculators – it is the economists who pretend to calculate, but their calculations are no more than a tautological logic of the choices that face a possible agent (and anyone who sees more in this rationalistic approach is in danger of imposing an intellectual construct onto the collectivity). The economic agent himself is confronted with fundamental uncertainty in a competitive situation. (Cf. further Venter, (b), pp. 231ff.)

In the light of these criticisms and of what was said above about the role of normativity and human dignity, one needs to have a look at a pre-modern idea of rationality, such as that of Anselm of Canterbury:

Finally to be rational for the rational creature is nothing but to be able to distinguish the just from the unjust, the true from the not true, the good from the not good, and the better from the less good. (*Monologion*, Chapter 68) For the rational nature is rational precisely to distinguish between just and unjust, good and evil, and between better and less good. Otherwise it would have been created rationally in vain. But God did not make it rational in vain. Thus it has undoubtedly been made rational for this purpose. (*Cur Deus homo*, II, Chapter 1)

Anselm followed an age-old tradition in his understanding of rationality as the logic of value distinctions – a tradition that goes back to Plato's views that the logical function of the intellect is to lead us to the Ideas of the Beautiful, the True and the Good. Some reminiscences of this are even to be found in Descartes' conception of rationality (cf. Venter, (a), p. 4). And Horkheimer points out that even in idealistic philosophy reason was understood as a liberation from the fetters of nature, and transcended narrow self-preservation of the elite, preserving also the life of the masses, bearing a true relation not only to one's own existence but rather to living as such, which is concomitant with self-preservation (i.e., "with obeying and adapting to objective ends"). It could recognize and denounce injustices and thus emancipate itself from them. Horkheimer does not believe that reason can keep itself aloof from history, intuiting the true order of things as ontological ideologies (such as that of Anselm and Plato), but he does find – even in idealistic conceptions of rationality – a sensitivity for the great values of liberation, life, and justice. It is exactly in the narrow competitive concept of rationality – self-preservation

– that the curtailment of reason is implied and brought out by skepticism, which spells the "end of reason" (Horkheimer, pp. 47–8).

I have argued that the neo-classical and Friedmannian free bartering society based on self-interested contracting is in itself a construct which does not exist, and that therefore the idealizations structured for calculative predictions have only limited predictive value (which may serve control to a certain extent but not care). I have also argued that economics cannot avoid normative basic concepts anyway, and suggested some norms, such as human dignity, to be taken into account. I have suggested that justice be understood as the claim of the person to a dignified participation in the expansion of culture. And since rationality in its older sense is not limited to self-preservation, but has been shown to be the logic of value distinction, we define economic rationality as the logic of use – and of care – values/norms which serve human dignity such that it serves everybody's participation in a culture in which care plays an important role.

*Potchefstroom University*
*Potchefstroom, South Africa*

## REFERENCES

Anselm of Canterbury. *Opera omnia*, ad fidem codicum recensuit F. S. Schmitt. Unverand. Photomech. Neudr. d. Ausg. Seckau (1938–1961). Stuttgart-Bad/Canstatt: Frommann (Holzboog), 1968.

Bérthoud, G. "Que nous dit l'économie?" in *L'économie dévoilée*. Ed. S. Latouche. Paris: Editions Autrement, 1995, pp. 61–73.

Butler et al. "Cover Story: Attention All Shoppers", *Time*, August 2, 1999, pp. 32–37.

Caillé, A. (a). "La science économique est-elle impérialiste?" in *L'économie dévoilée*. Ed. S. Latouche. Paris: Editions Autrement, 1995, pp. 52–60.

Caillé, A. (b). "Sortir de l'économie", in *L'économie dévoilée*. Ed. S. Latouche. Paris: Editions Autrement, 1995, pp. 177–189.

Clerc, D. "Vous avez dit 'économie'?" in *L'économie dévoilée*. Ed. S. Latouche. Paris: Editions Autrement, 1995, pp. 41–51.

Combenale, P. "L'hétérodoxie: une stratégie vouée a l'échec?" in *L'économie dévoilée*. Ed. S. Latouche. Paris: Editions Autrement, 1995, pp. 163–176.

Friedman, M. (a). *Essays in Positive Economics*. Chicago: University of Chicago Press, 1953.

Friedman, M. (b). *Milton Friedman in South Africa*. Ed. M. Feldberg, K. Jowell and S. Mulholland. Cape Town: University of Cape Town, 1976.

Goudzwaard, B. and De Lange, H. *Beyond Poverty and Affluence. Towards a Canadian Economy of Care*. Toronto, Buffalo, London: University of Toronto Press, 1995.

Horkheimer, M. "The End of Reason", in *The Essential Frankfurt School Reader*. Ed. A. Arato and E. Gebhardt. Oxford: Blackwell, 1978, pp. 26–48.

Klamer, A. *The New Classical Macroeconomics. Conversations with the New Classical Economists and Their Opponents*. Brighton, Sussex: Wheatsheaf Books (Harvester Press), 1985.

Latouche, S. (a). "Avant-propos", in *L'économie dévoilée*. Ed. S. Latouche. Paris: Editions Autrement, 1995, pp. 9–13.

Latouche, S. (b). "L'économie paradoxale", in *L'économie dévoilée*. Ed. S. Latouche. Paris: Editions Autrement, 1995, pp. 15–36.

Machlup, F. "Theories of the Firm: Marginalist, Behavioral, Managerial", *American Economic Review* 57 (1967), pp. 1–33.

Smith, A. *An Enquiry into the Nature and Causes of the Wealth of the Nations*. Ed. E. Cannan. London: Methuen, 1950.

Van der Kooy, T. P. (a). *Op het grensgebied van economie en religie*. Wageningen: Zomer en Keunings, 1953.

Van der Kooy, T. P. (b). *Om welvaart en geregtigheid*. Wageningen: Zomer & Keunings, 1954.

Venter, J. J. (a). "Rasionaliteit by Anselmus van Kantelberg", *Humanitas, Tydskrif vir navorsing in die geesteswetenskappe* 9: 1 (1983), pp. 1–13.

Venter, J. J. (b). "Conceiving Conflict/Competition – Gripped by a World Picture: C. Darwin, D. H. Lawrence and F. A. von Hayek", in *Life: In the Glory of Its Radiating Manifestations*. Ed. A-T. Tymieniecka. *Analecta Husserliana*, 48, pp. 205–248. Dordrecht: Kluwer Academic Publishers, 1996.

Venter, J. J. (c). "Mechanistic Individualism Versus Organismic Totalitarianism", *Ultimate Reality and Meaning* 20: 1 (1997), pp. 41–60.

# SECTION V

PIERO TRUPIA

# PEOPLEGRAM VS ORGANIZATION CHART: THE NEW MANAGEMENT OF HUMAN RESOURCES

## *TECHNEIN* AT WORK OUT OF WORK

Philosophers have always tried to explain the world, and that's their duty. Sometimes they have also tried to direct human life, and that's a worrisome extra task they are not paid for. But with just that in mind, they have looked at the hard sciences in order to borrow methods and models. The final result should have been a universal *technein*. This is a "reasoning without reason", in the words of Jurgen Habermas, that is a reasoning producing a set of relations embodied in a model, system or better, machinery that is made up of either material or human forces. Let us think only of Marxism and laissez-faire economism as practical expressions of Determinism and Positivism.

Politics, economics and labour management are nowadays the fields where a philosophically rooted *technein* has found the greatest application. This is today spread through the cultural world by the media industry and through the ethical field in the form of the business ethic, whereby both business and ethics are to be mechanically regulated, the first through the laws of the market, the second through a reward-punishment mechanism. Some totalitarian social systems formally aimed at people's welfare and even happiness have eventually grown to be a modern thraldom either in the form of a police state or in that of the consumer society. According to the philosophical views that embody these models, personal choice must be reduced to a minimum, every choice being a source of possible error or, better put, mismanagement.

The contemporary production system has mostly conformed to this model. As a matter of fact, it has adopted a form of strict organization where machines and individuals have a predetermined place, function and field of action, both to be directed according to an external, predetermined design. Within such an organization, individuals assigned to jobs are *organa*, that is, instrumental parts in a productive machine. This production structure is an *organization* and people working within it embody an *organization chart*, which has a logic of its own: a logic where *persons* are *organa* which, according to the Greek *ethymon*, means *implements*, firm or farm-implements.

This system is no longer working. The high cost of actual machinery, a customer orientation and the resulting emphasis on service over production,



A.-T. Tymieniecka (ed.), *Analecta Husserliana LXXVI*, 323–333.

together with the spur toward total quality, require that people at work exploit all their capacities, not only mental but even sentimental. These are capacities which are out of the reach of machines and of every sort of machinery.

Instrumental reason is going to collapse under its own weight, beginning just where it has had its greatest and most effective application. *Therefore, the time has come for passing from organization to the personalization of work systems.*

## TECHNOLOGY FOR EVERYBODY

Three factors have determined the long-term positive trend of the advanced economies: the standardization of products for reasons of mere convenience, the minimization of transportation costs, and the affirmation, at first within firms only and eventually between firms and their ambience, of a logistic culture.

Let's consider the first one of these trivializations.

Technological devices are the result, as everything else, of labour division.

There are people whose work is inventing things, and there are also major research institutions and macro-companies that produce new technologies. The global result is a massive production of inventions and prototypes that need to enter the market, a production that then obeys market laws. Nevertheless, unlike what the preachers of the free market believe, marketing does not operate automatically for the welfare of companies, consumers, and human beings. Nor does it always operate for the maximization of profit. It rather aims at short-term benefits that do not require too much trouble.

Many brilliant solutions, all of them being socially useful, technologically possible and economically sound and profitable, could be fully adopted, but not in the short run. They then remain unnoticed just because managers seek to avoid greater complications in their plans.

Industrial production is still largely "Fordist": it prefers simplicity and, sometimes, it falls into simple-mindedness.

The first example that comes to my mind is one I am usually a victim of. It is the aircraft mobile-boarding ramp that is not equipped to link with both doors of the aircraft. A second example is the dispute over the "pull-and-press tab" on soda cans, that is, the ring you pull and press back onto a soft drink can (23 billions cans of this kind circulate in Italy every year). It is totally unhygienic. As they say, if the can is clean, then the user's finger is not. Nor is it sufficient to add a "hide-cap", that is, a plastic cap that covers the sterilized can. You just need to have a tab that does not get into the beverage.

There are dozens of patents for this, the best one is an Italian one – the *rimansicura* has been rejected by all companies, Coca-Cola included, for two main reasons: factories would have to be converted, and so far that is legally not required. A third more substantial reason is that no direct relation can be established between infections and today's type of cans.

My conclusion: there is in the world today a surplus of technology compared with the will and capacity to apply it. If a technology is easily applicable, everybody applies it, and if it is not, once it is applied, it is immediately copied. Therefore, if you want to reach differentiated margins of productivity, you must aim at something else.

As a matter of fact, an entire area is basically left unattended and unexplored: that is people, whom, as somebody has said, we will soon do without, and that even before we ever learn how to use them just the way they are and not as machines or Taylorized monkeys.

## A RESERVE OF PRODUCTIVITY

"It's People, Stupid", was the title of an article in *The Economist*, talking about the rush for cutting employees or managers.

"It's modernization", they say with false empathy, before granting incentives for resigning which only losers can refuse, those who have no alternative. Then they hire young people who while they do cost less, also produce much less. That is what several daily newspapers are currently experiencing: they've "modernized," that is they have mechanized the process, forgetting, however, that if you want to make a good quality newspaper, capable of facing the overwhelming power of TV, you need experienced people who know their job well, hence people of a certain age: people who have just what a newspaper primarily needs – knowledge as input, information as output.

It is one thing to deal with machines and another to deal with persons. You only need to pay to get the best machines, and then they can work on their own. The more expensive and younger they are, the more they work. Of course, you also must pay to get the best people, but luckily they cannot work like a software program. And that is not because they are imperfect, but because they are too sophisticated to be considered as just machines or, to use the French word for computers, as "ordinateurs". The better they are, the less they can work automatically in response to objective impulses. You must motivate and involve them; you must accept the fact that they may make mistakes. That is the price you must pay for having personal initiative, for learning-by-doing practise. You must entertain people once in a while; you must take some time to deal with them. They are not donkeys, stubbornly simple and manageable with a sort of "carrot-and-stick" system, that is, the system of reward and punishment.

### THE BRILLIANT MIGRANTS

As a result of modernization, it often occurs that the most brilliant young people, with their degrees, as well as the most specialized technicians, once they get hired and inserted into the squalid daily routines of the company, tire in one year or two. They then leave and look for a new job; they migrate, after having grabbed and brought home experience and formation just as one might bring home some stationary. The more brilliant they are, the more frequently they migrate.

I know personally the cases of two such migrants. The first one, a 30-year-old mechanical engineer, after several job experiences, finally got to Detroit. Admittedly, Detroit is not a city with a great quality of life; nevertheless, he has found there a medium-size company where he is part of a team working on small diesel engine design, work which aims at obtaining all the merits of gasoline engines without the shortcomings of diesel engines. The second case is that of a brilliant computer science engineer who, after a three-year experience at some major companies, such as IBM, got a degree in Sociology, learned Spanish and moved to Spain. Today, he's living and working at a market research institute in Barcelona, a city that indeed has a great quality of life.

Those who do not migrate are the bureaucrats, those employees firmly attached to their position, to their function, to their competence, and to their acquired privileges, who are incapable of either working or socializing outside their well-known areas. They spend their entire life studying Rules, Collective Contracts and Company Integrative Clauses, reading the classified ads looking for a better job, a much better one if at a state-run or major private company. At those places, the Bureaucrat gets married to the Bureaucracy, and after that … off they go to live together – not happily but moderately unhappily – in a nice slot in the organization chart.

Paraphrasing *The Economist*, we could then say, "It's the environment, my friends!"

And what could we say about the will to fight of the most brilliant human resources? Being a fighter does not necessarily mean being aggressive. Aggressiveness arises from fear or social hyperphobia and becomes chronic as neuroses and tics. The will to fight arises instead from *prohairesis*, that is, the ability to see beyond. This is a will to fight in order to reach certain goals and overcome all obstacles, many of which often are simply inside ourselves.

I have noticed a higher tendency to migrate in the "brilliant-migrant-fighter." She or he migrates until she or he gets where the "good fight" can be fought, as Paul of Tarsus would say. Luckily, there are such places, and I have seen some of them. Frustration fuels aggressiveness, and, once it sets in, it destroys resources rather than creating them. This is the human exception to Darwinian logic. Aggressiveness also depends on the pressure originating in the centralized control developed within a formally hierarchical structure. All becomes a question, not of one's value, that is of stature (what you are), but of one's role in the organization chart (the slot one occupies in the company).

In the top positions, aggressiveness is open and clear, manifests itself as control and repression (Lacanian "surveillance and punishment"), and aims at expanding its sway by conquest and consolidation. In the lowest positions, aggressiveness is instead sly and tricky, manifests itself as sabotage, and aims at reinforcing borders out of fear. Lines and slots in the organization chart. What gets penalized is will, and eventually the capacity to move on the open field. What gets penalized – especially at the organizational level – is the innovation that requires the acceptance of the different, the challenging, and the troublesome. All this applies to single individuals, organizations, states and nations. But what is encouraged is sabotage of the kind illustrated in Jonathan Swift's pamphlet *Directions to Servants* (1745).

## IT'S HARD TO LIVE WITH OTHER PEOPLE

Western civilization expelled the Arabs out of fear, although they then represented the peak of humanist, scientific and technical development. It also expelled the Jews, always out of fear. Some history books say that the Sicilians expelled the Angevins during the notorious "Vespers," to defend honour (both that of a girl and that of the nation). The truth is that the Sicilians were afraid of losing their rural identity in an encounter with the modernizers coming from the North. They preferred to rely on the more stabilizing and conservative Aragonese. That is when the decay of the Italian south started.

A few conclusive remarks after this excursus. The peoplegram is refused out of fear, a generalized fear spreading at all levels.

The organization chart is a fortress made out of fortresses. It helps more the single individuals occupying its slots and moving suspiciously along its borders, than the company as a whole.

The company is a living, ever-changing organism. The organization chart is instead a rigid structure. It is at the same time an endo- and exoskeleton. Its slots allow people to say, "It's not up to me," at the lowest levels, and "It's not something in my competence," or "It's not strategically sound," at the highest levels. And what happens at the very top level? "Ladies and gentlemen, I'm done for today. Does anybody have something to add? No? Thanks a lot. See you on Monday."

All top meetings are totally useless in terms of strategic programming and control unless they're really and substantially convivial, that is, democratic and liberal. Otherwise, they are just like a theatre show where a faithful audience hails the boss of the day, no matter what she or he says.

True convivial occasions occurred in the Benedictine Chapter Hall, that is, the place where the monks used to discuss openly their conduct and the daily management of the abbey, indeed, a real multi-sectional company *ante litteram*. All monks were allowed to speak their opinions, just as the officers on 18th and 19th century ships did in their daily meetings with the Captain in the stern.

### THE COMPANY OF "BEING TOGETHER FOR"

The peoplegram calls for a return to this tradition. The main idea is that the individual assigned to some particular function in a company – and not in a pure bureaucracy – is not an "organ," which literally and logically means "tool," but is instead a "person," that is a centre of independent initiative and informed judgement on the different reasons and goals of both his/her work and the company as a whole.

I call this company a "being together for." This is a company of global quality and one where customer service is operational after the sale is made. No longer valid here is the old commercial principle according to which, "Once the goods are sold, no claims are accepted." Of course, you can always ignore claims but then not only do you run the risk of never seeing the client again, but you have also turned him/her into an agent of negative publicity.

A company truly serving its clients and taking care of quality in the product and service it offers cannot rely on organs.

The difference between Japanese and Italian motorbikes is not much of one in terms of price or quality. We have in Italy a glorious tradition in this field. Our motorbikes are competitive in world racing. Our engines and chassis have always been and still are market leaders. So, what constitutes the difference? Our motorbikes, after one year or two, start leaking on the garage floor; Japanese bikes simply do not leak.

Why? It is because of the care with which the Japanese companies work, a care that is a common attitude in the entire company, from the top managers to the last worker. Certainly, this is a kind of Confucian care, one deeply rooted in the five social relationships that lie at the base of that ethic: father-son; husband-wife; old-young; teacher-student; boss-employee: relationships of subordination and alienation. They do not work for us, and apparently they are no longer working for the Japanese either.

## THE FIRST LINK OF THE VALUE CHAIN

Our care can have other kinds of roots: those of the Renaissance "bottega," where excellent and unbeatable goods, such as the strings of Cremona's lute-makers, were handmade. In the "bottega" the young apprentice grinding the minerals to get the different colours, or planing the fir-tree wood with which to make the sound-box of a violin, was one important link in the value chain.

The highest glory for Tiepolo – and for the other schoolmasters of that time – is that in his works of art you cannot distinguish his hand from that of his apprentices. Such is, even today, a true "learning organization."

Who is the most important person in the value chain of a company operating in the field of wide distribution? The financial manager or the cashier? I'd say the latter, since she or he is on the frontline.

Let me give you one example. Having heard one day that a new supermarket was going to be opened in my neighbourhood, I decided to ask the cashier at the supermarket where I usually did my shopping for more information. So I started by saying, "I know you're going to open a new supermarket...." She did not even let me finish, turned around to the cashier next to her and asked, "Hey Patty, are you going to open a new site?" She then turned back to me and said, "No sir, you're wrong. We're not going to open any new site at all!"

These are the same cashiers who often, having no change, disappear to look for it while the line gets longer and longer and you wait and wait. They often have no answer (nor do they seem sorry about that) to the most elementary questions you ask them about the features or the availability of some particular products. "It's not up to me," they say; "It's not within their competence," the management reinforces. That is OK! But then, who is the person the client is in direct contact with? The sales manager or the cashier? The frontline at the supermarket is the cash-line, and that's what really counts. Finance is considered to be the core business of the supermarket, but it's at the cash register that the client's satisfaction or dissatisfaction gets expressed, and it is there that all the money for finance comes from. If a client does not come back – not just the statistical client, but *that* client – only the cashier knows why.

With the peoplegram you can avoid these negative behaviours. Where you do not risk adopting this courageous innovation, you might end up having disinterest, negligence, passivity and sabotage at the very base of the pyramid or at the periphery of the flat organizational system.

## THE SERVANT WARS IN THE ORGANIZED COMPANY

Let us go back to Jonathan Swift's pamphlet *Directions to Servants.*

Swift gives to servants the first self-defence and sabotage manual, one having strategies much more effective than any directive of Spartacus, and much more devastating than any slave rebellion.

"Don't get trapped by competition": that's the first recommendation Swift gives. And right after, "You can fight among yourselves as much as you like, but always remember you have a common enemy – your master – and a common cause to defend – yours."

His fellow masters he reminds of "the numberless tricks of these incorrigible rascals.... Be then ready to face your servants' defection, simulation, lies." Servants may also use terrifying chemical-bacteriological weapons, "If the butler has no chamber-pot within his reach, he might well be using the silver salad plate." He then concludes, hoping that his peers will take seriously this book he wrote for their instruction, that – "will improve their capacity as administrators and also save their patrimony and families from decline."

The lesson is clear: Where you have servants or passive employees there can be no collaboration nor loyalty, and there is no hierarchy or control capable of eliminating "defection."

There are much worse things than the "brilliant migrants." You have slander in the workplace that ends up damaging the company image: you have sabotage; you have people confined to their own competencies, hence paralyzing the company's dynamic life.

## MY ORGANIZATION CHART FOR AN *ENTENTE CORDIALE*

There are today places where people work "together for," where conviviality and personal initiative are common.

These are places where everybody constantly learns, and not just the objective techniques of an abstract, general and often generic recipe that has nothing to do with the concrete situation within a specific context and human condition. These are places where the peoplegram is the most natural solution, a sort of research laboratory where good ideas may come from the bottom: a rock band where the singer and the last electrician play the same role in the success of the performance; a theatre company where the straight man is as important as the main actor. Think of Toto without Castellani, of Dario Fo without Franca Rame. The latter are *simple* stooge-actors, but their performance is essential for that of the main actor. These are indivisible teams. Think also of the current trend where the stooge-actor takes over the role of the main actor: Renzo Arbore, Serena Dandini, Piero Chiambretti. Here is an entire tradition declining, or better, getting redefined. The main actor becomes the stooge-actor of his/her stooge-actor, and yet she or he never gives up her or his own role.

Similarly, in the companies where the peoplegram dominates, the system does not acquire objective and technical knowledge – which implies standard training and learning requirements – but the capacity to enhance the integration of varied knowledge and competencies and to find effective and operational solutions to the company's problems, always within some specific, peculiar and context-bound conditions and constraints.

Let us think of the watch industry in Switzerland. At the end of the 60s, it was facing the devastating rise of the Japanese digital watch industry. Here was a precise, indestructible, cheap, product, yet one "with no soul", as someone – at the bottom – commented. Therefore, traditional watch production, technically obsolete, was relaunched through the production of hand-made models using 18th-century technical devices such as the *tourbillon*, or the second counter *rattrapant*, or *complications astronomiques*. Significantly, the market-leader company Blancpain coined a slogan that declares, "Since 1735 no quartz Blancpain watch has been made, nor will one ever be."

When it was adopted for the first time, the idea of "innovative backwardness" did not involve objective technical knowledge. This was rather a totally unusual perspective; no consultant would have recommended it. In fact, the marketing gurus were at that time advocating abandoning old products that had no added value at all. History has instead demonstrated that the opposite was true.

## THE CONVIVIAL FAMILY-LIKE COMPANY

I mentioned earlier Cremona and its lute-makers. But that is not the only example. Italy is full of small companies where affability, convivial family-like relationships and rural wisdom prevail over organizational engineering and managerial bureaucracy. From the Alps to the Lilibeo.

The Marche – a region located in the central part of the country – is the perfect example of an area which, in four decades, has shifted from a rural, underdeveloped economy to mass industrialization. This development spread over all its territory: from the mountains, to the hills, to the seaside.

The founder of the Marche's industrial development is Aristide Merloni who, in the 30s, created a firm in a mountain town, Fabriano. It produced industrial domestic items. But why did Merloni establish a firm in the mountains with all the communication and transport difficulties that this choice implied at that time? Because he knew that over there he could easily find those human qualities and interpersonal relationships which, after the Taylorite and Fordist industrial phase of workers used as machines or machine appendices, were increasingly becoming a strategic factor of success. In other words, he had spotted the *genius loci.*

The elements dominating within this kind of company-laboratory were (and are) friendliness, convivial family-like relationships, rural wisdom. Contrary to Banfield's diagnosis of the *amoral familism* of southern people, that is, their substantial unsociability, Merloni's firms were family-like companies, where a convivial spirit was ever present. There was no need for extirpating the rural roots in order to industrialize and develop the area. Just the opposite was the case. The workers, who had been farmers for generations, were allowed to leave the firm for seasonal work in the countryside, that is, during the sowing or the grape and olive harvests. Indeed, a new figure was born, that of the "iron sharecropper."

The Marche used to be a land of sharecroppers, but after the end of World War II sharecropping was outlawed. Sharecroppers, however, did not accept the fact that they could no longer work on their own. Hence, a number of small family-run endeavours emerged all over the territory, ignoring all the common claims of the 50s about the need to create industrial poles of attraction, which were to be abstractly identified and designed on paper.

The entrepreneurs of the Marche have repeatedly fooled the gurus of development and marketing theory. They have invested in old products such as shoes, clothes, food, domestic items when the general idea was and is that "only high-tech products can give added value." Giuseppe de Rita, who has recently edited a book about the Marche's industrial experience, talks about "simple technology," that is, specific products strongly aiming at modernizing domestic life (household appliances, furniture). "A development" – as De Rita writes – "based on (social) cohesion rather than (interpersonal) competition."

## REFERENCES

Various authors. *Da Detroit a Lille Idee e Progetti per il Lavoro*. Rome: Ufficio Studi Telecom Italia, 1996.

Ardigò, Achille. *Crisi di Governabilità e Mondi Vitali*. Milano: Bomplani, 1960.

Bacharach, Samuel B. et al. *Il Pensiero Organizzativo Europeo*. Milano: Guerini e Associeti, 1995.

Barberis, Corrado. *Storia di un Uomo e di un'Impresa in Montagna*. Bologna: Il Mulino, 1987.

Beer, Michael et al. *Managing Human Assets*. New York: The Free Press, 1984.

Crozier, Michel. *The Bureaucratic Phenomenon*. Chicago: University of Chicago Press, 1964.

Deci, Edward L. "Paying People Doesn't Always Work the Way You Expect It To," *Human Resource Management* 12 (Summer 1973), pp. 28–32.

De Rita, Giuseppe (ed.). *La Lunga Progressione. I percorsi dell'Eredità Imprenditoriale di Aristide Merloni*. Febriano: Fondazione Merloni, 1998.

Ewing, David W. *"Do It My Way or You're Fired!" Employee Rights and the Changing Role of Management Prerogatives*. New York: John Wiley and Sons, 1983.

Nonaka, Ikujiro and Takeuchi, Hiro. *The Knowledge-Creating Company: How Japanese Companies Create the Dynamics of Innovation*. Oxford: Oxford University Press, 1995.

Ouchi, William G. "Markets Bureaucracies and Clans," *Administrative Science Quarterly* 25 (March 1980), pp. 129–141.

Simon, Herbert. *Il Comportamento Amministrativo*. Bologna: Il Mulino, 1958 (1947).

BRONISŁAW BOMBAŁA

# THE AUTOCREATION OF A MANAGER IN THE PROCESS OF TRANSFORMATIONAL LEADERSHIP

## 1. ORGANIZATION AND THE DEVELOPMENT OF THE PERSON

In the field of management sciences the opinion that dynamic and effective leadership constitutes the primary attribute of good organizations is more and more often voiced. As soon as it turned out that managers play a significant and direct role in the shaping of the organization's culture, interest in research on the personalities of great managers – organization leaders – has definitely been on the increase. This has been so because they are not only authors of the formal side of organization, but also creators of the cultural components of life within an organization. Problem solving and decision making takes place in the space of certain limitations resulting from the system of economic and financial aims, as well as from legislation and social norms. However, the strongest limitations are of psychological aspects and are connected with the system of views held by particular managers.

The system of views on the part of the managerial staff exerts a vital influence on decision making in organizations and influences the definition of the context of the activities for the remaining organization participants. The vision of the top management relating to the missions of an organization plays the fundamental role in the set of views. That vision influences the choice of the ways and methods of effecting the aims. Thus there may appear friction between the subjective views held by managers and related to the specialties and strengths in an organization and the objective pressures insisted upon by the environment. These issues are of foremost importance and require analysis of the so-called philosophy of management. That entails the system of beliefs manifesting themselves in the preferred patterns and methods of the effective operation. The assumptions relating to the nature of the human being – a participant in the organization, its aims, and norms indicating which methods are appropriate to achieve the aims – are important in particular. Shared values, such as effectiveness, innovation, order, social service, domination, conservatism and opinions on goodness, beauty and truth complement the managers' system of views.

The managers' system of views is shaped by many factors. It is made of family beliefs and values, religion, knowledge and basics solidified during the course of studies. The system of views is by no means static. It undergoes changes with the acquisition of new experiences, new ideas, together with



*A.-T. Tymieniecka (ed.), Analecta Husserliana LXXVI*, 335–346.

the organization's recruitment of new managers. The system of views guarantees the cohesion of the organization's activities, providing the bases for the so-called style of management.

The research on modern organizations shows that the best of them ground their functioning on the bureaucratic and technocratic ideology. These organizational conditions cause the vast majority of people to only marginally make use of their potential. What should be emphasized here is the rule for all the potential skills of the human being, that they develop only when a human being actually uses them. The same concerns the relations occurring in all organized human teams. Centralization affects all types of organization in a de-intellectualizing way. With the rise of centralization, the global total of creative thinking within the whole organization decreases. The mechanics of this process inevitably involves a partial loss of information, emotion, and, consequently, motivation on the way during the level-to-level information transfer. Some of the valid material useful for consideration, analysis and reflection can be irretrievably gone. Some of the decisions that should be made are never made. Those in turn which are made are grounded on the basis of scarce information and weakly rooted in reality. That process, multiplied by all the links of an organization, stands for one of the manifestations of the de-intellectualization and the de-personalization of bureaucratic administration. Such a state of affairs makes it viable to pose a thesis that a negative stereotype of the human being (labelled "theory X" by D. McGregor) still persists in managers' thinking in the bureaucratic pattern of organization, which becomes the cause of frequently occurring organizational pathologies.

Yet a moment of reflection on the work of a manager will do to realize that, on the level of management, the "subject" of work, i.e., the manager, affects the "object" of work, i.e., not an (inanimate) thing but a human being too. That means that each participant in an organization should be treated as the subject of work, and not as an object of work. For the essence (*eidos*) of a manager's work entails exposing horizons,[1] i.e., the future states of organization and the ways that will lead to them, or, in other words, possible future activities. However, it is not only the manager's activity that enables the attainment of new levels of organization. It is the activity and the involvement of all the participants in the organization. In order to attain such a state, the manager should allow, or even encourage unrestrained operations (*I am able to do it, I do it, I am able to it in a different way*).[2] These statements lead to the formulation of a thesis: The correct fulfillment of an organizational role requires a personalistic orientation on the part of the manager, which, in the theory of organization and management, is parallel to the idea of transformational leadership and to an organization culture oriented towards the person (personalist culture).[3]

To expand the above thought, one should make clear that the personalistic orientation does not entail a one-way, reformatory, or revolutionary transformation of an organization but a cooperation with all the participants in the organization, reshaping the organization from within, beginning with oneself as its integral element. That, in turn, requires one's realizing that activity is, to a great extent, the sphere connected with the shaping of the human life, both individual and social. This means not only direct production but also other factors connected with the whole of human life and the factors which transform, intensify and develop it, and influence the achievement of further levels of anthropogenesis. On the other hand, activities that are nonsensical, destructive, degrading for the human being, human co-existence, or nature can be called work only in the economic and social sense but not in the philosophical and the moral senses as well; they might just as well be called "antiwork."[4] The manager's activity is thus connected with a great deal of responsibility, multiplied by the number of workers directly under supervision and by the importance of the projects undertaken. This puts specific requirements on the managers; they are always obliged to act responsibly.[5]

## 2. FROM MANAGEMENT TO TRANSFORMATIONAL LEADERSHIP

Because of its significance in the effectiveness of an organization, leadership is still subject to intensive research and new theoretical judgements. Leadership is described as a process, but also as a property, not infrequently dubbed "the leadership gift," available to some people only. As a process, it involves the use of one's influence without reaching out for the tools of pressure, with a view to delineating the organization's aims, creating motivations that would give rise to action focused on achieving those aims, and helping to define the culture of the organization. As a property, leadership constitutes a set of features ascribed to people who are perceived as leaders. Hence the leaders are people who influence other people's behavior without having to use force. Leaders are those who are accepted in that role by others.

The role of leadership in the processes of organization management has been aptly defined by R. W. Griffin:

Organizations, so as to function effectively, need both management and leadership. Leadership is an indisposable condition facilitating changes, and management is an indisposable form of attaining systematic results. Management coupled with leadership can give rise to systematic changes, and leadership – in connection with management – allows the maintaining of the appropriate level of accord between an organization and its environment.[6]

Leadership and management, in spite of their correspondence, are not equivalent. The fundamental difference is that the manager will aspire to maintaining stability, whereas the leader will aspire to change.

To penetrate the essence of leadership, one should state that it does not entail the wielding of power.[7] Although – as with power – it is relational, collective and centred around aims, leaders do not ignore the motives of their followers. They may, however, arbitrarily give priority to certain needs. What is crucial is that they are in charge of other people, and not objects. Control over objects – tools, resources (also financial), energy, information – is an act of power, and not of leadership, for objects are devoid of motives. A ruler can treat people as objects, but a leader is not supposed to do so. Rulers, through their actions, take into primary consideration that which is valuable for ruling, accepting the values and needs of their subjects as long as they are indispensable for manipulation and exploitation.

On the contrary, one can speak about leadership when, owing to the leaders, their followers fulfill the aims which encompass the values and motivations (wishes, needs, aspirations, expectations, etc.) common to both. On the whole, the art of leadership is based on the ability to perceive and to attain common goals. The essence of the relationship between the leaders and their followers consists of the interaction of people who, aspiring to the achievement of common, or at least convergent, goals, are attuned to the higher levels of motivation. This interaction can principally assume two fundamentally contrasting forms. One of them can be called transactional leadership. It occurs when one person initiates contact with the others with the intention of exchanging something that is valuable for each and every side. Such an exchange can have an economic, political or psychological character. Each side in such a transaction is aware of its partner's attitude and resources, each one treats the other as a person. Their aims are connected at least to such a degree that they are woven into the process of the exchange and, thanks to that, they can be realized. Nonetheless, such a relation does not go beyond those boundaries. There is nothing here that would permanently bind the participants, and, accordingly, they may part. Although the act of leadership took place, it did not unite the leader and his followers in the common and continuous pursuit of a higher goal.

The second type of leadership is defined as charismatic, symbolic, or transformational leadership. It takes place when leaders and their followers enter mutual relationships which elevate them to a higher level of motivation and morality. Their aims, however they came to be united at the beginning, truly coalesce, and yet remain individual. The foundations of power no longer involve simple mutuality, but support a common goal. What is important is that the basically moral character of such a relationship involves elevating the behavior and the ethical aspirations of both the leader and of those who cooperate with him to a higher level. Consequently everyone undergoes transformation. Such leadership is dynamic, i.e., the leaders fully engage in relations with their followers who, inspired, become more active and often take up leadership themselves. It is (active) involvement that constitutes the essence of such leadership, which goes beyond the starting point. Mere ruling can have neither a transactional character, nor a transformational one. These are the possibilities open to leadership.

In the process of transformational leadership, attention is drawn to the communal character of the organization's operations. The organization becomes the object of collective decisions and actions, and it is dependent on the values professed by the participants in the organization. Observing moral norms and the dignity of person in the processes of management sanctions the definition of transformational leadership as personalistic leadership, and of the resultant culture as a personalistic culture.

## 3. ETHICAL DETERMINANTS OF TRANSFORMATIONAL-PERSONALISTIC LEADERSHIP

The starting point of thus perceived transformational leadership consists in the statement that human activity is marked by morality, and human actions are morally good or morally bad: "*(...) moral values – good and evil – constitute not only the inner feature of human activities, but they also make the human being – as a person – become good or evil himself through these morally good or bad actions.*"[8] Therefore, leadership, as a person's relationship towards a(nother) person, is one of the paths of ethical human behavior. Due to this, one can speak of the moral abuse of organizational power wherever those moral foundations of management are violated. "The essence of moral exploitation consists in the awareness of being exploited. The latter happens whenever work is abstracted from the ethical aims, which it should serve in substance and on the basis of a clear purpose."[9] The modern system of the advanced division of labor supports the rise of various forms of moral exploitation. Moral exploitation of work brings about an array of negative consequences. Primarily, it causes a moral disintegration of the collectiveness of work. This leads to lack of trust and opportunism, and it prevents the rise of a truly creative work community. Therefore, three dimensions of work should be selected:

- technical/economic: consisting in the reshaping of nature with a view to its rational utilization by man;
- psychological/moral: a set of human efforts aiming at man's psychosomatic perfection and at a fuller participation in transcendental values. In that sense work is made up of both physical and mental effort and involves gaining new skills and moral perfection.
- socio-cultural: understood as the creation of new cultural values as well as the perfection of human coexistence.[10]

Thus physical or biological activity alone does not constitute work. A definite completeness related to human life (purposefulness, rationality, freedom, etc.) is crucial. The reflective-conscious character of work constitutes its specific feature. Human work differs from the animal activity in that it is planned and consciously directed. It is a reflection of human nature, a creation of the human mind. Hence, in personalistic leadership, the differences between physical work and mental work are not accentuated. Moreover, it is highlighted that not only does work provide for physical human existence, but it also conditions the creation of culture. Work perceived in the exact sense (production) has the character of a "service"; the extra benefits resulting from it are assigned to education, art, and philosophy. Work, perceived broadly, embraces mental activity, too.

What this requires from the manager is respect for the natural aptitudes of each of the participants in the organization and the practical use of those management techniques which enable the development of creativity. For it is true that

> man is obliged to be active in such a manner that is allowed by his physical and mental abilities as well as by his aptitudes and interests which, needless to say, he should develop, as indicated in the Gospel parable of the talents (Mt 25: 14–30). Not only does the obligation to work relate to the fact of performing it, but also to its quality and our attitude towards it. If work is man's co-operation with God and fellowship with him, and if it aims at perfecting man and the world as well as at building an active community through co-operation with one's fellows, it should be treated as a kind of a mission, as man's vocation in the world.[11]

Thus, transformational leadership means providing work that will satisfy and give the opportunities for self-creation to all the participants in the organization. Work should facilitate the development of human energies and not their loss. And it absolutely should not endanger human health or life. The manager is obliged to try his best to do away with any risks. The principle that "*there are no economic arguments for which human life might be sacrificed or, ultimately, endangered*"[12] governs here.

Finally, work should not take our whole day. It should not stand in the way of doing one's other duties, such as participation in family, civil, religious, cultural and social life. Hence a valid tip for managers is for them to organize work in such a way that it will enable the worker to have a sufficient supply of time and energy to perform other duties. Thomas Merton writes that "activity is only one of the manifestations of life, and that life which finds its expression in it [i.e., activity] is all the more perfect, the more it is based on an orderly economy of activity. The latter requires the sensible assignment of labor and recreation. We will not reach a greater completeness of life by working more, by knowing more, by seeing more, by having more impressions, and by having more experiences...."[13]

The manager should also remember that work does not constitute the only, or the highest, value. The human being is more than a mere worker. It is not work that constitutes the most important factor and the ultimate meaning of life. There are also higher values such as Truth, Goodness and Beauty. Work is a value, too, but a value that is serviceable to other people. Therefore, *homo sapiens* cannot be whittled down to *homo faber* or *homo economicus*. Merton argues: "Human existence – out of necessity – aspires to its own expression through action, but this fact should not lead us to be convinced that we will cease to be the moment we stop being active. We do not live just to 'do something,' regardless of what it is that we do."[14] Man – according to Merton – should live consciously and discover the world and himself and thus reach higher levels of existence:

> Action and having an experience in themselves are not the only way of enriching our lives. Everything depends on the quality of our actions and experiences. A considerable number of poorly performed deeds and only half-felt experiences exhaust and impoverish our existence. By performing various duties in the wrong ways we progressively distance ourselves from reality. ... One needs to recover the possession of one's own existence first; only then can we act sensibly and go through different experiences in their human reality. As long as we do not possess ourselves, each action is futile.[15]

Reducing human life to sheer work leads to, in a sense, the objectification of the human being, for the integrally human dimension gets lost. Man is "called" to work, but also to the cognition of truth, to love and to reflection. The manager must remember that, although every designate of the word "human" has the same characteristic features, every human being is an individual person, leads an individual life, has an individual character, an individual fate, an individual appearance, and an individual historic role.

## 4. THE SELF-CREATION OF THE MANAGER IN THE PROCESSES OF MANAGEMENT

The personalistic approach uncovers the meaning of personal values in the processes of transformational leadership. The starting point is constituted by the assumption that not respecting the basic ethical values (the primary values of life)[16] while directing leads to the degradation of the manager as a person. The full realization of this fact's significance provides the foundation for one's stepping beyond the purely economic interpretation of management as a process whose only goal is economic effectiveness. Such an approach is a simplification distorting the truth about the manager, for it does not visualize the "richness" which is created by the fact of the manager's creating himself as the person. The technocratic approach analyzes all the events happening within the economy by narrowing them down to the category of profit-making, reducing the variety of human action to just that which has utilitarian motivation. This leads to a specific economic imperialism[17] which undermines the rationality of those choices which are not based on economic calculation. Economic imperialism entails not only the isolation of economic phenomena from their cultural context but also the economic interpretation of all social phenomena. That leads to the objective treatment of the economy as being solely that activity that entails the transformation of the material world. As has been noted by R. Wiśniewski, it is very seldom that the subjective character of economic activity is taken into consideration. Yet "*work, as a realization of values, is the medium of reflective processes, and of creativity; it is a domain of self-realization, exceptionally important for managerial environments which often treat difficulties, or any resistance by matter, as a challenge.... The example of business ethics shows that the ethics can indeed consider economic activity as a way of personal self-realization, or so-called vocation, self-fulfilment, and self-examination.*"[18] Following this track of reasoning, we reach the conclusion that on the basis of an ethics that is founded on the philosophy of the human person (personalism) one can build a personal model of the manager, who would possess besides the advantages of formal competence, those of moral competence, too.[19]

With reference to our earlier reflections, one could positively state that the subjectivity of persons, their self-creation, expresses itself in action: "*Through action the subject is capable of achieving something basically unusual, i.e., of becoming present in the 'matter' wherein the action takes place.*"[20] Here is the center of the person from which all the conscious deeds and actions originate. What needs to be highlighted is that the development of subjectivity proceeds differently in individual people. The rise and the subsequent development of subjectivity occurs whenever the human being breaks free from the structure of determinism, and begins to function within the structure of motivation. The less the scope of nonvoluntary action, and the more the nonrepetitive action facilitating the use of one's qualifications and creativity and allowing one's work to have its own "stamp," the fuller the development of the subjectivity of the human being.

The process of the manager's attaining subjectivity is connected with the broadening of the sphere of operation on which he leaves his imprint. This broadening is initiated by the manager, yet conditioned externally. As a result of this expansion the personal subject itself forms itself, gaining and solidifying new qualifications, owing to which it surpasses itself, and that means self-creation. This process is interwoven and integrated into the process of transformational leadership. It is through the process of transformational leadership that the self-creation of the manager's *person*, as well as both the creation and the self-creation of the remaining participants in the organization, come about. A specific role in the building of subjectivity is ascribed to moral values. Morally positive conduct introduces a trait of nobility, but also causes the integration, together with a certain simplification, of the personal subject. Life becomes a consistent, reasonable whole; and future choices and actions become comprehensible and anticipatable. It is then that the operation within the organization bears a clear moral stamp. This can be verified by a number of examples: transactional honesty, justice in the treatment of workers, ecological responsibility, and so on. Morally negative conduct exerts a diametrically different influence: it causes the inner disruption of the subject. In extreme cases it may lead to the manager's solidarity with the foundations of evil. Such a role can be played when the motive is that of becoming rich at any cost.

Unfortunately, a concentration of one's activity on the transformation of the outside world leads to one's ignorance of action oriented towards the inner world. The prospect of one's own spirituality easily escapes from managers' range of vision, because the history of Western civilization has overpraised the weight of outward achievements.[21] Utilitarian and practical values have dominated personal and moral ones. The situation of managers seems particularly difficult in that respect because of their extensive and intensive relationships with other people. Success in business is achieved by "submersion" in the economy rather than by pronounced individual isolation, such as not infrequently occurs in philosophy and art. Therefore, managers exposed to powerful social pressure are all the more bound to scrupulously protecting their own inegrity by giving priority to ethics over etiquette. "Remaining oneself," both for the manager and for every human being, means achieving personal maturity.

While analyzing the issue of self-creation we should remember the dangers of the reverse process, i.e., self-destruction (de-personalization). The change taking place in the subject sometimes assumes a backward direction, which entails a loss of positive self-definition and the emergence of negative self-definitions. In order to be able to develop, the subject must overcome that which disrupts it, or at least that which prevents the emergence of the positive values. Clearly, this is relevant to the manager, whose job hinders both the ability to distance himself from the external world and the assumption of the reflective stance with a focus on one's own inner life. This does not really have to lead to such a point that the market economy makes him a "bad man," although such cases when – under the pressure of the relentless competitive strife – a manager becomes greedy and uncompromising do occur. This may occur once a manager confines himself to superficial, materialistic achievements, which then come to stand in the way of his own access to intellectual and aesthetic values, and, most of all, to ethical ones. Then the potential development of his personality is dependent on the occurrence of a dramatic turning point which would entail the re-orientation of his purpose in life.

The above statements carry some essential implications relating to the choices made by managers. They (the managers) often have to make the dramatic choice between economic and moral values. Unjustified preference of economic values leads to the negative transformation of the subject and to the formation of an amoral personality. If an organization's influential managers are characterized by amoral personalities, this will lead to an immoral management and to an immoral organizational culture (anti-culture).

To sum up, it may be stated that the concept of transformational leadership, based on personalistic values, should constitute the pattern for future managerial models. At the moment this is becoming more and more popular, owing to both its theoretical attractiveness and its practical importance. Due to rapid changes, as well as to turbulent environments, transformational leaders are more and more often considered as a vital factor in an organization's success, for, as it turns out, not only are they capable of personal development (self-creation), but they also facilitate the development of the remaining participants in the organization. They leave their "stamp" on the organization, but, at the same time, they also solidify its identity and its integrity.

*Institute of Political and Sociophilosophical Sciences*
*Warmia and Masuria University in Olsztyn*

NOTES

[1] Cf. E. Husserl, *Méditations Cartésiennes, Introduction à la Phénoménologie* (Paris: Bibliothèque de la Société Française de Philosophie, 1931).

[2] Ibid.

[3] An in-depth analysis of the personalistic organizational culture is to be found in the author's doctoral dissertation entitled: *The Personalistic Vision of Organizational Culture.*

[4] Cf. C. S. Bartnik, *Theology of Human Work* (in Polish), (Warsaw: IW PAX, 1977), p. 15.

[5] The essence of responsible action is brilliantly handled by R. Ingarden: "*Responsible action is taken by its originator in a specific way. The originator undertakes it and performs it with a more or less complete aware understanding both of the situations resultant from his action, the situations being seen from the aspect of values (...), as well as of the value of the motives, which made him act. In all the phases of his action the originator realizes his own connection with the positive or the negative value of its outcome, and undertakes it, or continues it with a conscious acceptance of the outcome's value, and, accordingly, of the legitimacy and appropriateness of his action.*" (R. Ingarden, *A Little Book About Man* (Polish transl.), (Cracow: Wydawnictwo Literackie, 1987), p. 76.)

[6] R. W. Griffin, *Management* (Warsaw: PWN, 1998), p. 491.

[7] Cf. K. M. Bartol, D. C. Martin, *Management* (New York: McGraw-Hill, Inc., 1991), pp. 506–508.

[8] K. Wojtyla, *The Person and the Act* (in Polish), (Cracow: Polskie Towarzystwo Teologiczne, 1969), p. 16.

[9] J. Tischner, *The World of Human Hope* (in Polish), (Cracow: Wydawnictwo Znak, 1975), p. 86.

[10] J. Majka, *The Ethics of the Economic Life* (in Polish), (Warsaw: ODiSS, 1985), p. 140.

[11] Ibid., p. 146.

[12] Ibid., p. 157.

[13] T. Merton, *No Man Is an Island* (Polish transl.), (Cracow: IW ZNAK, 1982), p. 155.

[14] Ibid., p. 154.

[15] Ibid., pp. 155–156.

[16] This issue is very important and is connected with our sense of purpose in life. In our culture this is connected with a definite picture of oneself and of one's own actions. It entails the meeting of hopes and expectations in a way that would allow a man to consider his life satisfactory. These hopes and expectations can be included in the category of human needs. Because of their significance for contemporary man, they are called the primary values of life. They can be listed as follows: (a) the need to feel one's subjectivity in social acts, i.e., the conviction that in his actions, man has a considerable range of freedom of choice, that he is not manipulated and that his role is not reduced to a puppet-like function; (b) the need to be convinced that experienced

feelings – friendship, love, gratitude, readiness to help or to console – are "noble" and disinterested: that they have the weight of real altruism; that they are not attitudes perceived fundamentally as "cunning," or calculated; (c) the feeling of a membership in a definite human community, be it an ethnic, professional, or political one, together with pride or satisfaction in that (it is important that this not be an awareness of belonging to a community in any way inferior to other communities, or a despised one – awareness of a pariah condition); (d) the need to nurture hope for immortality, and the belief that death does not establish a complete and irrevocable elimination – *non omnis moriar* – which releases the aspiration to bequeath something of durable value (pieces of art, science, or organization); (e) the need to have the sensation of an Important Mission that we must accomplish in our lives, i.e., setting an aim so important that it would be worth the effort to organize one's own life as a strategy for achieving that aim (such a sensation may not only be shared by artists and leaders, but may also accompany the life of any "man in the street," and relate to a belief in service to the Important Cause (lack of awareness of such an aim is defined as a sensation of emptiness in life).

Unfortunately, many hypotheses and theories having their origins in the exact sciences (sociobiology, behaviorism) violate the primary values of human life. Cf. T. Bielicki, "The Scientific Outlook on Life and the Primary Values of Human Life: Harmony or Discordance?" in: *Views on Man and Society in Scientific Theory and Research*, a volume of studies in Polish edited by S. Nowak (Warsaw: PWN, 1984).

[17] Cf. K. Sosenko, the *Creation of Wealth in the Perspective of the Personal Subject's Self-creation*, (typescript), (in Polish).

[18] R. Wiśniewski, "Three Types of Ethical Theory vs. Business Ethics," in Business Ethics (in Polish), ed. J. Dietl and W. Gasparski (Warsaw: PWN 1997), p. 49.

[19] Building the pattern on the basis of the ethical values necessitates the introduction of the personal subject category (the person), which will define the individual human existence as a concrete, subsistent and possible whole which is responsible for the enacting of values and, at the same time, is the medium for some of these values. There are, naturally, those philosophical orientations (positivism), and corresponding ideological tendencies (scientism), as well as cultural trends (technocracy) which will not agree with the introduction of the notion of the personal subject, treating it as a metaphysical construct.

[20] A. Węgrzecki, *An Introduction to the Phenomenology of Subject* (in Polish), (Cracow: Ossolineum, 1996), p. 62.

[21] This phenomenon has been aptly described by H. Marcuse. According to him, Western civilization's dominant principle of productivity creates an artificial human nature supressing all that is not in accordance with the "technological mind." The transcending dimension of the existence, so typical of human nature, which manifests itself in the cognitive-aesthetic attitude towards the world, undergoes repression. Cf. H. Marcuse, *One-Dimensional Man* (in Polish transl.) (Warsaw: PWN. 1991).

MAREK PYKA

# BUSINESS AND ETHICS, A GENERAL APPROACH

One of the difficulties which we face when applying ethics to the actual problems and situations in our lives stems from the fact that ethical principles and values (as knowledge) must be of a general character, whereas our problems and situations in life are always individual and specific. This is an old and well-known difficulty. Another difficulty, which is emerging in the most developed countries of our times, is of a different nature. Traditional ethics appeals to an individual as such, to an individual in an abstract situation, or as if everything were taking place in the private sphere of life. In contrast to that, our present-day life is dominated by the structures of social institutions, among the most important of them being the institutions of the free market economy. Here in Poland we are receiving a few short, but very intensive lessons in the role and significance of these institutions and structures. It would be a great pity if ethics did not devote a considerable amount of attention to the problems arising in the recent market economy.

What is the impact of these institutions and structures on the situations in which an individual moral subject must act? Business ethics discusses different kinds of business activities and the respective moral issues arising within them. My perspective in this paper is more general: I will discuss three groups of factors that vitally influence the great number of situations in which a person must act in the present-day world. The impact of these factors constitutes some new difficulties in applying ethical principles to present-day organizational life, in addition to those mentioned in the opening paragraph. My discussion here is limited to the factors which I believe are the most important. All of them are rooted or present in the contemporary world of business. What are these three groups of factors? The first group stems from the fact that the place where a person must act is a place within an organizational structure and not an abstract situation. The second group of factors has its source in the limitations of the goals of a business organization in comparison to the richness of the goals and values of an individual. The last group of factors that will be discussed here comes from the complexities of present-day technological and management processes and from the complex and bureaucratic structures of large corporations. In fact, all these three groups of factors are interconnected, and here they are discussed separately for the purpose of clarity.



*A.-T. Tymieniecka (ed.), Analecta Husserliana LXXVI*, 347–354.

I.

The impact of the factors of the first group can best be observed in terms of responsibility. In the abstract perspective of traditional ethics a person acts on his/her own and is free from formal pressure to do or not to do something. In contrast to that, a person acting within an organizational framework is under constant pressure or even sometimes coercion to do this or that, and this pressure or coercion may be very powerful. This difference is difficult to overestimate.

The source of an individual's moral obligation to follow his/her organizational goals can be found in the theory of social agreement. In entering a job within an organization, a person has freely agreed to follow its goals and conform to its internal procedures, and this agreement constitutes also his/her moral duty to keep the promise. It is obvious that to refuse to do it or to do it improperly or in a negligent way is unethical. But what should people do if their organization requires them to do something which itself is ethically dubious or even unethical? Such a conflict is not very rare, especially when a corporation faces serious economic difficulties, the situation in the market changes dramatically, the competition is tough, and so on. Does the order of a superior justify a person's unethical act? Of course, it does not. Let us examine this point more carefully. Each manager has a moral and also legal duty to run a business for the owners according to the best of his knowledge and judgement. But it does not follow from this that, while fulfilling his duties, a manager is allowed to break the law or ethical principles, or to violate ethical values. On the contrary, he has to act in accordance with them. In ordering or suggesting an unethical action, a manager usually refers to "higher purposes" that follow from the goals of the firm, but what he actually has in mind is often his own position and interests. A different argument, from the basic perspective of the liberal tradition, leads to the same conclusion. Within that tradition each individual has to make his/her own decisions and to accept the consequences for his/her own life. The fundamental responsibility one has is, therefore, for oneself, and the shape of one's own life. Each of us has to decide which orders, instructions and so on to accept and follow, and which not to follow and refuse. External pressure cannot be a valid justification for unethical acts, and any person applying such pressure also becomes morally responsible for these acts. There is always a way out of the difficulty, which is open for an individual under pressure: to resign from the job and leave the firm. There is also, however, the other side of the coin. We can easily imagine that the person in question has a family to support, will not be able to find a comparable job in the market, becomes jobless, after having invested a lot of time and energy in work he/she loves and so on. Whatever the things might be, it is clear that to act ethically under real pressure from an institution is much more difficult than giving the correct answer in an abstract situation, and much more courage and determination is required.

Another phenomenon is related closely to those discussed above. This is the phenomenon of the dividing and dispersing of responsibility, which take place in a group. In the perspective of traditional ethics it is assumed that a person feels, or at least should feel, responsible for the final result of his/her acts. In contrast, when an act is performed by a group, the responsibility for its final results becomes divided among all the involved members of the group, which can be easily observed even in an informal, random group. It is a different thing to refuse to help somebody in the mountains or in a wasteland than to do the same thing in the downtown of a large city. In the first case, one has to face an entire responsibility, whereas in the second, only a small part of it, often too small to be noticed. When an action takes place in as large an organization as many corporations are, then an individual feels responsible for only a very small part of the final product or corporate activity, and quite often this small part can disappear from the horizon of his/her experienced world. In any case, the line between what is morally right and wrong is thinner and more easily crossed, particularly when accompanied by the pressure of superiors. For this reason it is easier there to hide information from workers about damaging effects on their lives caused by their work conditions than would be the case in private life. There is no need to mention that the former is a standard case of business ethics.

The division and dispersion of responsibility within a corporation are accompanied by a special vagueness and mobility of responsibility. When an organization or a division of one achieves a success, then the responsibility for this is assigned upwards in the organizational hierarchy of the firm. Then it may be ascribed as well to the people working on the lower levels of the organizational structure according to the will of managers. When an organization faces a failure or serious troubles, however, the responsibility for this is thrust down the organizational hierarchy of the firm. Abusing their power, managers too often try to make their people responsible for failure or problems. When a breach of the law or a morally unacceptable practice is disclosed, the pressure, instructions or encouragement of superiors are always carefully withdrawn and perfectly hidden.

The last problem of responsibility I would like to discuss here is the problem of the moral responsibility of a corporation as a whole. Of course, responsibility in terms of money obligations or in terms of the law can and should be ascribed to a corporation as a whole. But what about moral responsibility? Can we ascribe it also to a corporation as a whole? Some authors hold such a view, and this recalls the way in which Hegel explained the historical development of societies. It is true that inside an organization sometimes things happen that were nobody's intention, that nobody wanted, and so on. But the view that a corporation constitutes a kind of super-person, with its own intentions, goals or responsibility is obviously false. Who is, therefore, responsible for the morally wrong practices of a firm or a division of it? The basic way to solve this problem is to point out those members of the organizational structure who, acting in the name of the firm, actually did the things in question; those people are morally responsible for the wrongdoing of the firm. It is clear that the higher the position in the hierarchy of an organization, the wider the attached sphere of moral responsibility should be.

## II.

The second group of factors has its roots in the limited nature of the goals of a corporation and in the very nature of business itself. In the world of the values and goals of an individual there is, of course, an important place for economic values and aims. But even if a person is wholly devoted to achieving some economic goals, it would be difficult to say that these are the only goals he or she has. In contrast to that, the sphere of goals of a corporation is limited to economic goals alone. What is more, the achievement of these goals is the fundamental condition for the further existence and position of the corporation as a whole, its future, development and so on. There is no need to mention the personal positions and careers of the managers themselves. These are the most important reasons that make managers do or order things done that they would not do in their private lives. This is the case particularly in the face of serious economic difficulties for a firm, dramatic changes in the market, periods of crises or, as in Poland right now, a period of switching to the market economy. These are also the reasons to which some managers appeal in trying to justify their unethical actions or the pressure they exert on their inferiors to take such actions. Of course, such arguments cannot provide a valid justification in any case. The task of business ethics is not an easy one. It has to show how economic goals can be achieved without breaking ethical rules and values. The relevant ethical rules are not very numerous, and the difficulty lies rather in the ways in which they should be translated to the real situations of business activities. One of the fundamental claims of business ethics is that, in the long run, the ethical actions of the firm contribute to its economic success, or, in the weaker version of this claim, that they are not in opposition to it.

In his classic theory of the market economy, Adam Smith has shown that business activity must be an activity oriented to one's own profit. Smith's theory has been changed many times, and it has been argued that to define the goals of business in terms of profits is not enough. But it is impossible to ignore profit completely while thinking of business. On the other hand, what ethics demands from us is, roughly speaking, that we should care not only about our own goodness (which we do automatically) but about the goodness of other people as well. Is it, therefore, possible in psychological terms to be both ethical and successful in business? A few answers to this question can be found. The best known of them is perhaps that of Smith and following him classic liberalism: The motivation of each individual in business is egoistic but the final result of his/her actions for the whole of society is automatically the best possible. One of the weak points of this argument lies in the fact that in ethics we would like to deal with the intended results of the actions of the moral subject, if not with the very intentions. Besides, it is clear that we often face a practical dilemma: either to help somebody or to pursue our own interests and success. The last problem, as I believe, is most interesting not only for business ethics but for theoretical ethics as well.

## III.

The third and last group of factors that I would like to discuss here has its sources in the huge complexity of the recent procedures of management and administration, the complex structure of large and global corporations, and the complexity of the technological processes involved. In the perspective of axiological ethics, to act in a responsible way, a person must meet a few necessary requirements. First of all, one must judge correctly the situation in terms of the values that are at stake. Secondly, one must consider the actions possible in this situation in terms of the values which will be respected and realized or exposed to nullification. Thirdly, one must find the right action in the situation he or she faces, carry it out, and internally accept all the results it brings about. In the influential utilitarian perspective, in turn, to act in a responsible way requires knowledge of all the results for all relevant people, including remote results and remote people. In the light of traditional ethics to meet these requirements is not a difficult task for a person. In contrast to that, the situation of an individual acting in the complicated structures of the contemporary world of business may be quite different. This may particularly be the case when matched to the phenomena of divided and diminished responsibility discussed above. The conclusion that comes to mind is quite obvious: In order to act in a responsible and ethical way within a corporation, a new and necessary condition is required. And this condition reads: It is necessary to have a considerable amount of relevant knowledge. Without such knowledge one cannot adequately grasp the meanings of what he or she is doing and cannot properly judge the future results of one's actions. In the private sphere of life an analogous kind of knowledge is not necessary. In order to be on good terms with one's friends, one does not have to be a psychologist or sociologist. The meanings and results of our actions in private life are clearer than in the realm of complex social structures and interactions wherein they become difficult to see and ambiguous amid the mass of events, consequences and information. For this reason any lack of relevant knowledge, that is to say, any incompetence as such, is immoral, even if accompanied by the best moral intentions.

A spectacular case illustrating the role of the factors of the first and third groups can be found in the history of the sinking of the giant Soviet nuclear submarine *Komsomolec*.[1] The tragedy took place in the North Sea, 600 kilometers from the Norwegian coast, in 1989. Of course, there are significant differences between military and economic organizations, but some relevant factors (and types of behavior) are very similar. It had taken twelve years to design and five years to construct *Komsomolec*, which was the highest achievement of Soviet military technology and the best submarine in the world in the eighties. It could submerge twice as deep as any other ship, and from this depth it could launch its two missiles with nuclear warheads. The immediate cause of the catastrophe was the autocombustion of oil that somebody had spilt in the last compartment of the ship. Any present-day submarine should have an oxygen sensor in each of its compartments, for the oxygen concentration cannot be too low, with people unable to breathe, nor can it be too high, with inflammable substances self-igniting. Some time before the catastrophe the automatic sensor in the last compartment of *Komsomolec* worked faultily, and somebody decided to switch it off completely. At the same time an order was given to check the concentration of oxygen in this compartment with a manual meter every hour but, in fact, it was only checked every four hours. Then the manual meter was accidentally broken, so nobody did it at all. At the time of the fire, the concentration of oxygen was twice the acceptable level. When temperature sensors spotted the fire, the crew was not sure whether the reading was real. (A few days earlier, on April Fools' Day, some people had played jokes by heating up the temperature sensors with lighters.) Then when an order was given to surface, for some reason or other air was directed into the next two compartments instead of the ballast chamber. As a result those two compartments, including the power compartment, caught on fire. It should be mentioned that half of the total number of officers in the crew had no experience whatsoever, as they had just left naval colleges. When *Komsomolec* managed to surface, its crew was not able to put the fire out and after six hours of struggle the ship sank.

In comparison to the catastrophe of the *Titanic*, the *Komsomolec* catastrophe was not caused by any error in its construction or unfortuitous event. It was caused exclusively by the incremental coincidence of the small results of a few highly irresponsible acts. The builder of the ship spoke about the "untidiness" and "laziness" characterizing the Russian soul. From our perspective, we can say that a lack of practical knowledge (resulting from a lack of proper training) meant that some members of the crew did not understand the full implications and possible results of their actions and failures to act. Even if they understood these in a way, the division and dispersion of responsibility made them ignore them.

But this was not the end of the tragedy. The crew was not allowed to send an SOS signal, for that would be against military rules. The naval command in Murmansk did not allow any other ship in the vicinity to rescue the crew. As the ship sank, the crew tried to save themselves in automatic pneumatic dinghies. But they did not know how to operate the dinghies because of lack of training. In the end, naval command allowed the remaining crew to be rescued, but only after an hour; twenty-four people were rescued, but forty-two lost their lives.

To conclude: In the vast majority of situations in present-day organizational life, there are many factors influencing the acts of an individual and making them more difficult to decide. The three groups of factors discussed above constitute something that could be called the typical situation of an individual in the free market economy. Knowledge of these factors might help an individual to make the right decisions on his own despite all the difficulties to be faced. These factors and situations should be widely discussed in ethics in order to bridge the gap between abstract ethical debates and the vagueness and complexity of everyday life.

*Krakow University of Technology*

## NOTES

[1] Jacek Hugo-Bader, "Komsomolec z głębin": *Gazeta Wyborcza, Magazyn*, May 27, 1999, pp. 7–13 [in Polish].

## REFERENCES


Hartmann, Nicolai, *Ethics*, 3 vols. Trans. Stanton Coit. London: George Allen & Unwin Ltd.; New York: The Macmillan Company, 1932. Vol. III, pp. 143–196.

Ingarden, Roman. *Über die Verantworung. Ihre ontischen Fundamente*. Stuttgart: Philipp Reclam, Jun., 1970.

Jackall, Robert. "Moral Mazes: Bureaucracy and Managerial Work," *Harvard Business Review* 61: 5 (September–October 1983).

Pratley, Peter. *The Essence of Business Ethics*. London: Prentice-Hall International (UK) Ltd., 1995.

Shaw, W. H. *Business Ethics*, 3nd ed. Belmont CA: Wadsworth, 1999.

Stenberg, Elaine. *Just Business: Business Ethics in Action*. London: Little, Brown and Company, 1994.

Velasquez, Manuel G. *Business Ethics: Concept and Cases*, 4th ed. Upper Saddle River, New Jersey: Prentice-Hall Inc., 1998.

JIM I. UNAH

# INTELLECTUALS AND THE LEGITIMATION CRISIS: A PHENOMENOLOGICAL ONTOLOGY OF HUMAN RELATIONS

It is difficult to decide here, without any further ado, whether by "intellectuals" we mean those persons whose vocation includes teaching, research and consultancy in a tertiary school setting or whether the term can be stretched to embrace all those who, primarily and mainly, employ the intellect with dexterity and depth in their daily commerce with the world. Differently stated, by the term "intellectuals" do we mean academics, that is, intellectuals as scholars, or do we use the term to refer to all those persons who have, through education – private or public – sharpened their intellect in such a way that they can participate in reasoned social discourse and social criticism? Since classification is difficult to derive, we should explore, phenomenologically, the meaning of the term as not placed in this or that theory.

Phenomenology is a method of philosophy that refuses to privilege any position or subject and which by dint of self-reflexivity focuses its critical searchlight upon itself. It is an exercise in self-unmaking or an enterprise in the unmasking of layers of truth to get at the thing-in-itself which is the presumed target of traditional ontology. A critical *a priori* research, phenomenology disavows any "free floating accidental constructions" (Heidegger, 1962) and insists on a return to genesis, to prereflective consciousness, to the life-world in order to reconstitute the world anew. It is a "reconstruction after an initial demolition exercise" – an act of dissection and recombination and a paying of attention to every position and every point of view. In consequence of the foregoing, the way back to the essential reconstitution of the intellect and the role of those labeled as "intellectuals" in the legitimation crisis is to explore the concepts etymologically and phenomenologically. But, on the other hand, the concept of legitimation crisis is fairly well defined in available literature and its kernel resides in whether it is legally or socio-politically derived.

We have no intention here, as academics are alleged to do, to cause paralysis by analysis. Suffice it to state unambiguously that, since we speak from a background of ontology and phenomenology, the clarification of terms that we shall execute in this essay would show that in the last analysis a legitimation crisis arises fundamentally from the way and manner we intellectualize reality, from our conceptual and representational schemes, that is,



*A.-T. Tymieniecka (ed.), Analecta Husserliana LXXVI*, 355–367.

from the way our intellect, our reason, our ratiocinative faculty legislates for experience, for reality.

Consequently, this paper will focus on the role of the rationalistic intellect, that is, human pure reason in conceptualizing reality and how our orientation on the old Greek metaphysics and epistemology, upon which Western scholarship is essentially anchored, heightens the legitimation crisis, and how, finally, the mitigation of the crisis will depend on the radicalization of our ways of thinking, on the destabilization of the categories of Western thought (Lather, 1988), on a reorientation of the intellect to refuse to privilege any subject or position (Best and Kellner, 1991, 175), and on how those who exercise authority on behalf of the state (legal legitimation) can harmonize their aspirations through consensus-building (Habermas, 1981) with those of the civil society (social-political legitimation) to ameliorate human social conditions by the creation of adequate opportunities from which alone human transactions of all sorts can arise and be assured (Mannheim, 1940). Accordingly, the essay trifurcates into (i) the analysis of the structure of thought, that is, the rationalistic intellect; (ii) the analysis of the immanent tension between legally and socio-politically derived legitimations, and (iii) the analysis of the postmodern conditions which beckons on the intellect and, by extension, the intellectuals, to disavow authoritarian accounts of truth, the lust for "one true story" and the fruitless and deleterious search for a God's-eye perspective on reality, which have been responsible for all forms of rigidity, fixism, intolerance and the patent lack of a capacity for pragmatic compromise on the part of those who exercise the authority of state, on the one hand, and the resultant tension from the lack of opportunities which heightens social insecurity in the populace, on the other hand, and triggers political instability in consequence (Unah 1992), especially in developing, peripheral, societies. We begin our analysis with the intellect and intellectuals.

## THE INTELLECT AND INTELLECTUALS

First, the intellect is to be distinguished from intelligence. The *Encyclopaedia Britannica* (vol. 9) defines intelligence as a "hypothetical construction used to describe individual differences in an assumed latent variable that is by any direct means, unobservable and unmeasurable." The concept is employed in the context of variations in the ability to learn, to do well and to behave according to social expectations. Intelligence is a personal attribute and possession and is measured in the behavioral sciences through the

Intelligence Quotient (IQ) test. Accordingly, "the IQ score is an operational, manifest, observable and measurable representation of intelligence" ... and as such ... "the distribution of intelligence is theoretical and never can be known precisely ..." (p. 673).

Furthermore, an American author is quoted to have described intelligence as the native ability of the creature to achieve its ends by varying the use of its powers – living, as we may say, by its wits. Accordingly, we can "distinguish the intelligent from the stupid throughout the scale of sentient being" (cf. Fahm, 1978, 3). The point to be made from this is that "intelligence is an individual and private possession." It can never be passed on to a future generation, that is, it cannot be inherited by offspring.

The intellect, on the other hand, is a community property, which can be cultivated and stored up for generations. Etymologically, the world "intellect" is of Latin origin and it comes from the term "intellectus" signifying discernment or understanding. Accordingly, *The Oxford Dictionary of the English Language* [Vol. VII] defines the intellect as "That faculty, or sum of faculties, of the mind or soul by which one knows and reasons ... power of thought; understanding" [p. 1067]. The same dictionary defines the "intellectual" as "of, or belonging to, the intellect or understanding." Elsewhere, the intellect is defined as "the capitalized and communal form of live intelligence, it is intelligence stored up and made into habits of discipline, signs and symbols of meaning, chains of reasoning and ... a shorthand and a wireless by which the mind can skip connectives, recognize ability and communicate truth" [Cf. Fahm, p. 3]. The main difference between the intellect and intelligence is that while the latter is a private and individual possession, the former is a collective good, a community property. And whereas the latter is not transferable, the former is "the patrimony of all who care to cultivate it ... a product of social effort and acquirement" which presupposes concentration and continuity, self-awareness, articulate precision and fluency.

Having said that, it needs to be pointed out that the intellect is intimately related to intelligence. The intellect draws largely from intelligence and most times makes up for the lack of it. The intellect operates by accentuating "the force of intelligence by giving it quick recognition and apt embodiment." Hence the intellect is a communal form of live intelligence. But while we can talk of an intelligent child, an intelligent dog or an intelligent taxi-driver, we cannot speak of an intellectual child, an intellectual dog or an intellectual taxi-driver. This leads to the analysis of the intellectual. An intellectual, from the foregoing sketch of the intellect in its differentiation from intelligence, is one who is of the intellect or understanding, that is, one who cultivates the intellect or one who is devoted to matters of the intellect. In other words, an intellectual is one who vigorously cultivates the intellect and contributes to the pool or stock of the communal property. Professor Fahm understands the intellectual as a member of the clerisy, a *literatus* – a member of the class of learned men, a scholar.

Thus, it takes more than the possession of intelligence to attain the status of an intellectual. Surely, every human being is endowed with the faculty of reasoning – the intellect, but for the most part and for a great majority of people, it [the intellect] remains latent, hidden, uncultivated. That is the big difference and that is the point at issue. Everyone has a speck of intelligence and is endowed with the ratiocinative faculty, but not everyone has made it a point of duty and commitment to draw from the intelligence to build on and cultivate the intellect. Only those who have made it a vocation to develop the intellect with a view to sustaining it as a community property qualify to be described as the Literati, members of the clerisy, intellectuals. From this it follows that those who are domiciled in the universities and allied tertiary institutions do not exhaust the supply of intellectuals. In this respect the distinctions made by Antonio Gramsci in his prison notes become instructive and helpful. Although all human beings are "potentially intellectuals in the sense of having an intellect and using it, but not all are intellectuals by social function" [Gramsci, 1971, 1]. But even in the functional sense, there are two broad groups of intellectuals – those that inhabit society's institutions of higher learning – the traditional intellectuals; and the "organic" intellectuals who constitute the "thinking and organizing element of a particular fundamental social class." In Gramsci's opinion, organic intellectuals are distinguished more by their function in "directing the ideas and aspirations of the class to which they organically belong" such as a political party, a trade union, a ruling elite, than by their professional calling. What is common to intellectuals, whether they be those who inhabit the interstices of society or those who are organically committed to social class formation, is that they constitute the locomotive of world affairs. Perhaps, the concept of the "intellectual" is more carefully delineated by Adebayo Ninalowo in the distinctions which he makes between the "traditional" and "critical" intellectuals. Elucidating Gramsci [1971], Ninalowo [1990] makes the point that the notion of a nonintellectual class of *Homo sapiens* is an embarrassing nomenclature since thinking is one event that happens in all humans. To think is to intellectualize. Since all humans think, all humans are by nature intellectuals. But of course intellectuals are those who are so designated because they make thinking into a vocation. Thus, it would seem, every man is an intellectual animal. But the depth and profusion of intellectualism make the difference between the ordinary man and the intellectual.

Now the fine point which Ninalowo makes in his analysis is that "tradition" intellectuals are those who occupy positions in tertiary institutions of learning, research institutes and the various bureaucracies of state. This substratum of the intellectual elite group is fluid in its class character and usually lacking in ideological coherence, commitment and vitality. Then there are the "organic" intellectuals usually "linked to a class in terms of their practices" [Ninalowo, 1990, 117]. Even amongst those labeled as organic intellectuals there are two broad distinctions to be made between those who are clearly linked with the dominant, broad interests of the ruling oligarchy and those who identify with and defend the interests and champion the cause of the subaltern classes of society. The first subclass of organic intellectuals comprises, to use Ninalowo's example, "corporate lawyers, reactionary social scientists, technicians, technocrats ..." [p. 118]. The second subclass of organic intellectuals include "... those who, in their capacity as scholars, are humanistically and critically oriented. They comprise those who with their expert and specialized knowledge and training are actively engaged in formulating innovative ideas" [Ninalowo, 1984, 310]. In Ninalowo's view, the third group of intellectuals which is more or less an extension of the organic, is the critical intellectuals. Although they constitute a subgroup or subclass, critical intellectuals are "conceptually and practically similar with subaltern organic intellectuals" [Ninalowo, 1990] – a common worldview binds them together for they share similar aspirations and emancipatory values. Essentially, "critical intellectuals are of particular significance in that they help to highlight and draw attention to some far-reaching shortcomings and contradictions that are counter-productive to authentic developmental directions" [ibid. p. 119].

The first point to note in all this is that the classification of intellectuals into the traditional, organic and critical is not watertight. Organic intellectuals, whether they be bourgeois or liberal in orientation, share a family resemblance – they invariably have clearly defined economic, social and political interests to propagate and promote. In propagating and promoting these interests, organic intellectuals generate ideas regardless of whether or not the ideas in question are innovative or reactionary. Even the organic intellectuals of the ecclesiastical order who maintained a monopoly of religious theology in the medieval period had landed interests to promote. Consequently, the ideas generated by scholasticism were not at variance with the interests of the propertied aristocracy [Gramsci, 1971, 7].

The second point of interest is that the world thrives on ideas. Ideas, especially the sorts that reshape the world, are usually generated by cultivated intellects. This is where the university – the citadel of traditional intellectualism – emerges clearly into the picture. The primary responsibility of the university is to "cultivate the intellect, liquidate ignorance" [Fahm, 1978] and dispel illusions.

The third point which is closely aligned to the first is that corporate lawyers, technicians, technocrats, entrepreneurs of the managerial class, theologians, among others, of the organic class of intellectuals are largely produced by the universities. Thus the ideas they generate for the propagation and promotion of their class interests are ultimately procured from the traditional intellectuals, the *literati,* the clerisy. Thus it follows from this that those vested with the primary responsibility to cultivate the intellect are the traditional intellectuals of the academia, the "Dons," as they are called, of the Ivory Towers. The objective of the assigned role of cultivating the intellect is intended to liquidate ignorance, dispel illusions and generate ideas for the ordering or reordering of human affairs.

Consequently, the organic/critical intellectuals are secondary intellectuals, if you like, syncretic accretions of the clerisy of the university. They are extensions or derivations of what I choose to call the primary intellectuals of academia. Even where it can be demonstrated that some organic or secondary intellectuals did not train formally in a university setting, it can also be shown that they probably developed their intellectual potentials in parallel institutions – private or public – that are either forerunners of a conventional university or that ape the established tradition of the university. The ecclesiastical order, for example, has successfully created parallel theological institutions with the paraphernalia and trappings of a university. Even then, it can also be demonstrated that those who cultivate the intellect in such parallel institutions have themselves been trained in a normal university. The point is finally made, therefore, that organic intellectuals – whether they be critical or not, or whether they be of the emancipatory persuasion – are, in the last analysis, creations of the primary traditional intellectuals – the historical cultivators of the intellect. But then, what precisely is the role of the intellect? How do intellectual activities create the legitimation crisis and how can a radical employment of the intellect mitigate such a crisis?

## THE NATURE AND FUNCTION OF THE INTELLECT

Primarily, the function of the intellect is to receive, process, analyze and synthesize sense information, sense data, or sense presentations from phenomena. Experience, especially sense information, does not come to us comprehensively, that is, as an organized system of meaning. It comes to us rather fragmentally – in bits and pieces. The function of the intellect – the human faculty of understanding – is possible only because it (the intellect), even prior to any reflection, is in possession of pure knowledge *apriori*, that is cognition independently of experience. Thus apart from receiving, processing, analyzing and synthesizing sense data, which constitute cognition *aposteriori*, the faculty of understanding (the intellect) has a native capacity to form notions of connectedness, universality and necessity antecedently to experience.

Immanuel Kant makes the crucial distinction between empirical universality and absolute universality in relation to the ability of the intellect or human pure reason to design formal concepts or categories with which to order or reorder experience. Although they are inseparably connected with each other (which attests to the capacity of the human mind to legislate for experience), the distinction between loose and strict universality relate to cognition *apriori*, that is, judgements that apply to all without exception as, for instance, in the proposition, "all bodies are heavy" (Kant, 1934, 26). On the other hand, empirical universality deals with propositions that are "arbitrary extensions of validity," that is, from that "which may be predicated of propositions valid in most cases, to that which is asserted of propositions which holds good in all ..." [ibid].

The point to be made from this is twofold. There are two broad types of propositions and, by extension, two classes of concepts. These are analytic and synthetic propositions and nonempirical and empirical concepts respectively. Propositions are analytic if they are unaffected by spatio-temporal considerations, and are synthetic if they are subject to the vagaries of spatio-temporal considerations. By the same token, concepts are nonempirical if palpable experiential entities were not consulted before their manufacture by the intellect, and are empirical if the raw materials for their manufacture by the intellect were funded by entities in experience.

Now, in achieving strict universality, the intellect designs and employs nonempirical concepts woven in analytic propositions. A similar feat is performed by the intellect, this time through the creation and use of empirical concepts intertwined in synthetic propositions in achieving loose universality. What is of interest in both these cases is the fact that the world as presently constituted and assigned meanings is the product of the mind's active categorizing, that is, our active intellectualization of reality. Ontologically, the intellect can understand the world monistically and/or dualistically restrictively or pluralistically-dialectically unrestrictively. This is how and where law and politics creep into our intellectualization. The mind (i.e., the intellect) must in all cases legislate for experience. The mind must decide how experience is to be organized and by such decision legislates for it. The mind must design the pattern which experience, which reality should thread. It must define the course of reality. As the mind functions through thinking, if thinking's manner of procedure were monistic or dualistic then reality or experience would be restrictively organized. But if thought proceeds dialectically-pluralistically, then reality or experience would be unrestrictively organized. As politics is the struggle for supremacy or predominance, the monistic-dualistic and the pluralistic-dialectical modes of cognizing and organizing reality are in competition for supremacy and legitimacy. The immanent tension in this struggle for supremacy and legitimacy breeds the much-touted legitimation crisis in the social sciences.

Now, phenomenologically speaking, the procedure of thought, whether monistic-dualistic or pluralistic-dialectical, is always object-gravitational. Whenever the mind thinks, it thinks of something (Husserl, 1969, 261–262). But also, the content of thought is intentionally inexistent (Chisholm, 1972), something occupies the mind whether palpably real or imagined. Thought is relational to content and "True being ... whether real or ideal, has significance only as a particular correlate of our own intentionality" (Husserl, 1970, 23). Also, phenomenologically speaking, no one decrees the content of thought or what thought should be about. It comes freely. Thought is free. We cannot say that someone else should have been thinking our thought. From this (I presume there is no haste here), it would follow that the immanent tension between those who organize reality restrictively and those who do so unrestrictively is not merely contrived and so cannot be wished away. It is rooted fundamentally, ontologically, in our being. Differently stated, neither the restricted mode of cognizing being nor the unrestricted procedure of understanding the world is borne out of intellectual mischief. Both these modes of intellectualizing reality are structurally interconnected, that is, they are, ontologically with human life, equiprimordial. There is no one, straightforward way of looking at the world. And every way of looking at the world, of looking at reality, has a sound measure of legitimacy and, by extension, constitutes a means of contributing to the universal politics of making meaning. This state of affairs calls for understanding between those who cognize reality restrictively and those who intellectualize being unrestrictively. Such understanding is our way of access to mitigating the immanent tension between legally and sociopolitically derived legitimations in the social sphere.

## THE ORIGIN OF THE PROBLEM

A team of men and women who share common world-views, common values and aspirations, quite naturally pilots social regimes. If the world-views, values and aspirations are accretions of a restrictive intellect, then there is bound to be friction between popular interests and the interest of the ruling class. Usually, popular interests are accretions of unrestrictive intellectual orientation.

Consequently, the posers to explore phenomenologically in dealing with the legitimation crisis include the following:

i. Are we as intellectuals who design operational ideas or ordering principles open to alternative perspectives?
ii. Are we ready to accommodate uncomfortable reality?
iii. Are we tolerant of alien points of view and values or, better still, shouldn't we allow others to articulate their perceptions of issues even if such perceptions run counter to our cherished ways of looking at things?
iv. Do we cherish possibilities over against actualities since human life itself is ontologically rooted in possibilities?
v. As intellectuals do we in our quotidian decision-making processes truly seek consensus or rather allow dissensus (i.e. a permanent play of the crisis of legitimation) to guide and direct our affairs? The point to be made from these questions is that the world is an ontological scene (Unah, 1977) in which ideas of all sorts germinate and take root. It is the task of the essential thinker, the ontological researcher, to illuminate this perspective with a view to assisting all those who intellectualize reality with the prism of the old Greek metaphysics and epistemology to ascend the pedestal of authentic liberal humanism. The liberal temper, as we have had cause to emphasize elsewhere (Unah: 1996B), not only assures the blossoming of thought but radically mobilizes the human community to the point where those in control of social regimes take account of subaltern rank interests in the formulation and execution of social policies. The crisis of legitimation arises basically when the ruling oligarchy is not able to synchronize its world views, values and aspirations with those of the larger segment of society. Nonliberal policies could never conduce to the well-being of the rank

> and file of society. Nonliberal social policies are accretions of a restrictive intellectualization of experience.

All restrictive intellectualizations of reality take their rise from the old Greek metaphysical *contagion* that reality is one whose ramifications can be exhausted by the epistemological resources of a single knowing subject. Upon Greek orientation it is possible for mortal man to entertain a vision of eternal verities and maintain a God's eye perspective on reality. This orientation produced systems of transcendent truth with its controlling humanism, its instrumental rationality complex (Bertens, 1995, 21) that emasculates man in the mass and heightens social insecurity and tension in the populace. By its insistence on the absolutes of being, old Greek orientation produced authoritarian accounts of truth (as evidenced from traditional rationalism and empiricism) and introduced all sorts of fixism, dogmatism, rigidity, and intolerance into our social, legal and political transactions.

Fortunately, with the emergence of the phenomenological culture the archaic Greek orientation with its pretentious liberal humanism, which culminated in modernism, has vanished into history. With phenomenological hermeneutic replacing our "former aspirations to objectivity" and with the findings of quantum physics attesting to the anarchic character of reality, the old controlling humanism and its instrumental rationality complex have collapsed like a pack of cards. This is the reality of our postmodern conditions – that regimes of absolute truth no longer have any title to legitimacy.

## MITIGATING THE CRISIS

Truth nowadays, especially social truth is numerically and quantitatively *decidable*. There are many truths. This is what the intellectual of state power and authority should take account of in social policy formulation and implementation. But since intellectuals, whether they be apologists of state authority or of emancipatory persuasion, are products of our school systems, there is an urgent need for a reorientation of the trainer of intellectuals – the teacher. The teacher needs a retraining in theories of perspectives or first principles. Every perspective, every ordering principle, is a product of human pure reason and to that extent it is legitimate. However, an ordering principle or theory of *perspective delegitimizes* itself if it insists that it is absolute or that there are no alternatives to it. We should respect and accommodate alternative principles or possible perspectives. This does not imply that we should adopt conflicting ordering principles or perspectives simultaneously. An ordering principle or a combination of ordering principles (not all possible principles or perspectives) will effectively mobilize experience more comprehensively and thus assure stability and security in our social transactions. Only those should be employed at any given point in time. Other competing principles should be accepted but shelved until they become compelling for implementation.

Thus the phenomenological position that we should allow competing principles or possible perspectives to flower does not necessarily destroy our former aspirations to objectivity in the old Greek dispensation. It simply states that we replace our penchant for totality from an isolated knowing subject with intersubjective experience. Since every truth and every point of view imply a human setting and a human subjectivity (Sartre, 1977), in a community of participating *others* objectivity emerges from intersubjective consensus. In other words, social truth is intersubjectively arrived at (Unah, 1997) not by a single individual laying claim to an alleged vision of eternal verities.

Postmodern intellectual commerce demands that we seek and domicile intersubjective experience. The imperative is that we evolve nonimpositional ways of thinking and talking about our experiences which are "inherently culture bound and perspectival" (Lather, 1988); that we replace compliance with consensus, coercion with persuasion; and that while we may entertain our different points of view, we must surrender to superior logic.

As intellectuals of the interstices of society we should reorient thinking in the positive and humane direction even while accommodating negative thoughts and putting them in their rightful shelves. As intellectuals of the Ivory Tower we should never make believe that only when thinking and talking are of emancipatory persuasion or defensive of subaltern rank interests, is our intellectualizing really legitimate. As intellectuals we should allow the blossoming of thought in whichever direction. This attitude will be helpful in mitigating the legitimation crisis.

Be that as it may, intellectuals of state power and state authority need to do a lot more in mitigating the legitimation crisis. They need to tell those whose interest they defend that what legitimizes the right of command is not that the leader has a vision of eternal verities or that the ruler is radically autonomous but that he does genuine political deeds. Authentic political conduct consists in providing adequate opportunities from which alone human transactions of all sorts can arise. Once there are no adequate opportunities for citizens to realize their aspirations in accordance with their natural aptitude and training, social insecurity and tension are heightened in the populace. This is so because lack of opportunities destroys integrated labour relationships formed over the years, creates frustration and desperation and confers the right of rebellion on the populace.

The point to be made here by way of conclusion is that lack of adequate opportunities for the pursuit of enlightened self-interest and social rebellion are structurally interconnected. Thus, whoever is ambitious to reduce or expunge the tension between official thinking (legal legitimation) and popular interests (socio-political legitimation) has first of all to insist on the provision, by the State, of adequate opportunities for human transactions of all sorts.

*University of Lagos*

## REFERENCES

Bertens, H. (1995). *The Idea of the Postmodern: A History*. London: Routledge and Kegan Paul.

Best, S. and Douglas Kellner (1991). *Postmodern Theory: Critical Interpretations*. London: Macmillan.

Chisholm, R. M. (1972). *Intentionality, Mind and Language*. Ed. A. Marras. Chicago: University of Illinois Press.

Fahm, L. A. (1978). "The Responsibility of the University to Itself," *African Insight*, Vol. IV, pp. 1–6.

Gramsci, A. (1971). *Selections from the Prison Notebooks*. Trans. Quintin Hoare and G. N. Smith. New York: International Publishers.

Habermas, J. (1975). *Legitimation Crisis*. Trans. T. McCarthy. Boston: Beacon Press.

— (1981). "Modernity Versus Postmodernity," *New German Critique* 22 (Winter 1981), pp. 3–14.

Heidegger, M. (1962). *Being and Time*. Trans. J. Macquarrie and E. Robinson. Oxford: Basil Blackwell.

Husserl, E. (1969). *Ideas: General Introduction to Pure Phenomenology*. Trans. W. C. Boyce Gibson, London: George Allen and Unwin Limited.

— (1970). *The Paris Lectures*. Trans. Peter Koestenbaum. The Hague: Martinus Nijhoff.

Kant, I. (1934). *Critique of Pure Reason*. Trans. J. M. D. Meiklejohn. London: Everyman's Library.

Lather, P. (1988). "Feminist Perspectives on Empowering Research Methodologies," *Women's Studies International Forum*, Vol. II, No. 6, pp. 561–581.

Mannheim, K. (1940). *Man and Society in an Age of Reconstruction: Studies in Modern Social Structure*. London: Routledge and Kegan Paul.

Ninalowo, A. (1990). "Intellectuals, Possibilities and Ruptures," in *Bureaucracy and Social Change: Studies in Bureaucracy and Underdevelopment*. Ed. A. Ninalowo. Lagos: Pumark, pp. 115–131.

— (1996). "The State, Legitimation and Human-Centred Development," *Africa Development*, Vol. XXI, No. 4, pp. 55–73.

Sartre, J.-P. (1977). *Essays in Existentialism*. Secaucus, New Jersey: The Citadel Press.

Unah, J. I. (1993/94). "Prelude to Political Stability," *The Nigerian Journal of Philosophy* (a Publication of the Department of Philosophy, University of Lagos), Vol. 13, Nos. 1 and 2, pp. 46–54.

—— (1997). *Heidegger: Through Kant to Fundamental Ontology*. Ibadan: Hope Publications.
*The Encyclopaedia Britannica*, Vol. 9.
*The Oxford Dictionary of the English Language*, Vol. VII.

# INDEX OF NAMES


–A–

Adorno, T. 242
Ainsworth, M. W. 145
Alberti, L. B. 199
Allais, M. 306
St. Anselm 318, 319
Apel, K.-O. 118
Arbore, R. 331
Ardigó, A. 333
Aristotle ix, xv, 95, 104, 117, 119, 123, 125, 126, 156, 157, 159, 163, 195, 211, 224, 234, 238, 257, 262, 266
Assunto, R. 204, 212
St. Augustine 90, 121

–B–

Bacharach, S. B. 333
Bachelard, G. 261
Bacon, F. xv
Bacon, R. 220
Balthasar, H. U. von 90
Balzac, H. de 172
Banfield, E. C. 332
Barash, J. A. 166, 169
Barberis, C. 333
Barrios Lara, J. L. 261, 262
Bartnik, C. S. 345
Bartol, K. M. 345
Basho 48, 51
Basu, S. 24
Bataille, G. 247, 252, 259, 262
Baudrillard, J. 235, 246
de Beauvoir, S. 156, 157
Beckett, S. 171
Beer, M. 333
Beethoven, L. van 206, 207
Belting, H. 165, 169
Benjamin, W. 241–43, 246
Bergson, H. 91, 102, 115, 125, 126
Berkeley, G. 5, 17, 18, 197, 208
Bertens, H. 364, 366
Bérthoud, G. 290, 319
Best, S. 356, 366
Bhaumik, M. 24
Bhavana, R. 44
Bielicki, T. 346
Bion, W. R. 109
Blaga, L. 152, 157
Block, M. 159
Bogus, D. 246
Bombala, B. 345
Borges, J. L. 171, 261
Bossuet, J. B. 121
Bosteels, B. 238, 239, 246
Braudel, F. 162
Brilliant, R. 166, 168, 169
Bronowski, J. 215–221
Brookner 17
Brueghel, P. (the Elder) 173
Bruner, K. 290
Bruno, G. 220
Buber, M. 101, 102–104, 106, 114, 115
Buchanan, I. 240, 246
Buddha 47
Bultmann, R. 88
Burks, J. 215, 222
Butler, O. 310, 319
Butterworth, G. E. 12, 18

–C–

Caillé, A. 305, 315, 316, 319
Calabrese, O. 243, 245, 246
Calvino, I. 171
Carroll, L. 100
Cartier-Bresson, H. 163
Castellant 331
Cenini, C. 198
Cezanne, P. 14, 241
Chadwick, W. 146
Chakravarti, A. 28, 44, 46

Chapuis, J. 145
Chiambretti, P. 331
Chisholm, R. M. 362, 366
Chrétien, J. L. 247
Christiansen, K. 145
Clerc, D. 304, 319
Cochran, E. 12, 18
Combenale, P. 291, 294–96, 319
Confucius 328
Coomaraswamy, A. K. 33, 45
Copernicus 220, 242
Cortázar, J. 171
Cortés, J. M. G. 250, 254, 257, 258, 262
Critias of Athens 90
Croce, B. 205
Cromwell, O. 218
Cronenberg, D. 248, 250, 253, 254–56, 259
Crozier, M. 333
Cubie, D. 149

–D–

Dandini, R. 331
Dante 252
Danto, A. C. 237, 246
Darwin, C. 327
Da Vinci 173, 198
Debi, I. 35, 36
Debreu, G. 304
Deci, E. L. 333
De Duve, T. 235, 240, 246
De Lange, H. 310, 311, 313, 319
de León, L. 88
Deleuze, G. 233, 238–240, 244, 246, 248, 261, 262
della Francesca, P. 253
de Menil family 144
de Rita, G. 332, 333
Derrida, J. 159, 160, 168, 169, 233, 237, 238
Descartes, R. 4–6, 13, 18, 48, 177, 219, 222, 224, 318
Dewey, J. 139, 146
Dhanens, E. 143
Diels, H. 90
Dilthey, W. 117, 118, 120–25
Diogenes Laertius 154, 157
Dixon, L. S. 144
Dodd, G. 310
Dogen 48, 50, 52, 54–56
Doisneau, R. 163
Duchamp, M. 169, 235
Dumont, L. 292
Durand, G. 261
Dürer, A. 134, 144, 165
Duychaert, E. 244

–E–

Eckermann, J. P. 207
Edgeworth, F. X. 291
Einstein, A. 19–46, 219, 238
Elizondo, S. 171
Epicurus 156, 157, 292
Ewing, D. W. 333

–F–

Fahm, L. A. 357, 360, 366
Federico de Montefeltro 253
Fédida, P. 102, 115
Fichte, J. G. 197
Fitzgerald, P. 310
Fleming, W. 201, 202, 212
Fo, D. 331
Focillon, H. 200, 212
Foister, S. 169
Ford, H. 324, 332
Foster, H. 237, 240, 242, 246
Foucault, M. 235
Freud, S. 102, 104, 105, 108, 109, 113, 115
Friedman, K. 148
Friedman, M. 296–303, 309, 311, 315,319
Fromm, E. 95
Fry, R. 144

–G–

Gadamer, H.-G. 161, 180, 181
Galileo 220, 233, 234, 238, 239, 243, 245
García-Gómez, J. 88, 91–94
Gibbon, E. 201
Giotto 146, 147
Goethe, J. W. von 207
Goldberg, R. 148
Gottlieb, C. 144, 146, 147
Goudzwaard, B. 310–313, 319

Göz, G. B. 203
Gramsci, A. 358, 359, 366
Green, A. 106, 109, 115
Greenberg, C. 233, 235, 244
Grenier, C. 243
Griffin, R. W. 337, 345
Gruenwald, M. 143
Guattari, F. 233, 238, 239, 240, 246

–H–

Habermas, J. 118, 323, 356, 366
Hacking, I. 227, 231
Hainley, B. 147
Haldar, A. K. 41
Halévy, E. 313
Hanson, N. R. 228, 231
Hartmann, N. 93, 287, 354
Hayek, F. A. von 294, 318
Haywood, R. E. 137, 145, 148
Hegel, G. W. F. 118, 121, 171, 197,200, 212, 238, 241, 242
Heidegger, M. 80, 89, 90, 93, 94, 153, 167, 181, 233, 237, 238, 240–42, 246, 249, 355, 366
Helting, H. 88
Heraclitus 196
Herder, J. G. von 121
Hermann, I. 108
Hildebrand, D. von 168, 169
Hindmsan, J. T. 148
Hishiguro, K. 171,173, 175
Holbein, H. (the Younger) 164
Holloway, J. 67, 88
Horkheimer, M. 318, 319
Hugo-Bader, J. 354
Huineng 49, 52, 55
Huizinga, J. 137–39, 142, 143, 145
Hume, D. 119, 197
Husserl, E. ix, 3, 4, 17, 51–53, 55–57, 168, 176, 178, 181, 224, 233, 234, 237, 240, 242, 246, 265–279, 362, 366

–I–

Ihde, D. 149
Ingarden, R. 193, 251, 262, 281–287, 345, 354
Ishiguro, K. 171, 173, 175

–J–

Jackall, R. 354
Jauss, H. R. 257, 258, 261, 262
Jesus Christ 163, 218, 241, 304
John Paul II 103, 115
Jowett, B. 82
Jung, C. 242

–K–

Kafka, F. 175
Kalidasa 29
Kandinsky, W. 171
Kant, I. 91, 93, 117, 119–126, 197, 204, 206, 224, 228, 233–38, 241, 249, 265, 285, 286, 310, 311, 361, 366
Kaprow, A. 136–148
Kelley, J. 138, 142, 144–46
Keliner, D. 356, 366
Keynes, J. M. 290, 293–97, 299, 300, 306
Kinsley, C. 243
Kirby, M. 139, 145–47
Klamer, A. 290, 319
Koestler, A. 210, 212
Kosuth, J. 146
Koyré, A. 110, 111, 115
Krantz, W. 90
Kuhn, T. 305

–L–

Lacan, J. 242, 327
Lamo de Espinosa, E. 122
Laplace 285, 286
Lather, P 356, 365, 366
Latouche, S. 289, 291, 303–311, 313, 314, 317, 318, 320
Lauer, Q. 237, 246
Lavista, M. 171
Legge, J. 55
Lehnert, P. 146
Leibniz, G. W. 83–85, 96, 202–04, 261
Lévinas, E. 159, 161, 168, 169, 233, 238, 247, 260–62
Lévi-Strauss, C. 305
Lewes, J. 148
Ligeti, G. 171
Lippard, L. P. 147
Livingston, J. 147

Locke, J. 166
Luhmann, N. 230, 231

–M–

Machlup, F. 294, 319
MacIntyre, A. 17, 18
Majka, J. 345
Malthus, T. 291
Mannheim, K. 356, 366
Marcel, G. 220–22
Marcuse, H. 346
Marias, J. 67, 76, 80–82, 85, 88–97
Maritain, J. 93
Marrow, J. H. 145
Martin, D. C. 345
de Marval-McNair, N. 91
Marx, K. 241, 242, 291, 293, 305, 309, 323
McDarrah, F. W. 147
McGregor, D. 336
Medawar, P. B. 88
Dr. Mendel 21
Merleau-Ponty, M. 3, 13, 14, 16–18, 247, 261
Merloni, A. 332
Merton, T. 341, 345
Mill, J. S. 309
Montagu, J. 243
Montesquieu, C.–L. 308
Mullarkey, J. 239
Munschower, S. S. 144

–N–

Nagarjuna 54
Nasio, J.–D. 104, 106, 110, 113, 115
Natanson, M. 93
Némo, P. 147
Nevelson, L. 144
Newman, B. 144
Newton, I. 119, 120, 219, 229, 236, 239
Nietzsche, F. 125, 126
Ninalowo, A. 358, 359, 366
Nishitani, K. 51, 54, 55
Nonaka, I. 333
Nowak, S. 346
Nyojo 50

–O–

Oepke, A. 89
Olivier, L. 210
Ortega y Gasset, J. 67–97, 217–19, 222
Ouchi, W. G. 333

–P–

Parmenides xv, 89, 196
Parsons, B. 141
Pasolini, P. P. 248, 250–3, 256, 259, 262
St. Paul 327
Pelligrini, A. 146, 147
Penrose, R. 24
Philip II 136
Duke Philip the Good 135
Picasso, P. 164, 210
Pieper, J. 84, 96
Plato xv, 82, 83, 85–87, 90, 91, 97, 117, 125, 126, 151, 154, 156, 157, 159, 196, 245, 265, 304, 318
Pollock, J. 141, 147, 148
Popper, K. 119, 215, 220, 222, 305
Postman, N. 215, 222
Poussin, N. 205
Pratley, P. 354
Prigogine, I. 24, 120
Putnam, H. 17
Pythagoras 39, 196

–Q–

Quevedo, F. de 248
Quine, W. V. O. xv

–R–

Rame, F. 331
Rapaport, H. 238, 246
Raphael 241
Rauh, F. 153, 157
Reagan, R. 303
Reik, T. 106–08, 115
Reischuck, A. 149
Rembrandt 165, 166
Rheinberger, H. J. 227, 231
Ricardo, D. 291, 304, 305
Ricoeur, P. 159–169, 247, 259, 261, 262
Rock, I. 10, 11, 15, 16, 18

Rodrigues Huéscar, A. 74, 78–80, 81, 89–92, 94
Rosen, R. 226, 231
Rosenberg, H. 141, 146
Rothko, M. 144
Roy, A. 169
Rulfo, J. 173
Rush, M. 148

–S–

de Sade, D.-A.-F. 247, 248, 252
de Saint Phalle, N. 146
Salthe, S. N. 223, 231
Salus, C. 146
Saramago, J. 171, 173–75
Sartre, J.-P. 365, 366
De Saussure, F. 305
Say, J. B. 306
Schafer, R. M. 46
Schama, S. 166
Schapiro, M. 144, 145
Schechner, R. 142, 148
Scheler, M. 121, 168
Schelling, F. W. J. von 197
Schimmel, P. 147
Scholem, G. 242
Schönberg, A. 171
Schopenhauer, A. 151, 157
Schumpeter, J. 293
Schutz, A. 93
Schwartz, H. 146
Scillia, D. G. 144
Scrovegni Family 146
Sen, B. 24
Serota, N. 243
Shaner, D. 50, 55
Shaw, W. H. 354
Shinn, D. G. 145, 146, 148, 149
Sills, P. 146
Simon, H. 333
Smith, A. 290, 308, 313, 319, 351
Smith, B. C. 233, 239, 240, 246
Smith, J. C. 144
Socrates 99, 117
von Soden, H. F. 88
Sokolowski, R. 237, 246
Solow, R. M. 306
Sosenko, K. 346
Spartacus 330
Spicq, C. 88
Spinoza, B. 104
Stein, G. 169
Stenberg, E. 354
Stengers, I. 120
Stokstad, M. 147
Stumpf, S. E. 212
Summers, L. 313
Suzuki, D. T. 48, 55
Suzuki, S. 48, 55, 56
Sweeney, R. D. 149
Swift, J. 327, 330

–T–

Tagore, R. 19–46
Takemitsu, T. 171
Takeuchi, H. 333
Tarski, A. xv
Taylor, A. E. 95
Taylor, F. W. 325, 332
Thatcher, M. 303
St. Thomas Aquinas 95, 96, 119
Thumb, P. 203
Tiepolo, G. B. 329
Tinguely, J. 146
Tischner, J. 345
Tolstoy, L. 154, 157
Toto 331
Trajan 201
Tymieniecka, A-T. xi, 45

–U–

Uccello, P 199
Ultvelt, P. O. 146
Unah, J. I. 356, 363, 365–67

–V–

van der Goes, H. 143
van der Kooey, T. P. 312, 314, 315, 319
van der Marck, J. 145, 146
van der Weyden, R. 143
van Eyck, J. 135, 136, 140, 143, 165
Van Gogh, V. 233
Velasquez, M. G. 354
Venter, J. J. 290, 294, 295, 305, 318, 319

Vico, G. 121
Villoro, L. 224, 231

–W–

Wait, E. C. 18
Walras, L. 304
Walsh, R. J. 44
Warhol, A. 144
Weber, M. 118, 121
Wegrzecki, A. 346
Weingartner, C. 215, 222
White, H. V. 162
Wilkomerski, B. 167
Williams, L. 149
Wimsatt, W. C. 225, 229, 231
Wing-Tsit Chan 55
Wisch, B. 144
Wisniewski, R. 342, 346
Wojtyla, K. 345
Wyld, M. 169

–Z–

Zahavi, D. 113, 115
Zaner, R. 93
Zhuang Zhou 49
Zubiri, X. 89, 90

# Analecta Husserliana

## The Yearbook of Phenomenological Research

*Editor-in-Chief*

Anna-Teresa Tymieniecka

*The World Institute for Advanced Phenomenological Research and Learning, Belmont, Massachusetts, U.S.A.*

1. Tymieniecka, A-T. (ed.), *Volume 1 of Analecta Husserliana.* 1971 ISBN 90-277-0171-7
2. Tymieniecka, A-T. (ed.), *The Later Husserl and the Idea of Phenomenology.* Idealism – Realism, Historicity and Nature. 1972 ISBN 90-277-0223-3
3. Tymieniecka, A-T. (ed.), *The Phenomenological Realism of the Possible Worlds.* The "A Priori', Activity and Passivity of Consciousness, Phenomenology and Nature. 1974 ISBN 90-277-0426-0
4. Tymieniecka, A-T. (ed.), *Ingardeniana.* A Spectrum of Specialised Studies Establishing the Field of Research. 1976 ISBN 90-277-0628-X
5. Tymieniecka, A-T. (ed.), *The Crisis of Culture.* Steps to Reopen the Phenomenological Investigation of Man. 1976 ISBN 90-277-0632-8
6. Tymieniecka, A-T. (ed.), *The Self and the Other.* The Irreducible Element in Man, Part I. 1977 ISBN 90-277-0759-6
7. Tymieniecka, A-T. (ed.), *The Human Being in Action.* The Irreducible Element in Man, Part II. 1978 ISBN 90-277-0884-3
8. Nitta, Y. and Hirotaka Tatematsu (eds.), *Japanese Phenomenology.* Phenomenology as the Trans-cultural Philosophical Approach. 1979 ISBN 90-277-0924-6
9. Tymieniecka, A-T. (ed.), *The Teleologies in Husserlian Phenomenology.* The Irreducible Element in Man, Part III. 1979 ISBN 90-277-0981-5
10. Wojtyła, K., *The Acting Person.* Translated from Polish by A. Potocki. 1979 ISBN Hb 90-277-0969-6; Pb 90-277-0985-8
11. Ales Bello, A. (ed.), *The Great Chain of Being* and *Italian Phenomenology.* 1981 ISBN 90-277-1071-6
12. Tymieniecka, A-T. (ed.), *The Philosophical Reflection of Man in Literature.* Selected Papers from Several Conferences held by the International Society for Phenomenology and Literature in Cambridge, Massachusetts. Includes the essay by A-T. Tymieniecka, *Poetica Nova.* 1982 ISBN 90-277-1312-X
13. Kaelin, E. F., *The Unhappy Consciousness.* The Poetic Plight of Samuel Beckett. An Inquiry at the Intersection of Phenomenology and literature. 1981 ISBN 90-277-1313-8
14. Tymieniecka, A-T. (ed.), *The Phenomenology of Man and of the Human Condition.* Individualisation of Nature and the Human Being. (Part I:) Plotting the Territory for Interdisciplinary Communication. 1983 *Part II* see below under Volume 21. ISBN 90-277-1447-9

# Analecta Husserliana

15. Tymieniecka, A-T. and Calvin O. Schrag (eds.), *Foundations of Morality, Human Rights, and the Human Sciences.* Phenomenology in a Foundational Dialogue with Human Sciences. 1983 ISBN 90-277-1453-3
16. Tymieniecka, A-T. (ed.), *Soul and Body in Husserlian Phenomenology.* Man and Nature. 1983 ISBN 90-277-1518-1
17. Tymieniecka, A-T. (ed.), *Phenomenology of Life in a Dialogue Between Chinese and Occidental Philosophy.* 1984 ISBN 90-277-1620-X
18. Tymieniecka, A-T. (ed.), *The Existential Coordinates of the Human Condition: Poetic – Epic – Tragic.* The Literary Genre. 1984 ISBN 90-277-1702-8
19. Tymieniecka, A-T. (ed.), *Poetics of the Elements in the Human Condition.* (Part 1:) The Sea. From Elemental Stirrings to Symbolic Inspiration, Language, and Life-Significance in Literary Interpretation and Theory. 1985
For Part 2 and 3 *see below* under Volumes 23 and 28. ISBN 90-277-1906-3
20. Tymieniecka, A-T. (ed.), *The Moral Sense in the Communal Significance of Life.* Investigations in Phenomenological Praxeology: Psychiatric Therapeutics, Medical Ethics and Social Praxis within the Life- and Communal World. 1986
ISBN 90-277-2085-1
21. Tymieniecka, A-T. (ed.), *The Phenomenology of Man and of the Human Condition.* Part II: The Meeting Point Between Occidental and Oriental Philosophies. 1986 ISBN 90-277-2185-8
22. Tymieniecka, A-T. (ed.), *Morality within the Life- and Social World.* Interdisciplinary Phenomenology of the Authentic Life in the "Moral Sense'. 1987
*Sequel to Volumes 15 and 20.* ISBN 90-277-2411-3
23. Tymieniecka, A-T. (ed.), *Poetics of the Elements in the Human Condition.* Part 2: The Airy Elements in Poetic Imagination. Breath, Breeze, Wind, Tempest, Thunder, Snow, Flame, Fire, Volcano . . . 1988 ISBN 90-277-2569-1
24. Tymieniecka, A-T., *Logos and Life.* Book I: Creative Experience and the Critique of Reason. 1988 ISBN Hb 90-277-2539-X; Pb 90-277-2540-3
25. Tymieniecka, A-T., *Logos and Life.* Book II: The Three Movements of the Soul. 1988 ISBN Hb 90-277-2556-X; Pb 90-277-2557-8
26. Kaelin, E. F. and Calvin O. Schrag (eds.), *American Phenomenology.* Origins and Developments. 1989 ISBN 90-277-2690-6
27. Tymieniecka, A-T. (ed.), *Man within his Life-World.* Contributions to Phenomenology by Scholars from East-Central Europe. 1989 ISBN 90-277-2767-8
28. Tymieniecka, A-T. (ed.), *The Elemental Passions of the Soul.* Poetics of the Elements in the Human Condition, Part 3. 1990 ISBN 0-7923-0180-3
29. Tymieniecka, A-T. (ed.), *Man's Self-Interpretation-in-Existence.* Phenomenology and Philosophy of Life. – Introducing the Spanish Perspective. 1990
ISBN 0-7923-0324-5
30. Rudnick, H. H. (ed.), *Ingardeniana II.* New Studies in the Philosophy of Roman Ingarden. With a New International Ingarden Bibliography. 1990
ISBN 0-7923-0627-9

# Analecta Husserliana

31. Tymieniecka, A-T. (ed.), *The Moral Sense and Its Foundational Significance: Self, Person, Historicity, Community.* Phenomenological Praxeology and Psychiatry. 1990 ISBN 0-7923-0678-3
32. Kronegger, M. (ed.), *Phenomenology and Aesthetics.* Approaches to Comparative Literature and Other Arts. Homages to A-T. Tymieniecka. 1991 ISBN 0-7923-0738-0
33. Tymieniecka, A-T. (ed.), *Ingardeniana III.* Roman Ingarden's Aesthetics in a New Key and the Independent Approaches of Others: The Performing Arts, the Fine Arts, and Literature. 1991
*Sequel to Volumes 4 and 30* ISBN 0-7923-1014-4
34. Tymieniecka, A-T. (ed.), *The Turning Points of the New Phenomenological Era.* Husserl Research – Drawing upon the Full Extent of His Development. 1991 ISBN 0-7923-1134-5
35. Tymieniecka, A-T. (ed.), *Husserlian Phenomenology in a New Key.* Intersubjectivity, Ethos, the Societal Sphere, Human Encounter, Pathos. 1991 ISBN 0-7923-1146-9
36. Tymieniecka, A-T. (ed.), *Husserl's Legacy in Phenomenological Philosophies.* New Approaches to Reason, Language, Hermeneutics, the Human Condition. 1991 ISBN 0-7923-1178-7
37. Tymieniecka, A-T. (ed.), *New Queries in Aesthetics and Metaphysics.* Time, Historicity, Art, Culture, Metaphysics, the Transnatural. 1991 ISBN 0-7923-1195-7
38. Tymieniecka, A-T. (ed.), *The Elemental Dialectic of Light and Darkness.* The Passions of the Soul in the Onto-Poiesis of Life. 1992 ISBN 0-7923-1601-0
39. Tymieniecka, A-T. (ed.), *Reason, Life, Culture, Part I.* Phenomenology in the Baltics. 1993 ISBN 0-7923-1902-8
40. Tymieniecka, A-T. (ed.), *Manifestations of Reason: Life, Historicity, Culture.* Reason, Life, Culture, Part II. Phenomenology in the Adriatic Countries. 1993 ISBN 0-7923-2215-0
41. Tymieniecka, A-T. (ed.), *Allegory Revisited.* Ideals of Mankind. 1994 ISBN 0-7923-2312-2
42. Kronegger, M. and Tymieniecka, A-T. (eds.), *Allegory Old and New.* In Literature, the Fine Arts, Music and Theatre, and Its Continuity in Culture. 1994 ISBN 0-7923-2348-3
43. Tymieniecka, A-T. (ed.): *From the Sacred to the Divine.* A New Phenomenological Approach. 1994 ISBN 0-7923-2690-3
44. Tymieniecka, A-T. (ed.): *The Elemental Passion for Place in the Ontopoiesis of Life.* Passions of the Soul in the *Imaginatio Creatrix.* 1995 ISBN 0-7923-2749-7
45. Zhai, Z.: *The Radical Choice and Moral Theory.* Through Communicative Argumentation to Phenomenological Subjectivity. 1994 ISBN 0-7923-2891-4
46. Tymieniecka, A-T. (ed.): *The Logic of the Living Present.* Experience, Ordering, Onto-Poiesis of Culture. 1995 ISBN 0-7923-2930-9

# Analecta Husserliana

47. Tymieniecka, A-T. (ed.): *Heaven, Earth, and In-Between in the Harmony of Life.* Phenomenology in the Continuing Oriental/Occidental Dialogue. 1995 ISBN 0-7923-3373-X
48. Tymieniecka, A-T. (ed.): *Life. In the Glory of its Radiating Manifestations.* 25th Anniversary Publication. Book I. 1996 ISBN 0-7923-3825-1
49. Kronegger, M. and Tymieniecka, A-T. (eds.): *Life. The Human Quest for an Ideal.* 25th Anniversary Publication. Book II. 1996 ISBN 0-7923-3826-X
50. Tymieniecka, A-T. (ed.): *Life. Phenomenology of Life as the Starting Point of Philosophy.* 25th Anniversary Publication. Book III. 1997 ISBN 0-7923-4126-0
51. Tymieniecka, A-T. (ed.): *Passion for Place. Part II.* Between the Vital Spacing and the Creative Horizons of Fulfilment. 1997 ISBN 0-7923-4146-5
52. Tymieniecka, A-T. (ed.): *Phenomenology of Life and the Human Creative Condition.* Laying Down the Cornerstones of the Field. Book I. 1997 ISBN 0-7923-4445-6
53. Tymieniecka, A-T. (ed.): *The Reincarnating Mind, or the Ontopoietic Outburst in Creative Virtualities.* Harmonisations and Attunement in Cognition, the Fine Arts, Literature. Phenomenology of Life and the Human Creative Condition. Book II. 1997 ISBN 0-7923-4461-8
54. Tymieniecka, A-T. (ed.): *Ontopoietic Expansion in Human Self-Interpretation-in-Existence.* The I and the Other in their Creative Spacing of the Societal Circuits of Life. Phenomenology of Life and the Creative Condition. Book III. 1997 ISBN 0-7923-4462-6
55. Tymieniecka, A-T. (ed.): *Creative Virtualities in Human Self-Interpretation-in-Culture.* Phenomenology of Life and the Human Creative Condition. Book IV. 1997 ISBN 0-7923-4545-2
56. Tymieniecka, A-T. (ed.): *Enjoyment.* From Laughter to Delight in Philosophy, Literature, the Fine Arts and Aesthetics. 1998 ISBN 0-7923-4677-7
57. Kronegger M. and Tymieniecka, A-T. (eds.): *Life. Differentiation and Harmony... Vegetal, Animal, Human.* 1998 ISBN 0-7923-4887-7
58. Tymieniecka, A-T. and Matsuba, S. (eds.): *Immersing in the Concrete.* Maurice Merleau-Ponty in the Japanese Perspective. 1998 ISBN 0-7923-5093-6
59. Tymieniecka, A-T. (ed.): *Life - Scientific Philosophy/Phenomenology of Life and the Sciences of Life.* Ontopoiesis of Life and the Human Creative Condition. 1998 ISBN 0-7923-5141-X
60. Tymieniecka, A-T. (eds.): *Life - The Outburst of Life in the Human Sphere.* Scientific Philosophy / Phenomenology of Life and the Sciences of Life. Book II. 1998 ISBN 0-7923-5142-8
61. Tymieniecka, A-T. (ed.): *The Aesthetic Discourse of the Arts.* Breaking the Barriers. 2000 ISBN 0-7923-6006-0
62. Tymieniecka, A-T. (ed.): *Creative Mimesis of Emotion.* From Sorrow to Elation; Elegiac Virtuosity in Literature. 2000 ISBN 0-7923-6007-9

# Analecta Husserliana

63. Kronegger, M. (ed).: *The Orchestration of The Arts – A Creative Symbiosis of Existential Powers.* The Vibrating Interplay of Sound, Color, Image, Gesture, Movement, Rhythm, Fragrance, Word, Touch. 2000 ISBN 0-7923-6008-7
64. Tymieniecka, A-T. and Z. Zalewski (eds.): *Life - The Human Being Between Life and Death.* A Dialogue Between Medicine and Philosophy, Recurrent Issues and New Approaches. 2000 ISBN 0-7923-5962-3
65. Kronegger, M. and Tymieniecka, A-T. (eds.): *The Aesthetics of Enchantment in the Fine Arts.* 2000 ISBN 0-7923-6183-0
66. Tymieniecka, A-T. (ed.): *The Origins of Life, Volume I: The Primogenital Matrix of Life and Its Context.* 2000 ISBN 0-7923-6246-2; Set ISBN 0-7923-6446-5
67. Tymieniecka, A-T. (ed.): *The Origins of Life, Volume II: The Origins of the Existential Sharing-in-Life.* 2000 ISBN 0-7923-6276-4; Set ISBN 0-7923-6446-5
68. Tymieniecka, A-T. (ed.): *PAIDEIA.* Philosophy / Phenomenology of Life Inspiring Education of our Times. 2000 ISBN 0-7923-6319-1
69. Tymieniecka, A-T. (ed.): *The Poetry of Life in Literature.* 2000 ISBN 0-7923-6408-2
70. Tymieniecka, A-T. (ed.): *Impetus and Equipoise in the Life-Strategies of Reason.* Logos and Life, volume 4. 2000 ISBN 0-7923-6731-6; HB 0-7923-6730-8
71. Tymieniecka, A-T. (ed.): *Passions of the Earth in Human Existence, Creativity, and Literature.* 2001 ISBN 0-7923-6675-1
72. Tymieniecka, A-T. and E. Agazzi (eds.): *Life – Interpretation and the Sense of Illness within the Human Condition.* Medicine and Philosophy in a Dialogue. 2001 ISBN Hb 0-7923-6983-1; Pb 0-7923-6984-X
73. Tymieniecka, A-T. (ed.): *Life – The Play of Life on the Stage of the World in Fine Arts, Stage-Play, and Literature.* 2001 ISBN 0-7923-7032-5
74. to be published.
75. Tymieniecka, A-T. (ed.): *The Visible and the Invisible in the Interplay between Philosophy, Literature and Reality.* 2002 ISBN 1-4020-0070-7
76. Tymieniecka, A-T. (ed.): *Life – Truth in its Various Perspectives.* Cognition, Self-Knowledge, Creativity, Scientific Research, Sharing-in-Life, Economics...... 2002 ISBN 1-4020-0071-5

Kluwer Academic Publishers – Dordrecht / Boston / London